"In this hauntology of modern science, Jimena Canales performs a gentle exorcism: the corpus of technoscientific hyperrationalism is laid before the reader, and with a sure hand Canales brings forth its 'demons' one by one—Descartes's, Laplace's, Maxwell's. These and more flutter up from the history of science, which is here retold as an eternal return of the (barely) repressed. What reason will not allow, it again and again enlists to work in the shadowlands that edge the world as we know it."

—**D. Graham Burnett, Princeton University**

"Brilliantly conceived and written. Canales offers an entirely new perspective on well-known episodes in science, and on subjects as diverse as thermodynamics, evolution, neuroscience, and quantum mechanics. Readers will never look at demons the same way again."

—**Robert P. Crease, author of *The Workshop and the World: What Ten Thinkers Can Teach Us about Science and Authority***

"Thought provoking and entertaining. Canales casts the history of science in a new light, one in which an underworld of imaginary creatures features prominently. This wonderful book illustrates the fundamental role of imagination in science."

—**Oren Harman, author of *Evolutions: Fifteen Myths That Explain Our World***

"A rich and wide-ranging book on the intriguing topic of demons as they have figured in scientific imagination."

—**James Robert Brown, author of *Platonism, Naturalism, and Mathematical Knowledge***

BEDEVILED

BEDEVILED

A SHADOW HISTORY OF DEMONS IN SCIENCE

JIMENA CANALES

PRINCETON UNIVERSITY PRESS
PRINCETON AND OXFORD

Copyright © 2020 by Jimena Canales

Princeton University Press is committed to the protection of copyright and the intellectual property our authors entrust to us. Copyright promotes the progress and integrity of knowledge. Thank you for supporting free speech and the global exchange of ideas by purchasing an authorized edition of this book. If you wish to reproduce or distribute any part of it in any form, please obtain permission.

Requests for permission to reproduce material from this work should be sent to permissions@press.princeton.edu

Published by Princeton University Press
41 William Street, Princeton, New Jersey 08540
6 Oxford Street, Woodstock, Oxfordshire OX20 1TR

press.princeton.edu

All Rights Reserved
ISBN 978-0-691-17532-4
ISBN (e-book) 978-0-691-18607-8

British Library Cataloging-in-Publication Data is available

Editorial: Ingrid Gnerlich and Arthur Werneck
Production Editorial: Mark Bellis
Text Design: Carmina Alvarez
Jacket design by Sukutangan
Production: Jacqueline Poirier
Publicity: Sara Henning-Stout and Katie Lewis
Copyeditor: Cynthia Buck

This book has been composed in Sabon Next Pro

Printed on acid-free paper. ∞

Printed in the United States of America

1 3 5 7 9 10 8 6 4 2

For Billy, who opened up my imagination

Contents

	Preface	ix
	Introduction	1
1.	**Descartes's Evil Genius**	15
2.	**Laplace's Intelligence**	29
3.	**Maxwell's Demon**	49
4.	**Brownian Motion Demons**	79
5.	**Einstein's Ghosts**	93
6.	**Quantum Demons**	112
7.	**Cybernetic Metastable Demons**	157
8.	**Computer Daemons**	185
9.	**Biology's Demons**	246
10.	**Demons in the Global Economy**	277
	Conclusion: **The Audacity of Our Imagination**	298
	Postscript: **Philosophical Considerations**	313
	Notes	325
	Bibliography	361
	Index	383

Preface

On a train ride to Geneva, I finished reading yet another scientific paper that included the word "demon." I had brought with me a stack of articles and stashed them in a folder that I labeled "the demon papers." Inside it, I arranged chronologically copies of original documents from the seventeenth to the twentieth centuries.[1] The texts were not theological demonologies written by priests or inquisitors; nor were they illicit texts on black magic. They were not prurient necromancies detailing evil wonders; they were not written by poets or novelists; they were not authored by anthropologists interested in superstitious or backward cultures; and they were certainly not New Age texts on the supernatural. They were standard scientific articles.

Modern demons were not found in the old *grimoires* of magical spells and incantations. They appeared in classic texts of science and modern philosophy, authored by highly respected thinkers and scientists. From the nineteenth century onwards, they were published regularly in standard journals, such as in the prestigious academic venues *Nature* and the *American Journal of Physics*. Specialized science magazines, such as *Scientific American*, covered their adventures. Even mainstream news outlets, such as the *New York Times*, occasionally reported on them. Most of the demons were associated with the last name of the scientist who first speculated about their possible existence. Some were so influential that they became a fixture in standard textbooks. Most of the demon research papers were widely celebrated, and many of them pointed to key discoveries, such as thermodynamics, relativity theory, and quantum mechanics. I could barely put the material down before having to descend from my train.

The content of these texts seemed to contradict one of science's much-trumpeted virtues and its most lasting accomplishment: the elimination of imaginary or supernatural beings from this world, including witches, unicorns, mermaids, demons, and many others. The cosmologist Sean Carroll described the magnitude of this contradiction in his book surveying our current state of knowledge of the universe. "What is it with all the demons, anyway?" he asked, then noted, "it's beginning to look more like Dante's *Inferno* than a science book."[2] Carroll's apprehension is as rare as it is commendable. Most references to demons by scientists appear without remark.

Authors of demon papers often use that eerie designation simply for lack of a better word. For them, the term is a sort of placeholder for the unknown, a word used *faute de mieux*, to refer to a whatchamacallit they do not yet fully comprehend and for which they soon hope to find a more precise term. Some writers use the term casually and unsystematically. The informality behind their choice of terminology can be maddening. "The word demon should not be used lightly," I wrote in my travel notes.

The dossier under my eyes convinced me that scientists were not behaving *at all* as we commonly thought they did. Envisioning the uphill task of understanding these practices, my mind started wandering excitedly high into the Alps. I would follow my Frankenstein wherever it led me. Seminal texts on physics, biology, and beyond seemed to me like wizardry manuals. They were enchanting and magical, bordering on dangerous. I decided to reach a few decades back from 1666 and to peer into the decade past 1999 to learn about their most recent adventures. I started gathering materials to re-create the most precise and detailed picture possible of these mysterious beings.

Older texts, those that dated to the seventeenth century, were still marked by vestiges of medieval lore. They often referred to beings that had not yet been ruled out by the budding science of those days. Often designated by the Latin term *demonia*, those creatures were thought to be *possibly* real. They fascinated researchers because nobody could prove unequivocally and conclusively that they did not exist. Demons that appeared in recent texts were more often clearly designated as hypothetical and imaginary. Yet scientists were so fascinated by them that they explored ways to bring them into existence, sketching out new technologies to imitate their feats. These demons did not remain figments of the imagination for long.

Who or what were these beings concocted in the minds of scientists? How had they moved so swiftly from science to fiction to science fiction and back to science? "These categories must be kept separate," I wrote in my notes. An as-yet-unwritten book held me down like an iron ball and chain. Our new millennium, I concluded with excitement and trepidation, required *a modern demonology for the age of reason.*

BEDEVILED

Introduction

The glass of science is half empty. Researchers across the globe are fixated on all that we do not know *yet*. It was the same one hundred years ago, and more than one hundred years before then too. Every once in a while, progress arrives.

Ah ha! Something clicks in someone's head. Everything falls into place. The result is nothing short of magical. What had once been invisible suddenly seems to have been hiding in plain sight. Inspiration happens to all of us—writers, artists, scientists, as well as ordinary people. The gap between knowledge and imagination is not as inscrutable as it has been made to appear. It is consequential long after the moment when these first ideas evaporate and practical concerns take over.

By previewing a world of wonders long before the curtains are drawn and the show begins, we can sit in on the rehearsal of our own scientific and technological future. The antechamber of discovery is a place where ideas are forged before they see the light of day. It is the incubator that shapes science before it is tested. When the spectacle of our achievements includes the trials and tribulations that led to them, knowledge looks different.

How can we explain the trajectory of science and technology that has taken us from the steam engine to the microchip, or from the early automata of the Scientific Revolution to the artificial intelligence of today? Scientists wake up every morning, drive to the lab, write papers, teach courses, train colleagues, sometimes receive prizes and accolades, retire, and die. Sociologists and anthropologists have carefully followed them every step of the way. This path has a clear logic that works in piecemeal fashion, yet somewhere along the way something greater than the actors themselves seems to break in. Scholars have been fascinated by moments of discovery in science, when genius scientists have a brilliant eureka idea.

Breakthroughs often arrive when least expected. What was once impossible no longer is.

New experiments and technologies are first conceived in the minds of scientists. They are thinkable long before they become feasible. When scientists are hard at work, their minds are frequently up in the clouds.

The surprising nature of discovery and invention may lead us to suspect that something akin to an unconscious force connives behind the boundaries of reason and drives their development from outside. Discovery has its own twisted, fascinating, and at times terrifying history. It also has its own highly developed technical vocabulary. Scientists often use the word "demon" during the most preliminary phases of their research. It designates something that is not yet fully known or understood. These demons are not religious, supernatural, monstrous, or merely evil. They refer to something that defies rational explanation and may stump or break a hypothesis or a law of nature. Their function is not primarily metaphorical or figurative. They are technical terms whose definition can be found in almost any dictionary.

The *Oxford English Dictionary* defines "demons" in science as "any of various notional entities having special abilities, used in scientific thought experiments." They are frequently mentioned eponymously "with reference to the particular person associated with the experiment" and follow a pattern originating with René Descartes, the seventeenth-century thinker known for inaugurating the Age of Reason.[1]

Descartes's demon opened up the floodgates to many others, continuing up to this day. New names are added as soon as they become part of the argot of the laboratory. "Laplace's demon" followed on the heels of "Descartes's demon" and became a model for new calculating machines that could potentially determine the precise position and movement of all particles in the universe to know all of the past and even the future. These two demons soon faced stiff competition from the Victorian creature named "Maxwell's demon," who could wreak havoc with the usual course of nature. As science grew in prestige and complexity, many other demons were invoked and named after Charles Darwin, Albert Einstein, Max Planck, Richard Feynman, and others.

Figuring things out often involves invoking demons as a useful category to articulate and fill in the gaps of existing knowledge. When confronted with a particularly difficult problem, or when the universe is not working the way it should, scientists immediately start suspecting a perpetrator. Besides being given the last name of the scientist who first started thinking about the enigma, the culprits are often anthropomorphized as they become blueprints for future technologies. Researchers sometimes refer to them as *he*, other times as *she*, and often as *it*. As scientists imagine demons with competing abilities and picture them collaborating with or fighting against each other, they inspire the creation of ever-more-complex technological arrangements. Prototypes are constantly upgraded. New versions are right around the corner, soon to be released.

A variant spelling, "daemon," has yet another meaning in science. In the context of computing technology, it designates "a program (or part of a program)" running inside a computer. The term can be interpreted as an acronym, either for "Disk And Execution MONitor" or for "DEvice And MONitor." When you perform a search in your computer, lines of code called "daemons" are used to find the match you are looking for. When you log into the internet or use your smartphone, myriads of such daemons are put to work smoothing the process of communication between you, your device, and the devices of others. Today these daemons are central to the communication infrastructure around us.[2]

Such a *façon de parler* is eerily consequential. Dictionary entries reveal an open secret within a close-knit community: scientists are demon experts. Practitioners across fields agree that "science has not killed the demons" and that studying them can be extremely useful.[3] To know the world, to make it better, to overcome insurmountable difficulties and dead ends, scientists routinely look for them. How did they become part of scientists' vernacular? What broader consequences come with this mode of inquiry? What aftereffects do these practices have on the development of world history? What, if anything, relates these definitions to the original term, one derived from the ancient Greek δαιμόνιον? Is there any connection between them and the demons associated with hell and the devil?

Most dictionaries include similar entries. In them, demons no longer appear as opposite to angels. Nor are they interchangeable with any of the other creatures of religion or folklore. They are grouped with other similar creatures. The technical use of the word shows us why the religious, figurative, and literary understanding of demons remains so pertinent today.

The progress of science and technology has been marked by investigations into the possible existence of a fine and motley crew, a veritable troupe of colorful characters with recognizable outfits, proclivities, and abilities who can challenge established laws. To catch them, scientists think like them.

SCIENCE IS STRANGER THAN FICTION

Since ancient times, poets and literary authors have given us evocative narratives of demons. Some feature them as personifications of evil, while others associate them with benign forces, including at times our inner voice or our moral consciousness. Classical and modern literature, horror films and comic books are rife with demons and devils who travel indiscriminately from highbrow to lowbrow popular genres.

Lucifer, Beelzebub, and Sathanus are some of the most prominent demons of religion. Socrates's is one of the best-known demons of philosophy. Literature has many: Dante's Lucifer, Shakespeare's Prospero, Milton's Satan, Goethe's Mephistopheles, and Shelley's Frankenstein are some of the most well known. These demons share certain characteristics with science's demons, but not all. The latter no longer have any of the physical identifying marks that would connect them to the demons of old: they have nothing in common with those furnished with short horns, long tails, and cloven hoofs. The clichés associated with black magic and evildoers do not fit them. Their form is different. Nonetheless, science's demons share many underlying characteristics with the demons of old. While no longer *isomorphic* with them, they remain *isofunctional* in key respects. For this reason, they are daunting, outperforming their predecessors in unexpected ways.

By focusing almost exclusively on the demons of lore, legend, or religion, we have forgotten to watch for the demons in our midst. The nineteenth-century French poet Charles Baudelaire was exceptional for refusing to accept the demystification of the world by scientific and secular means. His work called on readers to remain attentive to the real power wielded by figures deemed to be largely symbolic. In a poem initially titled "Le Diable," he described the evil one's latest ruse: "The devil's finest trick is to persuade us that he does not exist."[4]

Technologies are frightfully diverse. What do x and y have in common? When thinking about all the things that get categorized under the label "technology," I am often reminded of the riddles that begin with that question. Only a few things so categorized have metallic gears and pistons. They may be organic or chemical, living or inert, tiny or huge, or they may not occupy fixed areas at all. Some are clearly useful, others not at all. What can a telescope possibly have in common with a calculator? Is there a basic characteristic that can be used to describe what steam engines, for example, share with lines of code?

Of the innumerable things and systems that we commonly group in the broad category "technology," many have been associated, at one time or another, with the demonic, the magical, or the fantastical. While the very idea of modern technology is one that is frequently at odds with a belief in the power of the supernatural, too many thinkers consider technology in those terms. How can we make sense of such contradictions? Something else in technology must give rise to these associations. That "something else" is the topic of this book.

THE DEMON OF TECHNOLOGY

"What have I done?" A stroll through the history of science and technology shows us that innovations often beget regret, determination can turn into hand-wringing, and initial exhilaration gives way to soul-searching. The literature of the history of science is full of retrospective memoirs written by scientists who all confronted the same question after they saw how their research had been put to use.

Knowledge gives us power, leaving us to cope with the additional complication that power by itself does not discriminate between good and evil. Even our most advanced technologies have not brought us all the benefits we hoped for. We live in fear that our most cherished innovations in science and technology might fall into the wrong hands and be used for the wrong ends. Even in the best-case scenarios, when science and technology are developed for virtuous and honorable purposes, new developments can be quickly adapted for destructive ones. All that is needed to turn something good into something horrible is a slightly larger dose, an incremental increase in quantity, or an imperceptible change of context. Pesticides have been used in gas chambers against innocent people, fertilizers can be used to build bombs, space rockets can deliver weapons of mass destruction, vaccines are easily adapted for biological warfare, the cure for genetic diseases can become the basis of eugenic interventions, the same implement can be used to heal or to hurt, and so on. What was once a solution can become a tool for perpetuating a crime. A dream can turn into a nightmare in a heartbeat.

The picture of technological development that emerges is not entirely good. The sword of knowledge cuts two ways. We have thought about the dangers of knowledge in this way since it first appeared as a concept in history. The biblical account of the expulsion of Adam and Eve from the Garden of Eden describes knowledge as something transgressive and even demonic. A creature associated with the Devil, craftier than any of the other wild animals, tempts Adam and Eve to bite into forbidden fruit.

> When the woman saw that the fruit of the tree was good for food and pleasing to the eye, and also desirable for gaining wisdom, she took some and ate it. She also gave some to her husband, who was with her, and he ate it.[5]

Since these words were first written down sometime in the fifth or sixth centuries before the Christian era, they have been repeated over and over again. They are especially central in Judeo-Christian traditions, yet their influence on other cultures has been profound.

To this day, an unbridled desire to acquire knowledge—to gain wisdom—continues to be considered transgressive and sometimes even sinful. In other translations of this famous passage, Adam and Eve are described as eating from "a tree to be desired to make one wise." The words used to describe the serpent have been variously translated from the Hebrew *arum* as "wise," "intelligent," "clever," "cunning," "shrewd," "subtil," "crafty," "astute," and "wiley." Why are intelligence and wisdom so directly tied to sinfulness and lawlessness in this biblical passage and beyond?

The biblical account of Adam and Eve was preceded by earlier myths with similar themes. The myths of Prometheus and Icarus are perhaps two of the best known from a list that goes on and on. The idea of technology as a double-edged sword was already explored in the myth of Hercules and his poisoned arrows. After these were used successfully against his enemies, they inadvertently returned to kill their unwitting creator. Yet another famous tale of ancient times that speaks to the dangers of technology is the Hebrew story of the Golem. In the story, a lump of clay was given life, and though it mostly behaved according to the wishes of its creator, one day it did not, leaving a trail of rampant destruction and ruin. Similar themes motivate the stories of Talos, an artificial soldier made of metal; Galatea, who was created by Pygmalion to be larger than life; and Pandora, who was responsible for opening Zeus's box of evils.

Stories exposing the moral dangers of science and technology used similar tropes in medieval times. Demons, devils, and contracts made with them became more prominent. In the sixth century, the example of the life of the cleric Theophilus of Adana was used to highlight the perils of exchanging one's soul for the promise of complete and total knowledge. The medieval legend of Faust reminded its listeners that signing a pact with the devil in exchange for unlimited knowledge could have dire consequences. The Elizabethan play of that name by Christopher Marlowe brought those themes to the theater. These kinds of stories frequently feature characters who, like Adam and Eve, are tempted to explore more and know more—sometimes learning too much, being fatefully attracted to forbidden or secret knowledge. In the nineteenth century, Johann Wolfgang von Goethe's celebrated *Faust* gave new life to old Christian and medieval myths. Goethe's novel soon became a sensation throughout a continent that was being rapidly transformed politically, scientifically, and technologically. *Frankenstein; Or, The Modern Prometheus* by Mary Shelley was so imbued with these antecedent themes that she even subtitled her work with a reference to the ancient myth. Less celebrated authors pursued similar themes, sometimes echoing unsophisticated, prosaic, and commonplace beliefs about the dangers of knowing too much.

Why have these themes persisted throughout millennia? The descriptions of the entrepreneur-inventor Elon Musk are typical of the genre. When speaking at the Centennial Symposium for MIT's Aeronautics and Astronautics Department in 2014, he described AI as a powerful means for "summoning the demon."[6] Is there something in it—or in science and technology—that is inherently dangerous and wonderful at the same time? Why do we think that curiosity killed the cat? In other words, is there something about the quest for knowledge that is almost always *demonic*?

If we look at the technologies that science's demons have inspired, we get a surprisingly coherent view of science's most celebrated successes. In the seventeenth century, the philosopher René Descartes was fascinated and terrified by a host of new innovations around him, such as automata, and by new entertainment techniques that blurred the boundary between reality and spectacle. In their context, he described a creature who could take over our senses to install an alternative reality and developed an entire philosophical school designed for defending ourselves against this being. Those early technologies are quaint compared to the ones of today, yet Descartes's demon still comes up in conversations among scientists and engineers who are interested in the challenges brought about by new virtual reality technologies or who are invested in this research area. A search for demons, even some quite old ones, still drives the development of ever more perfect models. Virtual reality is one example out of many.

The history of demons permits us to see something that most social or political histories miss: the arch of modern science and technology being raised across the world. Science's demons were typically first sought after in places we now count as significant sites of historical transformation. In the Dutch Golden Age, they provided lessons about the limitations of our senses and the power of reason. In Revolutionary France, they gave scientists hope that certain natural laws were ultimately immutable and stable. In Victorian England, they showed a growing number of practitioners how to cope with industrialization. Demons played key roles in Continental Europe during World War I, in Britain and America during World War II, and in a handful of American universities during the Cold War. By the end of the millennium, enterprises where they were studied were truly global, with research taking place in select laboratories from Helsinki to Tokyo. These studies were central to the development of mechanics, thermodynamics, relativity, quantum mechanics, and cosmology. The study of demons then spread to the life sciences, where they were seen as providing the necessary oomph that jump-started life itself from its lowly origins in brute matter. They then played key roles in evolutionary biology, molecular biology, and neuroscience. Eventually, they left the desk of theoretical physicists and the laboratory benches of experimentalists to affect economic theory and monetary policy.

Not every fork is a trident, nor every bowl a cauldron. Many technologies are considered magical and fantastical without being thought of as demonic. Some celebrated thought experiments do not feature demons at all. Most descriptions where one aspect of science or technology is seen as demonic typically stick only for a short period of time before being dropped and transferred promiscuously to describe something else entirely. It is only when research is new, innovative, mysterious, and potentially transformative across broad swaths of culture and society that it is described thusly. In the case of epoch-making, world-altering technologies, we are hard-pressed to find examples that have *not* been described as demonic, in one way or another, at one time or another.

IMAGINATION

Our imagination works wonders, and many scholars have dedicated themselves to studying it. Yet its role in science is often assumed to be secondary. It is traditionally considered to be a "private art," too unruly to study, off limits to rational inquiry, inchoate, slippery, obscure, and perhaps even unrecoverably unconscious.[7] While scholarship on thought experiments has grown in recent years, most scholars still consider them to be lesser than, or essentially distinct from, the "real" deal experiments performed in laboratories and research centers.[8] The role of the imagination in science continues to be portrayed as an inconvenient id hiding behind science's ego, as something that takes place primarily outside of the lab and slyly and occasionally sneaks in, as an embarrassing sibling or bastard child of the arts and the humanities showing up uninvited.[9] But its power does not stop when scientists enter the lab or write down their equations. The entire enterprise of science—from theory to experiment to public communication—is thoroughly permeated with our imagination. When we think, reason, and make decisions, we simultaneously think ahead, far and beyond.

From a distance, we can see just how much our imagination shapes technology. The great writer Victor Hugo excelled in seeing connections between the technologies of his era and imaginary creatures of yesteryear. He asked his reader to consider how steamboats had tamed the oceans much as Hercules had tamed the Hydra, how locomotives appeared to breathe fire like dragons, and how hot-air balloons were much like the griffins once imagined to roam through the air. "We have tamed the hydra, and he is called the steamer," he wrote in *Les Misérables* (1862), before continuing: "We have tamed the dragon, and he is called locomotive; we are on the point of taming the griffin, we have him already, and he is called the balloon." He envisioned future technologies as being shaped by these age-old myths. "The day when this Promethean work shall be finished," he continued, "when man shall have definitely harnessed to his will the triple chimera of the ancients, he will be the master of the water, the fire, and the air."[10]

Castles in the sky are rarely empty. A beautiful princess may be trapped in a tower, a hunchback may live in the bell tower, or a troll may be asleep under the bridge. Our imagination is almost limitless, but it is not infinitely so. "Even in the fairytale," the philosopher Ernst Bloch reminds us, "not everything runs smoothly."[11] Imaginary creatures cannot randomly break any and all norms and laws. They must stay in character. They cannot just go any which way and act in any way they please. Creatures of our imagination lead us into certain prescribed futures. Our fate might change if we choose to enter the dungeon, peer under the bridge, sleep in the princess's bed, climb the high tower, or summon a demon.

Not all imaginary creatures have been equally useful to science. Demons are by far the most common creature that populates the modern scientific imagination. References to them vastly outnumber allusions to monsters, ghosts, werewolves, zombies, fairies, witches, unicorns, elves, giants, dragons, sirens, basilisks, hippogriffs, dracs, exotica, and so many others. Like the others, they too are representatives of universal archetypes, symbolic figures who help us express universal feelings, such as dread and fear, that are prevalent across history and culture. Yet to understand the development of science and technology, it is necessary to distinguish them from other imaginary creatures more carefully. Demons' particular ancient lineage makes them valuable for thinking about the natural world. They cannot be placed in the same basket as any other creatures. For example, while unicorns have a recent use among venture capitalists to designate unusually successful startups, they are rarely mentioned in the technical literature of science. Elves and giants, which are mostly creations of the pre-Christian mythology of the Norse and other Germanic tribes, are sometimes invoked by scientists to describe what the world looks like at different scales. Their use in technical science literature, however, is sparse. The same can be said of vampires, which are mostly of nineteenth-century eastern European origin, or of the ghouls and goblins of European folklore. Although the general category of the monstrous was very important for the development of science during medieval times, its role in modern scientific practices is minor. None of these creatures feature as prominently in modern science as demons.

A DEMON-FREE WORLD

If it is unsurprising to see techno-science's critics highlight its demonic qualities, it is even less surprising to see that techno-science's advocates think about demons and the imagination differently. Science is often portrayed as a weapon against all sorts of pseudoscientific and superstitious beliefs that have been peddled by quacks or impostors and fanned by the forces of religion and superstition. Carl Sagan, famed cosmologist and popular science author, celebrated science for just this reason. His best-selling book *The Demon-Haunted World* (1996) described the scientific method as "the fine art of baloney detection" that permitted scientists to brush away irrational beliefs and other falsehoods from this world.[12]

Sagan was right. When the unreal suddenly appears to be real—or worse, when real and unreal appear to blur—our imagination can be tempered by putting it to the test. The laws of nature provide us with constraints we can apply to check our beliefs and corral our runaway imaginations. They hold us back. As tough as brick and mortar, the laws of nature limit our imaginings and force our most audacious plans to fall in line with practical realities. Experiments can help. If you think you have seen a demon, you better think twice. Were you agitated, delusional, or inebriated? If that impression is not dispelled after ruling out mental causes that might have fooled you into thinking you saw a demon, you can create an experiment to rule out other causes. Turn on the lights. Check the window. Look for suspicious footprints. Prepare to catch the culprit during a future visit. Spread flour on the floor of your room to see if anyone has tiptoed in. If you find no evidence ever again, then it is extremely unlikely that a bipedal being was the culprit.

Throughout the history of civilization, we have developed clear ways of testing our beliefs. By varying conditions to eliminate false hypotheses, sensible folk act just like scientists, using experimental techniques to get to the bottom of things and arrive at the truth. The trial-and-error reasoning that characterizes sound, rational thinking has been tremendously effective at eliminating a host of hypothetical beings whose existence is thus proven to be so improbable that we might as well scratch them off the list of things to search for. A scientist brandishing a telescope or microscope, holding a test tube or swan flask, or analyzing a petri dish, all to eliminate false hypotheses, is acting much like a valiant knight slaying a dragon or a demon.

Yet it is not so simple. Scientists routinely look for new particles, forces, materials, states of nature, laws, and new combinations thereof. Enthralled by the incredible and unbelievable, they set off on voyages of discovery. Among themselves, they often describe their enterprise as a search for demons that are not yet completely understood or eliminated by current experiments. "If we knew beforehand what we'd find, it would be unnecessary to go," admitted Sagan. "Surprises—even some of mythic proportions—are possible, maybe even likely," he concluded.[13] How can it be that scientific laws characterized by certainty, precision, and finality are improved upon, refined, and sometimes even overturned? How does new knowledge arise from determinate laws?

A contradiction lies at the heart of science. Our imagination is necessary for obtaining new knowledge. We can celebrate *homo sapiens* for having learned how to plan and calculate as no other species before it, and *homo faber* for having used tools better than any of its predecessors, yet we seem to have forgotten that both were initially motivated by the creator of creativity: *homo imaginor*. The back-and-forth commerce between the real and the imaginary is what permits us to create new knowledge. Scientific laws are sturdy, but they are not fixed, and our imagination is the best tool we have for extending and improving on them. Science grows when researchers push it to new limits, striving to become smarter than the smartest, bigger than the biggest, smaller than the smallest, slower than the slowest, and faster than the fastest.

Scientists know full well that the fact that something has not yet been found does not mean it will never be. To make this point, the philosopher A. J. Ayer felt authorized to invoke the search for the abominable snowman as an example. "One cannot say there are no abominable snowmen," he warned, because complete proof of their inexistence across all time and space is practically impossible to come by. "The fact that one had failed to find any would not prove conclusively that none existed," he concluded.[14] The gates to the Parthenon of the Real remain wide open.

The search for new entities is not blind. Trails run cold. Experienced scientists know where it is most profitable to look, what new discoveries might look like, what properties they might possess, and what they might be capable of. Well-funded research programs focus on topics that are most worthy of investigation. Luck, goes a well-known saying, favors only the prepared.[15] It takes years and years of education and training to become prepared, and hours after hours of study to master all the preexisting literature on a given topic. Before setting out to discover the fundamental laws of nature, scientists equip themselves carefully, much like voyagers sailing off on long journeys. But luck also favors those who dare to imagine. An essential part of the work of all young scientists consists in working hard to sharpen their imagination.

Where is our imagination taking us? The science of today, it is also commonly said, is the technology of tomorrow. Yet the relation of science to technology throughout history has not been so direct or transparent. Scientist themselves are often in the dark about the repercussions of their research. Sometimes the closer they are to the topic the further they are from understanding its broader impact.

The physicist Max Born gave us one of the most honest renditions of scientists' blind spots when it comes to the impact of their research. Reflecting on his own contributions, he admitted that "anyone who would have described the technical applications of this knowledge as we have them today would have been laughed at." The path taken by the development of technology in the last centuries has gone beyond anyone's wildest dreams. During Born's youth, "there were no automobiles, no airplanes, no wireless communication, no radio, no movies, no television, no assembly line, no mass production, and so on."[16] Scientists working in the fields most relevant to new technologies can be completely blind to the changes about to take place right under their noses. Writers of speculative science fiction who are intent on imagining future worlds miss future developments just as much. If a path cannot be traced back to scientists' conscious actions and intentions, how else can we understand the development of technological innovations? The interconnection between science and technology is so complex, and their development throughout history so confounding, that it quickly raises another question. What comes before both?

For centuries, scientists have been transfixed with studying a particular set of demons. By imagining what they can or cannot do, they have figured out some of the most important laws of the universe. When scientists developed the law of energy conservation, they imagined powerful demons that could break it. When developing the theories of thermodynamics, they imagined tiny demons who fiddled with individual atoms and could overturn entropy. When they developed the theory of relativity, they considered faster-than-light demons that could wreak havoc in the universe in unpredictable ways. When they looked deep into atoms at the level of the quantum, they considered whether demons might be interfering in the bizarre paths taken by photons or electrons that were affecting atomic decay, transmutation, and the release of previously unknown sources of energy. The demons that are still under investigation possess sufficiently credible characteristics that experts continue to consider how and if they might pass for real.

The jury is still out when it comes to some of the fundamental questions associated with these strange creatures. The most die-hard demons—those that have survived centuries of investigation—have so far stumped the cleverest elimination methods of resourceful researchers. Weak and clumsy demons have been culled from the batch, but strong and nimble ones slip like lucky fish through the holes of the most up-to-date experimental techniques. As science helps us sift illusions and irrational beliefs from the real laws of nature, scientists' search lists have grown as they explain what nature can do, where its limits lie, and how its boundaries might be pushed.

The nature of logic, virtual reality, thermodynamics, relativity theory, quantum mechanics, computing, cybernetics, artificial intelligence, information theory, origin-of-life biochemistry, molecular biology and evolutionary biology, DNA replication and transcription—all have been advanced by reference to demons. The discovery of seemingly unrelated things—molecules, atomic bombs, computers, DNA, neural networks, lines of code, quantum computers—was part of an epic effort to find and understand them.

Modern demons arrived with modern thought, which they made into their comfortable home. In some descriptions, demonkind has deft fingers and sharp eyesight. In others, demons hold photon-emitting torches or flashlights; some of them are capable of forming families, and yet others are described as organized in an army or a society. Some shriek wildly, and others are good-natured and polite. They lurk in a demondom that is often dark, chaotic, and well insulated, as is the inside of a computer. In all of their shapes, forms, and guises, these creatures share one consistent quality: they appear intent on either aiding us in living a good life or preventing us from doing so, an ideal often designated by the Greek term *eudemonia*. It no longer surprises me that the ancient term for "the good life" was made by combining the prefix *eu-*, for "good," with the word *demonia*, for "demons."

What follows is a history of science's demons, some imaginary and some real, some impossible and others less so, and through it a history of the universe *as we have come to know it*, filled with mystery and possibility.

1
Descartes's Evil Genius

Volumes have been written about a demon first imagined in 1641 by the French philosopher René Descartes. Descartes described a creature, using the Latin term malignum genium, *who could remove the world in front of us and provide us with an alternative reality. By reminding us that all we perceive might be a lie, or the bad joke of an intelligence higher than ours, Descartes's creation planted the seed of doubt firmly in our minds. Against it, we have developed tools and techniques to better understand the imperfection of our senses and to exploit the power of pure reason. Yet we also turn these insights against ourselves, when we use them to find ever more perfect ways of imitating reality to fool others.*

Descartes's demon is not alone—he inspired many other thinkers to search for others like him. Today he is the darling of magicians, advertisement executives, spin doctors, and the entertainment and media industry. He is presently most feared for his ability to spread fake news and create deepfakes. Descartes's creation saddled us with the responsibility of making reason our master. It challenged us to advance knowledge by questioning our dearest assumptions, everything and everyone, including social, religious, and political authorities.

Thanks to him, skepticism and doubt continue to be the most powerful tools of scientific discovery.

"I always had an extreme desire to learn to distinguish the true from the false," wrote the philosopher René Descartes in his *Discourse on the Method*.[1] The year was 1637. The task was not easy. Folk practices for judging reality and for distinguishing truth from superstition seemed perilously faulty, unsystematic, unreliable, and subject to error. Could they be replaced by a better, maybe even *entirely rational* method? Descartes worked hard to find ways to deliver us from a world of maddening illusions—to discover a method for distinguishing fact from fiction and separating sanity from insanity. He articulated the best techniques he could think of. The process he came up with earned him the distinction of being the founder of modern philosophy and the father of rationalism. Scholars usually associate the beginning of the Age of Reason with his work, which is widely considered to mark the beginning of the natural sciences and to have inspired the materialist and secular philosophies that would characterize the following centuries.

The Cartesian era was one of brain over brawn. It proved to be exciting and wonderfully creative. Descartes gave his readers detailed instructions for how to fend off the demon who would carry his name. His demon's weakness resided in his inability to fiddle with certain basic facts, beginning with the inescapable reality that if one thinks then one must be. "I think therefore I am," stressed the philosopher, writing for the first time the original Latin phrase *cogito ergo sum* that to this day is widely recognized.

Descartes's solution was to focus on the most certain truths he could find in an otherwise confusing and tricky world. "Two and three added together are five," he wrote excitedly in his *First Meditation*. "A square has no more than four sides," he continued.[2] From simple examples like these, he sought to build up a method for determining all other ineluctable "transparent" truths. These, he hoped, could serve as a firm foundation for understanding the world in a thoroughly rational manner. This analysis seemed to be going swimmingly, until the demon named after Descartes revealed just how hard it would be to draw a firm line between the real and the unreal. His tricks were sometimes so perfect that victims would remain unaware that what they perceived as ineluctable truth was fraught.

Descartes's demon could capture you by throwing a cloak over your head and, like a talented kidnapper, severing you from reality, before tossing you an alternative one. He could intercept all inputs leading to your brain, hijacking the source of your sensory impressions. In the nightmare scenario known as "the brain in the vat" thought experiment, which was widely used by neurophysiologists and philosophers to illustrate how thinking works, a thinking organ could be fooled into believing in an inexistent reality. Descartes's writings fueled many later speculations about what an isolated brain might think, feel, and sense. What might go on in a brain that is cut away and separated from its body and senses? Are we justified in fearing an evil being who might cultivate our worthy organ in a soup bowl? Philosophers today often explain the power of Descartes's manipulating demon by asking these questions. They wonder if a prankster or scientist could manipulate the input to the pink-grayish lump of neurons floating in some greenish-blue liquid and conceal from the victim the terrifying reality of their truly lamentable blobby condition.

After it was first invoked, Descartes's savvy illusionist became a symbol for the ultimate trickster: a trafficker between fiction and nonfiction, much like an ideal magician who can operate without smoke and mirrors. As the master of trompe l'oeil, he represents the promises and perils of virtual reality. Because of him, we have become increasingly aware that we can only know the world as if through a glass darkly. Descartes's demon offers the promises of virtual reality minus the headset or screen. More than the stuff of nightmares, this professional hoodwinker gives us daymares. He is a threat—and an inspiration—to scientists, artists, engineers, and con men. At any moment, the heavens, a landscape, or a seascape could become his simulacrum, his favorite playground. In Descartes's conceptualization, the elements constituting our universe might be nothing but props in a demon's fabulous show. Nature might simply be the most wonderful spectacle that could ever be, one practically indistinguishable from nonspectacle. How could anyone, even astronomers trying to uncover the secrets of our universe, resist being hypnotized by the beauty of the starry skies? Descartes fanned fears that perhaps we are all living in an immense production courtesy of our defective senses.

But the powers of Descartes's demon were found to be limited. He could reach only as far in as the retina. He could gaslight us only through our senses and did not mess with our brains directly. Faced with the power of our minds, his strength dwindled. His theatrical skills were indeed deep, but his knowledge of neurophysiology and chemistry was shallow. It would take years for scientists to conceive of another demon who could manipulate atoms, another one who could mess with photons, yet another who could control our bodies, and an even craftier one who could implant itself directly in our brains.

Descartes's demon was central to the foundation of *cerebral* personhood. As attention shifted to the power of our brains, our bodies were devalued as machines in its service. Descartes famously argued in his *Principles of Philosophy* (1644) that he could "not recognize any difference between artefacts and natural bodies."[3] In the Cartesian conception of the universe, sometimes referred to as the "Cartesian Theater," the universe was divided into mind and matter. This dualistic conceptualization—and the demon that led to it—arose in connection with the development of modern media, starting with early theater and print.

DON QUIXOTE'S WINDMILLS AND OTHER DEMONS

A well-known character in Descartes's time who was particularly confused about truth and falsehood was the famous Don Quixote de la Mancha. Scholars are quite certain that Descartes must have read *Don Quixote* by Miguel de Cervantes. It is likely that his exposure to the novel, alongside other works of his era exploring similar themes, played a role in fueling his obsession with drawing out and systematizing the laws of reason.[4] Descartes seemed to be concerned by how easily an unreal world had supplanted the hero's sense of reality. In the novel, the old geezer went off traveling on the plains of La Mancha with the confidence of a handsome young knight. Mounted on the feeble donkey Rocinante, he thought he was riding a beautiful stallion. Flirting with the rustic Aldonza, he was convinced that he was conquering the sweet princess Dulcinea. Charging violently at windmills, he was fighting giants. Taken into custody by well-meaning gentlemen, he was convinced that he was being kidnapped by demons. All the while, he and his faithful squire Sancho Panza famously disagreed about donkeys and horses, damsels and ladies, windmills and giants, gentlemen and demons.

Descartes warned against the dangers of reading novels such as those that fascinated Quixote and a growing public. He was specifically concerned about the "most accurate histories" of valiant knights. If Don Quixote was knocked off his rocker by reading too many chivalric histories, other readers could suffer a similar fate by following in his steps. "Those who regulate their conduct by examples drawn from these works," Descartes warned, "are liable to fall into the excesses of the knights-errant in our tales of chivalry, and conceive plans beyond their powers." "Fables" were just as dangerous, he cautioned, since they could also warp the sense of reality of gullible readers. They "make us imagine many events as possible when they are not," he explained.[5]

Miguel de Cervantes might have been responsible for these and other crimes. He took his readers along on a doubly perverse adventure. By writing a best-seller readers could barely put down, he got them hooked on a work of fiction about a man who had been permanently damaged by becoming hooked himself on fictions.

"They are demons that have taken fantastic shapes," exclaimed Quixote, caged and confused, facing a clear upset during his chivalric adventures.[6] The valiant knight was carted away from the vast expanses of La Mancha by a group of gentlemen who thought he was unhinged and perhaps a tad dangerous. His delicate mind told him that his captors were "all demons." But his general assessment of the dire situation was not all that clear. Why were they traveling so slowly in a rickety, uncomfortable, ox-driven cart? Storied accounts of such sequestrations tended to feature fancier means of transportation. Quixote expressed his surprise to Sancho. Why were they not whisked "away through the air with marvelous swiftness, enveloped in a dark thick cloud, or on a chariot of fire, or it may be on some hippogriff or other beast of the kind"?[7]

And who were those men—or from the perspective of the lanky master, those demons—who now seemed to control their fates? Cervantes's endearing Quixote sees them, although his faithful squire does not. A proclivity to see demons serves as a litmus test, a kind of barometer, through which readers can gauge the mental fitness of the two friends. In the novel, Sancho was a salutary counterpart to overzealous Torquemadas—believers who found evidence of the angelic or the demonic in every nook and crack, in every unpredictable event, and who felt justified in glorifying or violently persecuting every minor insinuation of otherworldly presences. Sancho is a sort of proto-scientist who brushed superstitions aside pragmatically; a no-nonsense commoner, his congenital simplicity led him to be more *in touch* with reality than his noble master, who confronted the world, not through his senses and fingers, but only indirectly by poking his lance (and other protuberances) where he should not.

The faithful servant perceived the men taking them away as regular flesh-and-blood mortals. Don Quixote thought otherwise. Could the discrepancy between them be resolved? Quixote urged Sancho to corroborate his thesis *experimentally*:

> And if you want to see this truth, touch them and feel them, and you will see how they do not have bodies but are air and do not consist of anything but appearances.[8]

In the novel, the act of testing Quixote's demon hypothesis by touching his captors *did not change the belief structure of either man*. Sancho responded to his master by saying that he had already touched them and smelled them, and that they were burly men who smelled of sweet amber perfume. If they were demons, they would smell of sulfur, Quixote insisted, and if these particular beings did not, it was only because of some clever ruse; perhaps they had disguised themselves with perfume.

Cervantes's story is thus very much the opposite of the biblical story in the Gospel of John, where the Apostle Thomas had his doubts about the resurrection dispelled when he touched Christ's wound. To Quixote, touching was no longer any good. The experiment made no difference. It only confirmed what Sancho already knew and what Quixote already believed. The Don was not brought any closer to the squire's views. There was no epistemic resolution. Cervantes offered readers no possible means through which his characters could be freed from their illusions. The story unravels as a comic tragedy that in the end leads readers to question their own sense of reality and even their own existence. By offering them a story within a story, he asked readers to consider whether perhaps we are all dupes of our own minds caught in an infinite hall of mirrors. Through the realistic dialogue between someone who saw demons and someone who did not, Cervantes invited us to laugh with devilish glee at every turn of the page as we question our own wits. Might we just be characters in a comic novel written by somebody else?

Shakespeare became fascinated by these same questions. He portrayed Hamlet, who was bewildered by the vision of a ghost appearing to be his deceased father, as someone who read too much. The habit had put his mental health in danger. Hamlet also read too much into the world around him, including the clouds in the sky. "Do you see yonder clowd in the shape of a camell?" he asked Polonius, who politely assented: "'Tis like a camell, indeed." Hamlet quickly changed his mind. "Now me thinkes it's like a weasel," and Polonius agreed once more: "'Tis back't like a weasel." "Or like a whale?" "Very like a whale," responded his obsequious friend. Like the vision of the ghost Hamlet could not shake from his mind, other things he saw confounded him as well. How could he rein in his imagination and regain clarity? The corroborations offered by his friends were not helpful. Groupthink led all of these young men astray.[9]

During Elizabethan times, Shakespeare and other dramatists honed their writing skills to fool us into taking in their theatrical creations as real. In the new brick-and-mortar venues such as London's Globe Theater, the stage became a stable home for innumerable fantastical creatures. Demons strutted brightly on the stage. Pyrotechnics, trapdoors, moving sets, and other stagecraft innovations portrayed demons, ghosts, witches, and other fantastic creatures in ever more credible ways. Well-oiled automata were designed to spit out real fire, arrive on the scene announced by thunderous tremors, and disappear into thin air trailed only by pungent smoke.[10] In *The Tempest*, Shakespeare's masterpiece written a few years after *Macbeth*, theatergoers witnessed devils descending upon unsuspecting voyagers during a violent sea storm. "Hell is empty, And all the Diuels are heere," exclaimed the King's son after his ship was struck by lightning.[11] With this phrase, Shakespeare set the stage for yet another chilling depiction of pandemonium on earth.

Demons in early modern theater were not simply *symbolic* fictions used to tell stories about the human condition more generally, nor were they considered to be representations of real demons. Playwrights and writers such as Cervantes and Shakespeare increasingly invoked these creatures to explore the porous boundary between the real and the unreal, the reasonable and the unreasonable, and the credible and the incredible, as well as to poke the bear of our imagination.

Truth and illusion became even harder to untangle with the rise of theatrical and literary technologies. Fiction writers confronted some of the same questions that would later concern philosophers and scientists: Should we trust the testimony of our senses? What should we do when confronted by something that appears to be real but is so unusual that it seems incredible? What is the difference between reality and simulation, between life and theater? Can the latter be made so perfect that it will match the former? Complicated plots turned on slippages between life, imagination, and simulation. Macbeth famously reflected that "Life's but a walking Shadow, a poore Player, that struts and frets his hour upon the Stage."[12] A concern with exploring the limits of the trustworthiness of our senses and the reasonableness of our minds marked the arts as much as philosophy. These genres were all ideal petri dishes for exploring these questions, offering useful lessons about who we should trust, how we should go on with our lives, and how to understand the universe of possibilities before us in all of their complexity.

The works created under the patronage of King James I of England, including Shakespeare's, were marked by the uncomfortable fact that the expert on witches, demons, and spirits sat on the throne. King James's *Daemonologie* (1597) contained descriptions of clever creatures who were to be feared. The gullibility of commoners and their proclivity to attribute supernatural causes to almost anything could lead to exaggerated views about demons. Ignorance could sometimes lead to more fear than was justified. James was horrified by how easily townsfolk fell for fraudsters who claimed to speak for or to have come from the other world. Believing that many accounts of demons were mere fabrications, he and his inquisitors disputed them by carefully analyzing individual cases of apparitions and possession. Only those vouchsafed by the ecclesiastical and regal authorities were considered to be legitimate; all the other ones were to be excluded as folk superstitions or lies. Demonologists detailed the ways and wiles of real demons and gave readers advice for separating these from fraudulent impostors. Inquisitors and theologians routinely performed rigorous tests and interviews to determine which accounts of miracles and saints, which sightings of strange beings, should be considered veridical. Establishing their precise nature, as well as that of angels and archangels, was a central concern of the theological tradition from which European science emerged. A generalized belief in the actual existence of most demons started to be associated not only with ignorance but with backwardness, superstition, irrationality, and even mental instability.

King James I described the devil as powerful, but only because the devil was worldly and wise. The king explained that, in contrast to the Almighty, who had awesome powers, the devil had to *learn* everything through education. "Taught by continual experience ever since the creation," James wrote, "he judges by the likelihood of things to come *according to the like that has passed before*, and the natural causes in respect of the vicissitude of all things worldly."[13] The monarch's description of the devil as worldly proved fascinating far beyond the confines of religion. Tracking and understanding how the devil acted could be immensely useful for anyone interested in developing a sound theory of knowledge. With texts such as James's, the faithful learned that the devil was a lowly practical operator who did not possess the foresight and powers of his highly superior archrival. For better or for worse, we mortals were much like James's devil: we did not know everything automatically, nor were we omniscient. Like him, we too had to work hard. Like him, we learned slowly, in fits and starts, from experience. For those interested in trying to learn more, from ambitious leaders like King James I himself to the budding philosophers of his era, the techniques of the devil and his underlings proved to be illuminating. Understanding his *modus operandi* was and would continue to be useful.

As playwrights and writers populated fictional texts with imaginary creatures, their purpose became increasingly distinct from that of the demon experts. These practices became secular over time, with difficulty, and only partially. Demonologists focused on understanding real demons and their actions; playwrights and novelists, in contrast, investigated the complicated ways in which people made judgments about truth and falsehood in a wider variety of cases and contexts. Their craft consisted in manipulating spectators' imagination to suspend disbelief. Demonologists, instead, were concerned with untangling the real from the imaginary.

FABLES, TALES OF CHIVALRY, AND A *MALIGNUM GENIUM*

In Descartes's time as now, distinguishing the real from the fake was politically and philosophically pertinent. Although a belief in the existence of most real demons started to fall into ill repute in elite circles, there was one particular demon that Descartes and his followers were unable to entirely disprove, no matter how hard they tried. It would be known as Descartes's demon.[14]

The years around the publication of Descartes's *Discourse* were marked by the violent prosecution of witches and necromancers. Although most of those accused of these crimes proclaimed their innocence, a few willingly confessed to commerce with the devil. They too had to be put to the test. In a famous case, a peasant girl with a penchant for disappearing at witching hours claimed to have been acquiring supernatural healing skills during her outings. After strict interrogation, the ecclesiastic authorities determined that she was lying. No matter how loudly some of the accused advertised their connections to their evil master, in the cases involving fraud the impersonators were pardoned and spared from torture or execution.

Simply mentioning beings with supernatural powers was enough to get Descartes in serious trouble with the authorities. The last paragraph of his *First Meditation*, a document widely acknowledged as having opened the door to modern rational thought, was singled out by two Protestant theologians, Jacobus Revius and Jacques Triglandius. In Descartes's original Latin text, the paragraph described a *malignum genium*, which can be literally translated as an "evil genius."[15] Was this someone who could rival the powers of God? If so, a mere mention of that suspicious personage could count as heresy. And what right did Descartes have to discuss topics that traditionally concerned theologians? These questions led the philosopher into a detailed excursus on what he had really meant when he mentioned an "evil genius." The "Genie," a word connected to the Arabic term *Jinn*, shared many characteristics with a demon. The frequent attribution of genius and intelligence applied to both, and plots involving them frequently turned on the theme of outsmarting an opponent.

In the contentious paragraph, Descartes imagined an "arch-deceiver" who could be fooling with our sense of reality: "I shall then suppose, not that God who is supremely good and the fountain of truth, but some evil genius [*genium aliquem malignum*] not less powerful than deceitful [*summe potens & callidus*], has employed his whole energies in deceiving me." The "arch-deceiver" had the ability to alter our sense of the external world by supplanting it with another reality: "I shall consider that the heavens, the earth, colours, figures, sound, and all other external things are nought but the illusions and dreams of which this genius has availed himself in order to lay traps for my credulity." He could take control of all our sensations, even when we looked at our own bodies, affecting our perception of our very own flesh and blood: "I shall consider myself as having no hands, no eyes, no flesh, no blood, nor any senses, yet falsely believing myself to possess all these things."[16] In Descartes's descriptions, this demon was capable of usurping the role of any dramaturge previously running the show.

The terrifying line of attack of such evil genius was not infallible. Descartes continued his *Meditations* by teaching us how we could wake up from this "slumber" and escape our "captivity." It would be hard work—"laborious," no doubt—but necessary.

With these words, Descartes invited a growing number of thinkers to question the role of sensations vis-à-vis ratiocination in their understanding of reality. Numerous philosophers and scientists would use his example to investigate the relation between body and soul—and later between brain and mind—to explore the possibility of virtual reality.

Descartes's enemies in Leiden quickly accused him publicly of heresy and blasphemy, putting his reputation—and his life—in danger. He could be imprisoned, driven away from the Netherlands to God-knows-where, or executed. Could it be that his words had implied that God could be such an evil deceiver?

The philosopher clarified his intentions in an apologetic letter sent to the theological faculty at the university.[17] His accusers had argued that the "evil genius" described by Descartes could be interpreted as all-powerful, and therefore as equal to God. Such an equation would be heretical. No, responded Descartes in his defense. What he described in the first meditation was something more akin to a demon. Descartes turned the tables on his attackers and accused the two theologians of slandering him.

OMNIA DAEMONIA

The crux of the accusations centered on Descartes's use of the Latin term *summe potens*, which means "all powerful." In his response to the curators of the University of Leiden, the philosopher argued that this phrase did not attribute to such a deceiver a power equal to God's. What Descartes had in mind, he clarified, was far from it. It was actually more akin to "all demons, all idols or all pagan powers [*omnia daemonia, omnia idola, omnia Gentilium numina*]" who had comparably modest abilities. There was neither heresy nor blasphemy in that claim, he protested: "But I will merely say that since the context demanded the supposition of an extremely powerful deceiver, I distinguished the good God from the evil genius, and thought that if *per impossibile* there were such an extremely powerful deceiver, it would not be the good God . . . and could only be regarded as some malicious genius." By claiming that this being was not at all God-like, his accusers—and not him—could be found guilty of heresy for the sole reason of having elevated this example to such high status in their wrongful interpretation: "Following that line of argument they must hold that all demons, idols or pagan powers are the true God or gods, because the description of any one of them will contain some attribute that in reality belongs only to God."[18] These clarifications were necessary to avoid further confusion and to protect the philosopher from accusations of calumny or heresy. In their wake, Descartes's example would become widely known and widely translated as "demon." Descartes was convincing in his explanation that the creature he described was like a pagan demon with limited powers. It was similar to those that predated the Christian era.

The common association of demons with the devil, understood as a rival to God, was a distinctly Christian practice. The association of the demonic with the sinful was a relatively contained and short episode within the much longer history of demons. Demons were not systematically considered "fallen Angels" who had rebelled against God until the end of the New Testament period, around the first century BC. The figure of the devil as a malfeasant, the *maleficus maximus*, the Prince of Darkness, the manipulative kingpin of innumerable minions and servants, the universe's top villain, emerged only in reaction to the pantheism of Greek, pagan, and folk traditions based on the actions of various creatures, some of whom were not evil at all. Pagan and folk *daemonia* and other *exotica*, unlike Christian ones, were often quite benevolent. In fairy tales and myths, they adopted sundry and malleable roles in which they frequently traveled between the lurid and the pious, the immoral and the just, and the imaginary and the concrete. Operating from a demimondaine territory between heaven and earth, they inspired laughter as much as fear. Less powerful than the devil, they were widely considered his derivative evildoers, servile minions charged with doing his dirty work by tempting potential victims. Although in early modern times a belief in them was increasingly associated with backwardness and idolatry, the demons vouchsafed by the Christian Church often inherited some of the features and abilities of their older kin. In science, they continued to hark back to their ancient lineage by being both good and bad.

DESCARTES'S SOLUTION

In his *Second Meditation*, Descartes continued his discussion of "a deceiver of supreme power and cunning who is deliberately and constantly deceiving me."[19] Such a deceiver would fail at one thing. It could never keep its victims from knowing the essential truth of their being: *cogito ergo sum*, or "I think therefore I am." Since then, this phrase has become widely used to validate the power of our minds and remains central to our understanding of human subjectivity to this day.

> But there is a deceiver of supreme power and cunning who is deliberately and constantly deceiving me. In that case I too undoubtedly exist if he is deceiving me; and let him deceive me as much as he can, he will never bring it about that I am nothing so long as I think that I think I am something. So after considering everything very thoroughly, I must finally conclude that the proposition "I am, I exist" is necessarily true whenever it is put forward by me or conceived in my mind.[20]

Our ability to think critically, to doubt, and to question the reality before us, could circumvent a demon's manipulating ruses. Along with the truth of our existence came the truth of God's existence and a handful of other truths; for instance, the "idea of a triangle includes the equality of its three angles to two right angles, or the idea of a sphere includes the equidistance from the centre of all the points on the surface."²¹ None of these truths could be touched by the manipulating trickster Descartes described.

Sense-manipulating demons were to be feared most in light of Descartes's conception of the world as *a kind of theater to be taken in by human spectators*. "So far I have been a spectator in this theater which is the world, but I am now about to mount the stage," he wrote. Descartes made a commitment to live a public life, to become a player in the *theatrum mundi*, and to offer to the world a new philosophical understanding of the universe as something like a very large theatrical production.

By the time Descartes was in his forties, he was systematically investigating the possibility that the world around us might be an illusion—all of it, including the everyday and the mundane. This line of thought led him to another: he wondered what the world might be like for us if we were completely cut off from our bodies and our senses. "I shall consider myself as having no hands, no eyes, no flesh, no blood, nor any senses," he continued.²² What would it feel like to be not only deaf and blind, and unable to access the other senses of taste, smell, or touch?

THE WORLD AS THEATER

As *scientia* became the preferred method for distinguishing between the true and the false, more and more cases of demonic apparitions were disproven, and expertise in the manners and customs of demons started shifting away from theology. New musings about disconnected brains or machine-brain chimeras shed light on a more mundane line of questions: How was our sense of reality affected by the media we consumed? Was it harmful to read stories, such as those of medieval errant knights? How could they mess with our minds and sense of reality? What were the limits to mind manipulation or indoctrination? These same questions continue to concern us. What powers do advertisements and propaganda have, and how do they affect our thoughts? What are the risks of focusing too much on the pages or screens before us? The discipline of psychology became obsessed with answering those questions.

The need to understand technologies designed to close in on our minds and make independent thought difficult or downright impossible became more pertinent than ever. The powerful phantasmagoric tricks of Descartes's evil genius led more and more people to be convinced of the wisdom of not taking things at face value. If what reaches our senses was only phenomena that might be completely shielding the noumena behind it, then perhaps absolute reality was ultimately impossible to grasp. The philosopher's defense of skepticism as a cure for superstition and erroneous beliefs would characterize Enlightenment thought for years to come. It is a useful precedent for understanding why an extreme distrust of things as simple as things (such as the things-in-themselves considered by the philosopher Immanuel Kant) became central to philosophy. Given the feats of legerdemain and skullduggery Descartes's demon was known to be capable of, such caution was—and continues to be—hardly paranoid.

2

Laplace's Intelligence

Laplace's demon was born in revolutionary times. Her powers were formidable: by knowing the exact atomic configuration of the present state of nature and submitting the data to analysis, she could know it all—including all that came before and all that was about to happen next. This creature was thought to possess a perfect record of all of history. Her existence was tightly coupled with the idea of a "universal formula" that described the universe in its entirety.

The mathematician Pierre-Simon Laplace at first called this demon "une intelligence," but in most translations of Laplace's work the feminine pronoun was dropped and the term "intelligence" was translated as "intellect." Laplace's intelligence gained the special designation of "demon" only late in life, in the 1920s, but once she did, the moniker stuck.

Laplace's idea fueled the dream that by combining big data with simple mechanical rules we might soon solve all the problems of this world. Perhaps past was prologue. Perhaps free will did not exist.

Charles Babbage was inspired by Laplace's idea to build one of the world's first computers. Those close to him, including Ada Lovelace and Charles Darwin, started new speculations about what these machines told us about the universe. Lovelace wondered if instruments possessing intelligence could ever be built. Darwin asked if perhaps evolution itself, and the origin of the species, were in fact a mechanical, rather than God-driven, process. Laplace's expert calculator motivated the construction of ever more powerful calculating machines, first using gears and levers, then vacuum tubes, then semiconductors and microchips. When individual supercomputers were not deemed powerful enough, scientists networked them together to build the large-scale data farms of today.

This demon's powers of omniscience, foresight, and prediction were highly coveted. At the same time, they elicited fears that we might calculate everything and ourselves to death without emerging from all the data-mining any wiser in the end.

At the venerable Académie française in Paris, about a decade before the French Revolution exploded, the statistician Pierre-Simon Laplace was hard at work demolishing long-held myths and superstitions, felling them one by one. The strategy behind the method he developed was simple yet brilliant: attach statistical weights to events in order to determine their chances of occurring and their frequency. In contrast to Descartes, who focused on finding simple truths in cases involving one or two calculations, Laplace found counterintuitive lessons hiding in giant data sets by using highly complex mathematical analysis. Thanks to him, statistics emerged as a powerful technique for distinguishing the true from the false.

The mathematician expressed grand hopes for his theory of probability that went far beyond math and astronomy and impinged on religion, superstition, and reason. His work served to further a secular view of the universe as functioning mechanically and independently of the intervention of religious beings, including God.

Laplace's contributions stemmed in part from his willingness to think about the abilities of a curious know-it-all creature. He first mentioned this being in an article on the theory of calculus that he published in 1773. The description was buried in the middle of the text. "If we imagine an intelligence [*une intelligence*] who, for a given instant, embraces all the relationships of the beings of this universe," wrote Laplace, "she [*elle*] could determine for any time taken in the past or in the future the respective position, the movements, and generally the attachments [*affectations*] of all these beings."[1] Another reference to this "intelligence" appeared later in the introduction to his groundbreaking book titled *Philosophical Essay on Probabilities* (1814). In that book, statistics left the comfortable nest of specialized mathematics and entered the universe of ideas, politics, and culture. It helped change how ordinary citizens looked at the world, including their savings, their marriages, their life spans, and their opportunities or the lack thereof. By studying human behavior statistically and in aggregate, Laplace inaugurated the field of "social physics," which led to the innumerable quantitative studies of human societies that would later characterize the discipline of quantitative sociology.

Starting off from initial conditions, Laplace's hypothetical being could calculate the movement of each and every particle in our universe throughout space and time. All she needed was a sufficiently large brain and knowledge of basic physics:

> An intellect which at any given moment knew all of the forces that animate nature and the mutual positions of the beings that compose it, if this intellect were vast enough to submit the data to analysis, could condense into a single formula the movement of the greatest bodies of the universe and that of the lightest atom; for such an intellect nothing could be uncertain and the future just like the past would be present before its eyes.[2]

The importance of this creature for science cannot be underestimated. "Canonical is the word for his image of an infinite intelligence that recalls the past and predicts the future state of all things from knowledge of the position and motion of every particle at any given moment," stated one of Laplace's biographers.[3] A philosopher explained that "this intelligence has been called 'Laplace's demon,' and it has become the patron saint of determinism."[4] "This statement, or part of it," claimed a physicist in the 1970s, referring to the famous lines written by Laplace, "has often been quoted as the gospel of the deterministic view of the world."[5] "Almost every scholar and philosopher of the first half of the nineteenth century," claimed a renowned economist during those same years, was "fascinated by the spectacular successes of the science of mechanics in astronomy and accepted Laplace's famous apotheosis of mechanics as the evangel of ultimate scientific knowledge."[6] From the time she was first conjured, Laplace's demon lorded over the universe, guaranteeing it would tick predictably according to fixed laws and embodying the very "essence of causality."[7] She faced strong competition at the end of the nineteenth century when scientists became enamored by a very different being from Laplace's, one who, although quite tiny and marginally bright, had other qualities. Nevertheless, despite her death having been announced a number of times, Laplace's demon continues to enjoy a long life.

Would science advance to the point that one day we might know it all? That fantasy—or nightmare—was tightly coupled with investigations of who this demon really was. Writers after Laplace would refer to his demon— the ultimate soothsayer—in various ways. In German, she was almost always called a *Geist*, which can be translated as "spirit" or "ghost." Elsewhere, she was referred to as an "intellect," a "mind," a "prophet," and a "superhuman" or "superman." In the twentieth century, she gained the designation of "demon," but would never fit entirely under that single label, which represented an ideal much larger than Laplace's creation. Sometimes she was even referred to as the "Laplacean God."[8] Other times she was just called a "calculator" or a "supercomputer."

When Laplace wrote "know all," he meant it literally. Just as astronomers could predict the path of a planet at a future moment, he explained, it would be possible to follow the lightest "molecule of air or vapor" on earth every step of its way.[9] Laplace's intelligence at first motivated scientists to produce more precise almanacs and build ever more powerful forecasting and calculating devices. Mechanization, determinism, inevitability, and the sense that humankind might be going down a path that was uncontrollable and unpredictable was frightening. To this day, we can contribute to this stability, or we can fight it. When we sit in front of our computers setting off algorithms every time we note our plans in our calendar to make sure we meet our commitments and deadlines, when we pay our insurance and mortgage bills regularly, and when we act predictably at the office or in the boudoir, we are helping the cause of Laplace's demon.

Who could possess such abilities? Laplace's revolutionary idea consisted in thinking that mathematicians might. Traditionally, it was thought that only higher spirits could gain such intellectual powers. The Enlightenment philosopher John Locke had made this clear in his *Of the Conduct of the Understanding* (1706). "We may imagine a vast and almost infinite advantage that angels and separate spirits may have over us," he wrote, who "having perfect and exact views of all finite beings that come under their consideration, can, as it were, in the twinkling of an eye, collect together all their scattered and almost boundless relations."[10] Locke had started to think about what it might take to have a human "mind so furnished." It was "an extravagant conjecture" to think of such possibilities.[11] For Laplace, there was nothing extravagant about it.

By using statistics, Laplace also solved some of the problems that had been posed earlier by the philosopher David Hume. Hume had tried to evaluate the common technique for judging truths by weighing their likelihood against our previous experience. When we presupposed a certain regularity and stability in our lives and in the universe, were our assumptions scientifically justified? Our assessment that something might be unlikely might only be based on the fact that, well, *so far* it had been unlikely. Aware of the circularity behind this mode of reasoning, Hume thought of a way of escaping it. "All inferences from experience," he wrote, "suppose, as their foundation, that the future will resemble the past."[12] To evaluate which inferences were justified would require investigating the actual regularity of natural phenomena, taking note of all those instances when the future did indeed resemble the past and those when it did not. Hume lamented that "the *avidum genus auricularum*, the gazing populace, receive greedily, without examination, whatever sooths superstition, and promotes wonder."[13] The Scottish philosopher sought a better method for figuring out which miracles were real and which ones were not. The philosopher famously stressed that assessing the credibility of a particular claim required not only thinking about the trustworthiness of the person making it (whether the person was generally honest or not), but also looking at the likelihood of the event in question. In addition to inviting us to ask *who* we should trust, Hume asked us to think about *what* we should trust. "Man," Hume concluded optimistically, had the ability to "raise up to himself imaginary enemies, the demons of his fancy, who haunt him with superstitious terrors."[14] Hume was ambitious in proposing such a goal, yet it was Laplace who figured out the math that might achieve it.

ENTER LAPLACE

The son of a syndic of the parish and a family of well-to-do farmers from the provinces, Laplace displayed such talent in mathematics that few would compete with him. Napoleon, as a young army cadet, was examined by Laplace at the École militaire. The soon-to-be emperor would never forget his teacher. In five hefty volumes, Laplace laid out the principles of *celestial mechanics*—producing a mathematical treatise with no room for God, angels, or demons; it became the most comprehensive account of the universe known to humankind. In 1788, one year before the French Revolution broke out, he proposed a model of the universe that was highly stable and predictable. "This stability in the system of the world, which assures its duration, is one of the most notable among all phenomena," he wrote.[15] Despite the social, political, and cultural bouleversements quickly spreading throughout Europe, Laplace's heavenly oeuvre was a testament to the general stability of all things, including the universe.

Some of Laplace's topics concerned widely revered omens and miracles. He applauded the work of a clever astronomer named Edmond Halley, who found clear patterns in the arrival of a comet that would be named after him. During the year 1456, terror had spread across Europe as the long tail of a comet seemed to indicate the wrath of the heavens taking their revenge on earth. After the fall of Constantine, the Roman Empire had been dangerously losing ground against the Turks, and the strange light in the sky appeared to be an omen signaling worse things to come. The comet's portentousness started to be tempered when it appeared again in 1531, and then once again in 1607 and in 1682. Halley noted a pattern. He predicted that the comet would appear again by the end of 1758, or the beginning of 1759 at the latest. After his estimates were revised by others, the timing of the comet's return was set to sometime in early April 1759. The event was eagerly anticipated in learned circles. When the comet arrived at the right time and place, it proved the soundness of the prediction method.

Laplace also examined a miraculous cure that drew hundreds of pilgrims to the site where it had occurred. In the year 1656, during the reign of Louis XIV, a little girl known as *la petite Perrier* suffered from a lacrimal fistula on her left eye. It was so pernicious that her nose and throat grew deformed. But when she touched a relic containing one of the thorns from the crown of Christ, she was instantly cured. Word of the miracle spread, crowds flocked to the Abbey of Port-Royal, and allegedly many more miracles started descending upon the faithful.

Laplace remained unconvinced by this miraculous event and considered an alternative explanation. It was probable that the monks from the Abbey of Port-Royal needed to defend the religious doctrine they were developing because it was under attack by the Jesuits. If only we understood somebody else's circumstances in their entirety, explained Laplace, not only would we understand the world in a better way, but we might even learn why others held beliefs that differed from ours. In other words, there were not only at least two sides to every story, but possibly many more.

In addition to reexamining such miracles and omens, Laplace offered advice to gambling addicts by showing why it was almost always more profitable to stay away from casinos and gambling dens. Then he went even further and showed that many aspects of life were skewed in ways similar to the odds of a national lottery. The weather's effects on crops, the ratio of male births to female births, and the length of life spans could all be explained statistically. Those in risky professions, such as mariners, had lower average life spans than farmers. Deaths at sea were surely tragic, but they were predictable too.

Probability theory could show us how to distinguish truth from quackery by identifying liars, no matter how charming they were. Instead of blindly following the authorities or running with the crowd on matters of crucial importance, his mathematical methods offered ways to temper the "influence of the opinion of those who the multitude considers most informed" by testing whether they were really in the know.[16] Laplace's was a work of—in every sense of the word—Enlightenment. By invoking our mathematical abilities, Laplace explained why we were superior to animals and how we could control irrational fears. Would the sun rise tomorrow? The chances that it would not could be calculated with precision to 1 in 1,826,214.[17]

Anyone entering a print shop, Laplace recounted, and seeing typeset characters on a table spelling the word "Constantinople" might normally assume that the arrangement of these words was not due to mere chance. People tend to make general inferences from particular cases. But why? Only because it would seem "extraordinary" if it were otherwise. This inference was natural, and Laplace agreed wholeheartedly with it. A textbook on logic published the previous century had already explained why these inferences were justified. "It would be foolish," warned the book, "to play twenty sols against ten millions of livres, or against a kingdom, on the condition that one could win it if a child randomly arranged letters from a printing press to compose all at once the first twenty verses of the Aeneid of Virgil."[18] Laplace took these investigations one step further, not only by teaching his readers how to calculate the exact probability odds for such conditions, but by explaining how erroneous conclusions were often drawn from those cases. Extraordinary events, merely by virtue of their rarity, often compelled people to look for causes that explained them. But probability showed that truly extraordinary events were sometimes just that: extraordinary. They were due to freak causes and therefore were not worthy of much consideration: freaks were just freaks. Yet they were not to be entirely discounted, since they *could* happen. So the more extraordinary something was, the more reasons one had to suspect that when explanations for it were adduced, they were most likely sheer fabrications or lies, no matter how plausible they seemed.

Inspired by Laplace and his associates, secularism spread, diminishing the power of the Church. The French astronomer François Arago, taking his cue from Laplace, lambasted Isaac Newton for leaving to "a powerful hand" some of the most important work required to keep the universe in order.

"Newton," he argued, "believed that a powerful hand should intervene every once in a while to fix the disorder" of the universe.[19] A rumor soon spread among the scientific elites that when Napoleon once asked Laplace about the place of God in his universe, the mathematician responded: "Sir, I have no need of that hypothesis."

Laplace's work quickly became admired across Europe, symbolizing in the eyes of many a progressive scientific spirit. In the eyes of his enemies, however, he represented a dangerous philosophical materialism that they associated with the secularism and crimes of the French Revolution, with its political and moral debauchery. It was up to men like Laplace to show young cadets their way into the Age of Enlightenment. The challenge that lay in front of him was to educate young men to become modern citizens—to teach them how to think as forward-looking engineers rather than as backward-looking parishioners. An education in science and math would be the basis of this new form of secular reasonableness that left superstition, religion, and blind allegiance to the authority of church and king behind.

A UNIVERSE OF ATOMS IN MOTION

Revolutionaries attacked the clergy, pillaging and confiscating church lands, yet it would take many more decades for everyday citizens to replace theological explanations with scientific ones. Nevertheless, even as the cult of reason advanced only by fits and spurts, Laplace labored on. He had personal, political, and professional reasons to prove that one thing caused the next. Even "free will" actions, he argued, were "caused" by something prior to them.

After the success of the Eighteenth Brumaire coup on November 9, 1799, Napoleon seized power. He named Laplace minister of the interior and put him in charge of rearranging nearly everything in France, except for public finances and the police. Laplace lasted only six weeks in his new position. After only a little more than a month in office, Napoleon realized that outside of the ivory tower that brilliant mind was in fact quite hopeless. Laplace, he lamented, could never "get a grasp on any question in its true significance; he sought subtleties everywhere, had only problematic ideas, and in short carries the spirit of the infinitesimal into administration."[20] The scientist would serve him better "as an ornament, though not an instrument, of state."[21] Napoleon would give Laplace a good salary and make him a count of the empire. Louis XVIII would elevate his peerage to that of marquis.

A few decades after Laplace described this powerful "intelligence," the modern era as we know it—the era of trains, telegraphs, photography, statistics—started to come into existence. New machines were built—mechanical calculators with rotating gears and drums—that offered the possibility of reducing the labor involved in carrying out some of the most complicated calculations outlined by Laplace. Blaise Pascal's calculator, known as "Pascaline," was quickly followed by "Leibniz's wheel," a calculating machine invented by Gottfried Leibniz. More and more scientists used these machines to crunch more and more numbers with greater ease. Bolstered by these advances, Laplace's speculations took off.

Some of Laplace's younger peers developed even more powerful mathematical techniques. Chief among them was the mathematician Joseph Fourier, who developed mathematical methods for predicting the future—or at least that part of it that had to do with the distribution of heat—using differential equations. At first blush, a mathematical analysis of heat distribution could seem like a minor, merely technical contribution. But it was much more than that. Fourier explained the relevance of his research in broader economic and cosmological terms: "It is easy to see how much this research interests the physical sciences and the civil economy, and what influence they may have on the progress of the arts, which require the employment and distribution of fire."[22] His research promised to advance knowledge of one of the most important elements of the national and global economy, as "there is no mathematical theory that is more related to the public economy than this, since it can serve to illuminate and perfect the use of the numerous arts which are based on the employment of the heat." It could give scientists insights across time of universal proportions. Anyone who "wishes to know the spectacle of the heavens for successive epochs separated by a great number of ages" could even use his equations to combat the limitations of our own limited life spans, as mathematical reasoning of this kind revealed "a faculty of human understanding that seems to be destined to supplement the brevity of life."[23]

BUILDING LAPLACE'S DEMON

The revolutionary potential of French mathematics was soon felt across the channel. Laplace's brainchild took England by storm. In 1831, the translation of Laplace's *Traité de mécanique céleste* appeared to public acclaim. A brilliant scientist and engineer set out to build an actual machine that could do nearly automatically what Laplace's mind could do with pen, paper, and printed tables. Like some other radical Brits of his generation, he was both seduced and alarmed by the revolutionary French politics and by the new math that had emerged from that environment.

His name was Charles Babbage, and he would be known as one of the inventors of the computer. How could one man begin to process all the seemingly innumerable marks on Earth that might contain hidden clues about our past and future? Babbage wasted no time asking the British government for funding to build his machine. Calling it the "difference engine," he armed it with mechanical buttons, gears, cranks, and levers to crunch numbers of this magnitude. Babbage's large "calculating engine" was left unfinished during his lifetime, but its blueprints inspired many others to build better, faster, and more powerful such devices.

What had Babbage done? Was his machine a kind of artificial intelligence that could rival, or perhaps even surpass, the intelligence of humans? How did it compare against a thinking being?

"I am myself astonished," wrote Babbage to his colleagues gathered at the general meeting of the Royal Academy of Science in Brussels in the spring of 1835, "at the power I have been enabled to give this machine; a year ago I should not have believed this result possible."[24] A few years later, when he described his machine once again, he cited Laplace directly. "Let us imagine a being, invested with such knowledge," he urged.[25] To introduce readers to his ideas, he included an extract of the relevant passage of Laplace's *Theorie analytique de probabilités* in the "Appendix Note C" to his *Ninth Bridgewater Treatise* (1837), an unauthorized response to eight treatises on natural theology published previously by other authors.[26]

Babbage's treatise detailed the significance of the new calculating engines. He explained that the superior "being" described by Laplace had to be powerful, but not infinitely so. It only needed to master one area of science, namely mathematics: "If man enjoyed a larger command over mathematical analysis, his knowledge of these motions would be more extensive; but a being possessed of unbounded knowledge of that science, could trace every the minutest consequence of that primary impulse." This "being," according to Babbage, would be racially superior to the Englishman: "Such a being, however far exalted above our race, would still be immeasurably below even our conception of infinite intelligence."[27]

In Babbage's optimistic view, nothing would ever be permanently lost in the universe, not even something that sank to the bottom of the ocean.

Moreover, troubled waters would serve as repositories and ocean waves as telling messengers:

> The ripple on the ocean's surface caused by a gentle breeze, or the still water which marks the more immediate track of a ponderous vessel gliding with scarcely expanded sails over its bosom, are equally indelible. The momentary waves raised by the passing breeze, apparently born but to die on the spot which saw their birth, leave behind them an endless progeny, which, reviving with diminished energy in other seas, visiting a thousand shores, reflected from each and perhaps again partially concentrated, will pursue their ceaseless course till ocean be itself annihilated.
>
> The track of every canoe, of every vessel which has yet disturbed the surface of the ocean, whether impelled by manual force or elemental power, remains for ever registered in the future movement of all succeeding particles which may occupy its place. The furrow which it left is, indeed, instantly filled up by the closing waters; but they draw after them other and larger portions of the surrounding element, and these again once moved, communicate motion to others in endless succession.[28]

Knowing everything was interesting for science, but what was *most* exciting about this possibility was its potential for uncovering hidden secrets and past crimes. "The air itself is one vast library, on whose pages are for ever written all that man has ever said or woman whispered." No deed could be hidden forever: "But if the air we breathe is the never-failing historian of the sentiments we have uttered, earth, air, and ocean, are the eternal witnesses of the acts we have done."[29] Babbage showed the world how to create an engine to crunch such potentially revelatory numbers.

In a second edition of his treatise published the following year, Babbage's claims about the powers of this being were even stronger. It could "distinctly foresee and might absolutely predict for any, even the remotest period of time, the circumstances and future history of every particle of that atmosphere."[30]

In May 1837, off the coast of Africa, the slave-trading vessel *Adalia*, with 409 slaves onboard, was captured by Captain R. Wauchope. During the chase, "she threw overboard upwards of 150 of the poor wretches who were on board, besides almost all her heavy stores."[31] Babbage read the fascinating account in the *Western Luminary*. What if the ship had eluded capture through such an inhuman strategy? Could a scientist such as he conceive of a way to bring it to justice? Babbage started to think that the work of Laplace could be of use. Even if the "Christian master" of such a slave vessel "might escape the limited justice at length assigned by civilized man to crimes whose profit had long gilded their atrocity," the crime would not escape scrutiny forever.[32] The moral conscience of Europe depended on the very possibility. Babbage hoped to prove mathematically that what goes around comes around.

NOT A THINKING BEING—YET

Laplace's translator, the mathematician Mary Somerville, was none other than the private tutor of the British aristocrat Ada Lovelace, daughter of Lord Byron. Lovelace was one of the first thinkers to fully appreciate the potential of computers. As part of her education, Somerville introduced Lovelace to Laplace's works. She also introduced her to Babbage, who was then immersed in building the world's most powerful computer. "We frequently went to see Mr Babbage while he was making his calculating machines," wrote Somerville, recalling her days tutoring Lord Byron's daughter.[33] Lovelace fully appreciated the creation of her new acquaintance. In a letter to Babbage, she told him about her excitement for the project. "I am working very hard for you; like the Devil in fact; (which perhaps I am)," she explained.[34]

Decades earlier, Lovelace had been left out of a trip to Geneva organized by her father, Lord Byron, with his friends and lovers. Lovelace's mother had recently separated from him, taking baby Ada away from him. She would always make sure her daughter stayed far away from her philandering father and his hard-partying coterie. A rainy day during that trip resulted in the production of two literary classics. Mary Godwin (later Shelley) started writing *Frankenstein; or, The Modern Prometheus* and John Polidori, Byron's friend and personal physician, produced the initial fragment of *The Vampyre: A Tale* after Ada's deadbeat father initiated a late-night fireside contest in writing ghost stories. Unlike those who went along with Lord Byron and engaged in his literary or libertine challenges, Lovelace would later write about computers.

The monster in *Frankenstein* was unambiguously defined by its creator as a demon. "The monster whom I had created," explained Victor Frankenstein in the novel, the "miserable daemon whom I had sent abroad into the world for my destruction" would haunt his maker for the rest of his life. Victor had obsessed about bringing the dead to life. Inspired by necromancers and occultists skilled in the art of summoning demons, he confessed in his recollections that "the raising of ghosts or devils was a promise liberally accorded by my favorite authors." Victor's creation recognized himself in those terms. "I considered Satan as the fitter emblem of my condition," he explained, "for often, like him, when I viewed the bliss of my protectors, the bitter gall of envy rose within me." When the monster taught himself to read, one of his favorite writers was Johann Wolfgang von Goethe, author of *Faust*.

Goethe's and Shelley's cautionary tales about the hubris of research run amok anticipated one of the century's most enduring themes, and in some respects these authors were ahead of their time. In other respects, however, especially when compared to those actually involved in the cutting-age scientific research of their time, they were far behind. Selling one's soul to gain total knowledge (as in *Faust*) and using electricity to give life to hybrids made up from pieces of cadavers (as in *Frankenstein*) were obviously morally perilous. The dangers of the machine-based economy of the Industrial Revolution were also widely recognized. Many thinkers attributed demonic qualities to steam engines. In the *Communist Manifesto* (1848), the political economist Karl Marx and his collaborator Friedrich Engels compared the capitalist in charge of these new industrial machines to a "sorcerer [*Hexenmeister*] who is no longer able to control the powers of the underworld whom he has called up by his spells."[35] Later, in *Capital*, Marx described the new industrial machines that altered old labor relations as being driven by a "demonic power [*dämonische Kraft*]."[36]

Thomas Carlyle, the great historian of the Industrial Revolution, also described new industry as being driven by a "Steam-demon." Like many other British thinkers, he too was captivated by the English translation of Laplace's work, yet he was much less concerned about it than he was about steam engines. Carlyle cited Laplace's "Book on the Stars" with humor and derision, crafting a curious character who was amused about the ripple effects of throwing even the tiniest of pebbles.[37] "It is a mathematical fact that the casting of this pebble from my hand alters the centre of gravity of the universe," he said.[38] The determinism of Laplace was frightfully complex. Even local events could be significantly affected by faraway, seemingly unrelated ones: "I say, there is not a red Indian, hunting by Lake Winnipic, can quarrel with his squaw, but the whole world must smart for it: will not the price of beaver rise?"[39] Such tales motivated many minds to start thinking about the possibility that their economy was part of a much larger global system, affected by everything from the transatlantic slave trade to the Indian politics of remote territories in North America. They also started to think more carefully about the potential for slight interventions that could radically change the course of history. The more scientists calculated, the more afraid they became about the consequences for their results of a slight change or mistake, even at the nth decimal. Using colorful examples such as Carlyle's, many other writers considered the consequences of living in a universe where potentially all could be calculated. Some made a mockery of their era's greatest mathematical accomplishments, of its attempts at taming chance and faith in a "universal formula," but others embraced them.

Speculations about the amazing calculating capacities had been raised before, mainly by the German mathematician and philosopher Gottfried Leibniz and the Jesuit polymath Roger Boscovich. But before Laplace, those ideas had, for the most part, been the subject of a great deal of ridicule. In musing about the explosive additive effects that small changes could have across space and time in his *Pensées* (1778 edition), Pascal reached some absurd conclusions. A minute change in the size of Cleopatra's nose could have had ripple effects throughout history: "If it had been shorter," wrote Pascal, it "would have changed the face of the earth."[40] Could there be any point in exploring causal connections quantitatively given how their proliferation could lead to bizarre combinations? It might simply be pointless to try to trace causes back to their humble origins, let alone to try to calculate their relation mathematically. Human passions could compound with the mysteries of the physical universe, spreading causal effects throughout the cosmos in myriad ways.

Critical thinkers such as Voltaire were quick to call attention to these complications. To highlight the problems that arose from this causal mode of thinking, he jumped on the opportunity to criticize the royals. A petty fight behind closed doors could change the entire course of European history without the populace ever knowing the real cause. Concentrating too much power in a few individuals could have dramatic consequences, setting in motion a set of domino or butterfly effects with world historical consequences. With keen and ironic wit, Voltaire pointed his finger at the goings-on between Queen Anne and her favorite in court, Baroness Masham, to which only a few were privy (rumors circulated that they were lesbian lovers).[41] A combinatorial explosion could be too much to handle, effectively forestalling the possibility of understanding historical development in reasonable, let alone quantitative ways. Even for the most optimistic philosophers of this period, the world seemed a hopelessly complex place where the awesome power of mathematics diminished as it descended from an abstract realm to the concrete world.

But what were the real dangers and promises associated with the new mathematical and mechanical innovations sweeping across Europe and soon the world? Few noticed them. While intellectuals and the general public increasingly considered the potential negative effects of scientific research in terms of the fictional stories of Mephistopheles's tempting offers, Frankenstein's unbridled ambitions, or old demonic steam engines, Laplace's followers boldly pushed on with their research. Fictional demons stole the public's imagination during the first decades of the nineteenth century, distracting readers from the nonfictional demons that were being created at the same time. Scholars, literati, historians, social critics, and others, including Marx, largely ignored the new calculating machines that were being built around them—but not Ada Lovelace.

Could we—or should we—calculate everything? What were the limits and dangers of knowing by calculation? These were the questions posed by Lovelace. In 1842, she published a translation of the "Sketch on the Analytical Engine Invented by Charles Babbage." The main text was written by Luigi Federico Menabrea, a military engineer and general who had recently met Babbage in Turin and who would go on to become prime minister of Italy. His essay had already been published in French in Geneva.[42] In it Menabrea explained that "the imagination is at first astounded at the idea of such an undertaking" as Babbage's, "but the more calm reflection we bestow on it, the less impossible does success appear." At various points in the essay, the author insisted that "the machine is not a thinking being," yet he also argued that "although it is not itself the being that thinks, it may yet be considered as the being which executes the conceptions of intelligence."[43]

Could it think? Lovelace responded with hefty "notes by the translator," which she signed only with her initials. These notes would later be viewed as containing some of the most profound insights on early computers. In her published annotations, Lovelace fended off the common impression that a computer could be considered a "thinking being."[44] Yet even after raising this qualification, she remained in awe of the possibility of creating a "thinking or reasoning machine."[45] Lovelace and Menabrea were only partially successful in countering the impression that programmable calculating engines could indeed be a kind of intelligent being, and that they could be so improved in the future to perhaps even become a rival primate. For the next two centuries, scientists and philosophers reexamined their claims, working hard to improve on these technologies as they tried to bring Laplace's being to life.

DARWIN: THE LAPLACE OF BIOLOGY

The great naturalist Charles Darwin at first resisted Babbage's generous invitations to attend the lively dinner soirees he hosted at his London home. Babbage, Lovelace, and their friends were not the only ones wondering about how machines compared to living, and perhaps even thinking, beings. Darwin started considering these themes too—but in a diametrically opposite direction. If Babbage could build a machine from gears and punch cards that appeared somewhat like a "thinking being," could it be that actual living beings could be similarly mechanical? If machines seemed to act like living beings in some ways, he naturally wondered whether living beings could be mechanical in some ways. Darwin started thinking in detail about simple, perhaps even mechanical, processes that could give rise to living species.

"I am very much obliged to you for sending me cards for your parties," he wrote to Babbage, "but I am afraid of accepting them, for I should meet some people there, to whom I have sworn by all the saints in Heaven, I never go out, & should, therefore, be ashamed to meet them."[46] As excitement around Babbage's new engines increased, however, Darwin's curiosity was piqued. He gave up his initial misgivings and became a regular guest at Babbage's, bringing friends and family with him and sending them along when he could not attend. At Babbage's home, he wrote in awe, one "may see the World."[47] A few years had passed since Darwin had returned from his now-famous around-the-world trip aboard the *HMS Beagle*. Back in London, he had to settle for studying the mating rituals of the local fauna at Babbage's London parties. A comparison with what he had seen going down with the plants and animals in the Galápagos Islands would soon be unavoidable.

When jotting down some of the main ideas that would become central to his *On the Origin of Species*, Darwin mused about a "selecting" being who could direct evolution, scribbling about him in his notebooks and describing him in his correspondence. He speculated about the possible existence of "a being infinitely more sagacious than man" who had "foresight" and who, "during thousands and thousands of years," was able "to select all the variations which tended towards certain ends." In parentheses, he made sure to note that this creature should not be confused with God: "(not an omniscient creator)." Could this being breed a new race? Could it breed what it willed? In his notes (known as the *Sketches*), Darwin wondered whether, for instance, it foresaw that "a canine animal would be better off, owing to the country producing more hares, if he were longer legged and [had] keener sight," and so produced a greyhound. How would the work of this being compare to that of the breeders he saw around him? "What blind foolish man has done in a few years" will be nothing compared to what "an all-seeing being in thousands of years could effect."[48] Would it be able to produce better animals or more succulent produce? This being, speculated Darwin, would be capable of producing "a new race."

> Let us now suppose a Being with penetration sufficient to perceive differences in the outer and innermost organization quite imperceptible to man, and with forethought extending over future centuries to watch with unerring care and select for any object the offspring of an organism produced under the foregoing circumstances: I can see no conceivable reason why he could not form a new race (or several were he to separate the stock of the original organism and work on several islands) adapted to new ends.[49]

Only a few years before completing his magnum opus, Darwin returned to thinking about such a creature. In a letter to the American naturalist and Harvard professor Asa Gray, he asked Gray to "suppose there was a being who did not judge by mere external appearances, but who could study the whole internal organization, who was never capricious, and who would go on selecting for one object for millions of generations." The possibilities of these imaginings were endless. "Who will say what he might not effect?" he asked. This being could produce creations even more wonderful than those so far created by European breeders, who "by this power of accumulating variations, adapts living beings to his wants—may be said to make the wool of one sheep good for carpets, of another for cloth, &c." Darwin explained to Gray that "even breeders have been astounded by their results," before concluding his letter with an apology for offering such a speculative theory: "Your imagination must fill up very wide blanks."[50]

The "wise and perceptive Being" mentioned by Darwin in his 1840s notes and 1850s correspondence played "a dramatic role" leading to the publication of *On the Origin of Species*. But in the published tome, Darwin's imaginary being was nowhere to be found. When the editor of his *Sketches* looked for it, he only found "a corresponding passage" in *On the Origin* "where however Nature takes the place of the selecting Being."[51] Other historians have noted that it "never appears onstage." Neither did it appear in any of the "subsequent versions of the theory." Did it simply vanish? On the contrary, it "went underground, whence he rumbled his presence like the thunder machine in a Renaissance theater."[52] Darwin was intent on explaining nature without recourse to "miraculous additions." "I would give absolutely nothing for the theory of Natural Selection," he wrote to his friend the geologist Charles Lyell, "if it requires miraculous additions at any one stage of descent." On this point, he was unsparing: "If I were convinced that I required such additions to the theory of natural selection, I would reject it as rubbish."[53] Despite Darwin's determination to send miraculous causes to the cutting room floor, a general sense remained among many of his peers that other actions were at play in evolution that should not be entirely eliminated from discussions of evolutionary selection theories. Critics pointed out that there was a lot of old wine in the new bottle of "natural" selection.

The projects of Laplace and Darwin were comparable in many ways. Some of Darwin's staunchest supporters saw his contributions as extending a mechanistic view of the universe into the realm of life—in other words, doing for biology what Laplace did for physics. Thomas Huxley, one of Darwin's staunchest "bulldogs," explained that the value of Darwin's work—"the fundamental proposition of Evolution"—consisted in conceiving that "the whole world, living and not living, is the result of the mutual interaction, according to definite laws, of the forces possessed by the molecules of which the primitive nebulosity of the universe was composed."[54] The usefulness of Laplace's being for understanding the physical universe had already proven itself, and Darwin, according to Huxley, extended these lessons into the realm of biology.

"A sufficient intelligence," wrote Huxley, "could, from a knowledge of the properties of the molecules of that [cosmic] vapour, have predicted, say the state of the Fauna of Britain in 1869, with as much certainty as one can say what will happen to the vapour of the breath in a cold winter's day."[55] Such an intelligent being could trace all molecules dancing in the universe—even those constituting living organisms—in an eternal "Struggle for Existence."

Just as certain atoms went left or right and others up or down, joining together and forming new combinations, so certain species mated and survived while others did not.[56] Huxley argued that "multitudes of these [molecules], having diverse tendencies, are competing with one another for opportunity to exist and multiply; and the organism, as a whole, is as much the product of the molecules which are victorious as the Fauna, or Flora, of a country is the product of the victorious organic beings in it."[57] While Huxley was convinced that the predictable movement of molecules could explain all of life, others—including most notably the British biologist Alfred Russel Wallace—still noted that other forces seemed to be at work behind the development of living beings.

The split between those who believed, with Darwin, that natural selection could explain the emergence of life mechanically and those who did not, such as Wallace, became one of the most enduring splits of the next two centuries. Wallace had almost beaten Darwin to the discovery of natural selection. When Darwin first caught wind of Wallace's research, he grew alarmed. He had been working on a similar theory but had not yet published anything. His friends helped him catch up with the younger upstart, who did not have Darwin's high-society connections. Darwin quickly polished what he had and published it alongside Wallace's text, effectively preventing his competitor from beating him. As a result, Darwin's name would become associated with evolution and natural selection, while Wallace would become known to history only as co-discoverer. Even though for the most part Darwin and Wallace remained cordial with each other, their interpretation of what they had just discovered diverged greatly as the years passed by. At stake in their disagreement was the possible existence of agents that could guide evolution.

Wallace criticized Darwin for using "natural" selection as a cover for forces, some possibly supernatural, that affected the development of species. After the publication of *On the Origin of Species*, Wallace got into an argument with Darwin about the matter, accusing his competitor of tacitly using these forces in his theory. Wallace was convinced that some more powerful agency lay behind the magnificent results that Darwin was explaining away as mere "natural selection." In a letter to his well-connected colleague, he asked for clarification pertaining to his use of the term "natural selection." Wallace charged Darwin with suffering from "something like blindness, in your not seeing that 'Natural Selection' *requires the constant watching of an intelligent 'chooser'* like man's selection to which you so often compare it," he wrote.[58] Who was doing the *choosing*? Nature? If so, then Darwin's conception of nature also had to be driven by some kind of directive agent. Perceptive readers, according to Wallace, had similarly concluded that "choice and direction" were needed to produce the effects of "natural selection," even though the naturalist himself tried to downplay these aspects of his theory.[59] Wallace mentioned that the French scientist and philosopher Paul Janet had also noted such a deficiency in Darwin's work and had leveled similar criticisms. "Your so frequently personifying Nature as 'selecting' as 'preferring' as 'seeking only the good of the species' &c. &c.," protested Wallace, was a fatal weakness in Darwin's mode of explanation.[60] Darwin's proclivity for describing nature as animate betrayed his desire to offer an explanation of evolution that did not include the intervention of God or other otherworldly agencies. Coming from a poorer background and lacking the connections of the other, Wallace would remain much less successful than his erstwhile competitor, and he increasingly voiced anti-Victorian-establishment views. It did not help that he was a socialist and interested in studying spiritualist phenomena.

In *The Action of Natural Selection on Man* (1871), Wallace pointed out the limitations of Darwin's approach, which he had previously only broached privately. He explicitly stated that, when it came to the evolution of man, "some *other power* than Natural Selection has been engaged in his production."[61] What could this other power be? Just as humans had domesticated animals and vegetables, someone else had created us, he reasoned. "A superior intelligence," he surmised, "has guided the development of man in a definite direction, and for a special purpose, just as man guides the development of many animal and vegetable forms."[62] Readers of Wallace came back to him with a witty criticism: they accused him of offering a theory that portrayed man as "God's domestic animal."[63]

"Angels, and archangels, spirits and demons have been so long banished from our belief as to have become actually unthinkable as actual existences," explained Wallace in notes added to the second edition of his *Contributions to the Theory of Natural Selection* (1871). They were absent in accounts by most modern thinkers, including Darwin. "Nothing in modern philosophy takes their place." But the complete elimination of these beings from our thinking left "an infinite chasm between man and the Great Mind of the universe," which, Wallace concluded, was to "the highest degree improbable."[64]

3

Maxwell's Demon

Maxwell's demon is the most famous of the lot.[1] Still active today, he is known for working slowly but surely, nearly effortlessly, without much expenditure, as tirelessly and efficiently as a perpetual motion machine. Although tiny, his small size inversely reflects his strength. He is suspected of being present where tiny causes can accumulate to create much larger effects—such as in rapidly reproducing viruses, in replicating strands of DNA, in enzymes that set off chemical chain reactions, in certain subatomic particles, and in the forces that arrange snowflakes into wonderful shapes. Originally a product of the sprawling British Empire, he was born in a Victorian world of combustion engines and expanding railway, electric, and telegraph networks. Maxwell's demon is also known as the "sorting demon," because his special ability resides in his power to deftly control tiny amounts, from individual atoms and molecules to digital bits of information.

He is more dangerous than Descartes's demon, since he can act directly on the natural world and has no need to deceive anyone. He can stump Laplace's because he has the power to change the course of history midway. He is a control freak who intervenes occasionally to prevent nature from running its course. Because of him, the future may not follow as it ordinarily would. He works on the fly and adjusts his behavior by the seat of his pants, able to act suddenly and strategically to upset forces of equilibrium. He is much like a fish who can eat a whale, a David who can beat a Goliath, or the straw that breaks the camel's back. Like a miniature Katechon, the biblical "restrainer" who can delay the end of the world, he can stop entropy, put an end to decay, and make the world run in reverse.

Maxwell's demon is a model for all sorts of mechanisms that function as one-way conduits (such as orifices, valves, pumps, filters, semipermeable membranes, mechanical ratchets, electrical rectifiers, microprocessors, and mechanisms that permit the selection and transference of genetic material), as well as for more complex machines that create work at the expense of the environment, such as engines and refrigerators. We behave very much like Maxwell's demon when we, or parts of us, act intelligently by performing highly selective and consequential acts of discernment. He is the ultimate discriminator.

Mechanization elicited many fears, some of them connected to the possibility that God was not the main force behind the central operations of the universe. But the idea that there were to be absolute limits to mechanization provoked near-panic. The stability of nature described by Laplace was called into question during the Victorian era with the rise not only of evolutionary theory but also of modern physics. In light of this possibility, the end of the world—the actual death of the universe—seemed inescapable and perhaps not so far off. Another being, this one named after the brilliant Scottish scientist James Clerk Maxwell, appeared willing and able to stop Laplace's demon in his tracks. His specialty was altering the regular course of nature, for better or for worse.

If the history of the physical sciences was to be divided into the contribution of only three men, most historians would place Maxwell after Isaac Newton and before Albert Einstein. Maxwell is known mainly for his contributions to electrodynamics and remembered for having predicted the existence of electromagnetic waves. In a series of renowned scientific papers, he established that electricity, light, and magnetic forces were of the same nature, and he wrote down a set of famous equations that described them. These equations were retrospectively seen as prophetic: Heinrich Hertz's detection of electromagnetic waves much later, in 1888, paved the way for the development of the radio, television, and wireless telecommunication technologies on which the visual and communication networks of the modern world would be based. X-rays, radar, Wi-Fi, electric motors, and microwave ovens are all still understood by reference to Maxwell's equations. A recent biography of him was aptly titled *The Man Who Changed Everything*.[2]

The science of Maxwell's time was characterized by a dual concern with motors and electricity. On the centennial anniversary of Maxwell's birth, Einstein described the Scottish scientist's work as having brought about the "greatest alteration" that physics had experienced "since the foundation of theoretical physics by Newton."[3] The physicist Richard Feynman, who devoted one of his best-known articles to studying Maxwell's demon, claimed that "from a long view of the history of mankind . . . there can be little doubt that the most significant event of the 19th century will be judged as Maxwell's discovery of the laws of thermodynamics." Even the American Civil War, Feynman argued, "will pale into provincial insignificance."[4] How did the work of a Scottish mathematician come to be regarded by later generations as arguably more important than the war that brought about the abolition of slavery in America?

Maxwell disputed accounts of nature's simple regularity. He took aim at the view that nature was like a book in which one page led to the next. Instead, he argued that nature should be considered more like a magazine. At the very outset of his career, he envisioned this possibility. "Perhaps the 'book,' as it was called, of nature is regularly paged," he wrote, "but if it is not a 'book' at all, but a magazine, nothing is more foolish to suppose that one part can throw light on another."[5]

The demon that would be named after Maxwell is undoubtedly the best-known demon of modern times. A scientist recently described the nineteenth century as an era "hypostatized by twin hypothetical ideal beings, Laplace's Demon and Maxwell's Demon." Of the former he said that it "could predict in Newtonian, cause-and-effect, action-reaction fashion the complete state of the universe, given knowledge of the positions and velocities of all its particles for a single instant."[6] The powers of the latter were no less impressive. That demon could temporarily violate the second law of thermodynamics and, in so doing, almost rivaled a perpetual motion machine. A literary scholar described his widespread impact on our modern worldview: "Men had trusted in the Great Watchmaker to keep things running; but they had forgotten the Great Stoker, and now, it appeared, he had fallen down on the job. Our universe was mortal, and we have been anxiously reading its temperature ever since."[7]

Maxwell's demon is now "perhaps more well known than its creator."[8] Numerous books and articles have been written about him.[9] He has dedicated biographers and a stellar artistic career—theatrical, cinematic, literary, and even musical.[10] The most important articles discussing him have been reprinted. Two comprehensive edited volumes compiling some of the most important research on him predicted that his "continuing active life is likely" and concluded that "we expect Maxwell's demon to remain a potent teacher for many years to come!"[11] Thousands of students learn about him, and his image graces the pages of many a scientific journal and textbook. His life has been carefully documented in comprehensive bibliographies tracking his every appearance and archived by librarians driven

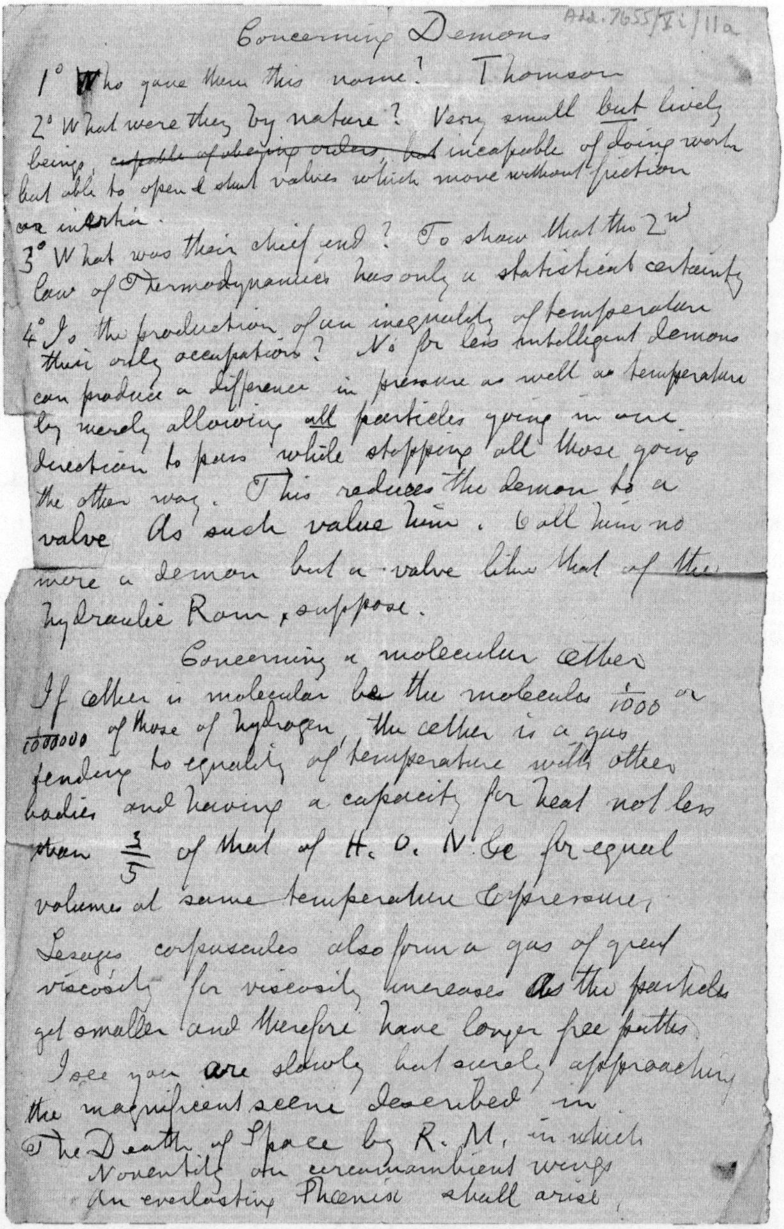
Add. 7655/Vi/11a

Concerning Demons

1° Who gave them this name? Thomson

2° What were they by nature? Very small but lively beings ~~capable of obeying orders but~~ incapable of doing work but able to open & shut valves which move without friction or inertia.

3° What was their chief end? To show that the 2nd Law of Thermodynamics has only a statistical certainty

4° Is the production of an inequality of temperature their only occupation? No for less intelligent demons can produce a difference in pressure as well as temperature by merely allowing all particles going in one direction to pass while stopping all those going the other way. This reduces the demon to a valve As such value him. Call him no more a demon but a valve like that of the hydraulic Ram, suppose.

Concerning a molecular aether

If aether is molecular be the molecules $\frac{1}{1000}$ or $\frac{1}{1000000}$ of those of hydrogen the aether is a gas tending to equality of temperature with other bodies and having a capacity for heat not less than $\frac{3}{5}$ of that of H. O. N. &c for equal volumes at same temperature & pressure.

Lesages corpuscules also form a gas of great viscosity for viscosity increases as the particles get smaller and therefore have longer free paths.

I see you are slowly but surely approaching the magnificent scene described in The Death of Space by R. M. in which

Noventily on circumambient wings
An everlasting Phoenix shall arise,

Figure 1. "Concerning Demons," undated manuscript, in Maxwell's handwriting, found in Peter Guthrie Tait's papers after his death. Ms. [Add. 7655 Vi/11a], James Clerk Maxwell Papers, Cambridge University Library. Reproduced by kind permission of the Syndics of Cambridge University Library.

by paparazzi-like zeal. The original letter in which he was first mentioned and a note related to it titled "Concerning Demons" are kept in the library at Cambridge University as a prized possession.[12]

NOT YET NAMED

Maxwell imagined this creature first. In early December 1867, he wrote a letter to his colleague Peter Guthrie Tait in which he discussed the laws of thermodynamics. He referred to a "finite being" who was "very observant and neat fingered." Tait then forwarded Maxwell's letter to his colleague William Thomson, one of the most revered scientists of his era; he would be knighted as Lord Kelvin, the first scientist to be elevated to the House of Lords. Thomson was responsible for calling the creature "Maxwell's demon" and would investigate the demon's abilities at the same time that he was developing valuable patents that made him rich and famous.

Tait considered Maxwell a man "of *originality*, and *fertility*, and *leisure*," and told him so in his letter. He wrote to Maxwell because he wanted to ask his opinion about a book he was writing on the history of thermodynamics: "Are you sufficiently up to the history of Thermodynamics to critically examine & put right a little treatise I am about to print—and will you kindly apply your critical powers to it?"[13] Apply his critical powers Maxwell did. By introducing his "finite being" in his reply to Tait, Maxwell was able to "pick a hole" in one of the most foundational theories of his era.

Maxwell's contributions to electrodynamics and the "hole" he picked in thermodynamics are now legion. His demon is used as proof for why we cannot predict everything perfectly all the time and why we need to think of the second law of thermodynamics—also known as the law of entropy—as valid only *statistically*.

According to the first law of thermodynamics, energy can be neither created nor destroyed but merely transformed. According to the second law (the law of entropy), heat flows from hot to cold and perpetual motion machines can never be found. Any cup of coffee will eventually cool down, but the reverse just does not happen. Hot patches in your soup or in nature do not last long. Temperature differences are very hard to maintain, and even the best insulators dissipate heat.

The two main laws of energy are some of the most comprehensive laws of science; running the gamut from cosmology to biology, they bring under their purview elements as basic and diverse as time, heat, food, information, and communication. The second law explains why we grow old, why we need to eat in order to replenish the nutrients in our bodies, why plants require light and water to thrive, why the future differs from the past, why we sense an arrow of time, why we need to invest so much work and effort into putting things back in order. The law of entropy tells us why all the king's horses and all the king's men were unable to put Humpty-Dumpty back together again. The law even has implications for how we think about justice and fairness, as well as chance and fate. Such comprehensiveness leaves room for exceptions—which are often described by scientists in terms of the actions of Maxwell's demon. Maxwell's demon explains why, despite all odds to the contrary, nature sometimes does follow the path less traveled.

Consider that we fight entropy all the time—and what is more, we are constantly creating and improving technologies to aid us in that endeavor. The method of statistical prediction inaugurated by Laplace was not flawless. Could Maxwell's creature take advantage of the statistical nature of thermodynamics to obtain a different outcome? Could we? In his original letter to Tait, Maxwell certainly did not think so. "Only we can't, not being clever enough," he had concluded.[14] Understanding our role in reducing entropy and that of the innovations we create for this purpose proved to be a complex riddle.

WHY DEMONS?

An undated note in Maxwell's handwriting titled "Concerning Demons" was found among Tait's papers after his death. Experts believe that it was written around 1876. Its structure followed closely that of a catechism. Like prayers and recitations, catechisms typically have a catchy pattern based on short questions followed by pithy answers. They are meant to be repeated over and over again, serving a mnemonic function. They are ideal pedagogic tools used to inculcate beliefs. The mysterious note read:

1. Who gave them this name? Thomson.
2. What were they by nature? Very small BUT lively beings incapable of doing work but able to open and shut valves which move without friction or inertia.
3. What was their chief end? To show that the 2nd Law of Thermodynamics has only a statistical certainty.

4. Is the production of the inequality of temperature their only occupation? No, for less intelligent demons can produce a difference in pressure as well as temperature by merely allowing all particles going in one direction while stopping all those going the other way. This reduces the demon to a valve. As such value him. Call him no more a demon but a valve like that of the hydraulic ram, suppose.[15]

The fourth item in Maxwell's list broke with the pattern of the first three items. It showed Maxwell disowning his own creation. "Call him no more a demon," pleaded the physicist. A lowly "valve," even one as simple as a hydraulic pump, he would argue on various occasions, could perform its main functions. But for how long? Was such a mechanism able to defy the second law of thermodynamics? If so, could this mechanism be used to build one of the most sought-after machines of all times: a perpetual motion machine?

MAXWELL'S THEORY OF HEAT

Where did the powers of Maxwell's demon come from? The demon could alter the course of individual molecules by acting selectively on them, reversing temperature equilibrium. Instead of having mixtures of unequal temperature even out as mixtures cooled down, he could make one part of the sample hotter as the other became colder. All he needed to do was to stop fast atoms (which produce hot temperatures) from mixing with slow ones (which bring temperature down).

Maxwell started discussing the idea directly with Thomson in another letter. The next winter, in a letter to another physicist, he described himself as a humble "pointsman on a railway."[16] The comparison to such a profession was as descriptive as it was evocative in Victorian England, where a single lever could be used to redirect a train coming from London to Cambridge or Oxford, and where a simple timing mistake could cause the deaths of hundreds of passengers. Could a demon working on molecules produce similar effects by simply redirecting a few of them away from their preordained paths? If such a minor action took place during a tipping point, could it change the fate of the universe?

Maxwell's now-famous character first appeared in print in the last section of his book *Theory of Heat* (originally published in 1871), appropriately titled "Limitation of the Second Law of Thermodynamics." "Before I conclude," he wrote, "I wish to direct attention to an aspect of the molecular theory which deserves consideration." This pertained to "one of the best-established facts in thermodynamics"—that it is impossible "to produce any inequality of temperature or of pressure without the expenditure of work." Maxwell explained: "This is the second law of thermodynamics, and it is undoubtedly true as long as we can deal with bodies only in mass, and have no power of perceiving or handling the separate molecules of which they are made up." This last point immediately left open a very interesting possibility. "But if we conceive *a being whose faculties are so sharpened* that he can follow every molecule in its course," he wrote, "such a being, whose attributes are still as essentially finite as our own, would be able to do what is at present impossible to us."

Maxwell went on to describe his being in the way it would be frequently pictured by the many authors who became fascinated by it, as working with two insulated containers attached to each other with a single opening between them to allow certain molecules to pass through:

> Now let us suppose that such a vessel is divided into two portions, A and B, by a division in which there is a small hole, and that a being, who can see the individual molecules, opens and closes this hole, so as to allow only the swifter molecules to pass from A to B, and only the slower ones to pass from B to A.

With his delicate movements, this being could "contradict" the second law. And he could do it nearly effortlessly: "He will thus, without expenditure of work, raise the temperature of B and lower that of A, in contradiction to the second law of thermodynamics."[17]

In his *Theory of Heat*, Maxwell also reviewed the contributions of Fourier, who had successfully expanded Laplace's work to better understand the transmission of heat. He had good reasons to pay close attention to the work of the Frenchman. As the world economy increasingly depended on coal, studies on heat dissipation became more pertinent than ever. Maxwell marveled at how, thanks to Fourier, "we are now able, by means of a thorough knowledge of the thermal state of the body at a given instant, to determine the rate at which the temperature of every part must be changing, and therefore we are able to predict its state in the succeeding instant. Knowing this, we can predict its state in the next instant following, and so on." A few measurements "will afford sufficient data for calculating the temperature of every point during all time to come."[18]

Scientists could not—would not—stop talking about Maxwell's miniature being, whose fame quickly increased after he appeared in Maxwell's *Theory of Heat*. But the official public debut at which he was named "Maxwell's demon" came almost half a decade later, when Thomson read a paper to the Royal Society of Edinburgh on February 16, 1874. His electrifying words were subsequently published twice, in the society's official *Proceedings* and in the journal *Nature*.[19] The local newspaper, the *Edinburgh Courant*, covered the news of this intriguing character in an article that ran the following Tuesday. In his articles and presentations, Thomson explained the tendency toward equilibrium in terms of an absent demon. "If no selective influence, such as that of the ideal 'demon,' guides individual molecules, the average result of their free motions and collisions must be to equalize the distribution of energy among them in the gross," he explained.[20]

The demon then appeared once again as a character in the controversial *The Unseen Universe*, a book coauthored by the physicist Balfour Stewart, who was the director of the Kew Observatory, and by Tait. First published anonymously in 1875, *The Unseen Universe* was republished numerous times. When the mathematician William Kingdon Clifford read it, he was taken aback by the numerous references in the book to supernatural creatures. "There is room not only for deities to preside over their properties and functions, existence, energy and life, but all other machinery of Christian mythology—spiritual bodies replete with energy, angels, archangels, incarnation, molecular demons, miracles and 'universal gehenna.'"[21] Controversy surrounding the creature and its possible existence ensued.

REVERSING TIME

If we could only turn back time! Why not? Maxwell's demon could. He could do it by working against the regular causal direction of physical particles. In a universe comprising molecules of motion, such a rearrangement amounted to a reversal of time.

Thomson was one of the first scientists to articulate the consequences that exceptions to the law of entropy would have for time. In his landmark lecture published in *Nature* on April 9, 1874, "The Kinetic Theory of the

Dissipation of Energy," he described how the world would look like if it suddenly started running in reverse:

> The bursting bubble of foam at the foot of a waterfall would reunite and descend into the water.... Boulders would recover from the mud the materials required to build them into their previous jagged forms, and would become reunited to the mountain peak from which they had formerly broken away... living creatures would grow backwards, with conscious knowledge of the future, but no memory of the past, and would become again unborn.[22]

Thomson then speculated that an "army of Maxwell's 'intelligent demons'*" could be employed to produce those effects. The asterisk after his evocative phrase pointed to a footnote explaining what was meant by the term: "The definition of 'demon' according to the use of this word by Maxwell, is an intelligent being endowed with free will, and [of] fine enough tactile and perceptive organization to give him the faculty of observing and influencing individual molecules of matter."[23]

Thomson's lecture dealt with energy, work, and heat in relation to the movement of underlying molecules. Like Maxwell before him, he investigated regular and exceptional cases, focusing on diffusion processes that led to temperature equilibrium. Although diffusion was the norm, he explained, "this process of diffusion could be perfectly prevented by an army of Maxwell's 'intelligent demons,' stationed at the surface, or interface as we may call it." How would this demon-army operate? "The duty of each demon is to guard his allotment, turning molecules back, or allowing them to pass through from either side, according to certain definite orders." The army would be properly equipped: he suggested that one "suppose the weapon of the ideal army to be a club, or, as it were, a molecular cricket bat." The army would stand guard at designated positions: "Each demon is to keep as nearly as possible to a certain station, making only such excursions from it as the execution of his orders requires." Good timing along with excellent hand-eye coordination and statuesque steadfastness were essential for this soldier or sportsman: "He may so time his strokes that he shall never move to any considerable distance from his station."[24]

The effects of Thomson's invisible demon-army could be actually measured and calculated. To illustrate them, Thomson chose the example of an insulated iron bar heated at one end. "By operating selectively on individual atoms he can reverse the natural dissipation of energy, can cause one-half of a closed jar of air, or a bar of iron, to become glowingly hot and the other ice-cold."[25] In the case of a spoon dipped in hot soup, the handle gets hotter while the soup gets colder. Eventually both reach the same temperature. In the case of an insulated iron bar heated at one end, the temperature of the entire bar eventually reaches equilibrium: opposite sides end up with the same temperature. But how probable would it be for heat to run in the other direction? It was hardly likely to happen, but it was not completely impossible either. "It is very improbable that in the course of 1000 years one-half of the bar of iron shall of itself become warmer by a degree than the other half," he explained. Chances were slim, but real. The "probability of this happening before 1,000,000 years pass is 1000 times as great as that it will happen in the course of 1000 years." Thomson was certain that such an effect would occur (at some point). "It certainly will happen in the course of some very long time," he concluded.[26] The actions of his demon-army were as real and palpable as anything else on this earth, even if unlikely and rare at human scale.

Thomson continued to think and lecture about Maxwell's demon for years. In a Friday evening lecture before the Royal Institution of Great Britain on February 28, 1879, he urged listeners to think about his choice of the term in positive terms, stressing that the "word 'demon,' which originally in Greek meant a supernatural being, has never been properly used as signifying a real or ideal personification of malignity." For him, Maxwell's demon was "a creature of imagination." But if purely imaginary, what was its purpose? "The conception of the 'sorting demon' . . . is of great value in purely physical science," he explained, clarifying that "it was not invented to help us to deal with questions regarding the influence of life and of mind on the motions of matter, questions essentially beyond the range of mere dynamics." His particular value was due to how he could "help us to understand the 'Dissipation of Energy'" in nature. Thomson described the demon as animalesque—"a being with no preternatural qualities, and differs from real living animals only in extreme smallness and agility." He also considered it to be "purely mechanical."[27]

Thomson explained that because this being was similar to us, possessing "attributes" that "are still as essentially finite as our own," we could learn from him.[28] His understanding of this creature as someone from whom we could learn something echoed Maxwell's own explanation of the utility of his being. When first imagining him, Maxwell underscored that he "would be able to do what is at present impossible to us," but which might be possible in the future.

MINUSCULE LEVIATHANS

Maxwell's demon was small, but milquetoast he was not. The molecules he could divert traveled at tremendous velocities, swirling by at speeds greater than one thousand miles per hour with more energy than the mythical Leviathan. "The molecules of the air in this hall," lectured Maxwell, were "flying about in all directions, at a rate of about seventeen miles in a minute." This was so "very great as compared with that of bullets," and no lesser than the velocity "which proceeds from the mouth of a cannon."[29] At the very least, it was "greater than any velocity obtained in artillery practice." We do not feel the impact of these minuscule bullets because they fly in all directions and collide with each other, canceling out their great forces in the aggregate. Scientists calculated that every hydrogen particle underwent an average 17,700,000,000 collisions per second and changed its direction about 17,700,000,000 times during that interval.

In Thomson's account, the demon could do even more. He could grab two atoms, combine them, and "press as it were on his two hands." In this way, he could create new molecular compounds. "Then let him apply the two hands" to "tear them asunder," and he could decompose them. What could he do if, instead of belonging to an army equipped with bats and clubs, he could pick and pull delicately and with care? "He can at pleasure stop, or strike, or push, or pull any single atom of matter, and so moderate its natural course of motion." What if he had the hands and skill of a piano player? "Endowed ideally with hands and fingers—two hands and ten fingers suffice—he can do as much for atoms as a pianoforte player can do for the keys of a piano—just a little more, he can push or pull each individual atom *in any direction*."[30]

With these attributes, Thomson's demon could transform himself into a master chemist with the ability to create explosive mixtures and combustibles. Give him water and he could give you in return "the explosive mixture of oxygen and hydrogen." Give him the carbonic acid we regularly breathe out and he could provide in exchange "the separated mutual combustibles, carbon and oxygen." Such talents permitted him to produce coveted fuels and explosives and solve the growing problem of pollution in cities. He could turn chemical waste into fuel at no cost over and over again, giving life to the dream of renewable energy sources—recapturing lost energy and reusing spent fuel.

To maximize the efficiency of engines in the mid-nineteenth century, they needed to be watched by a stoker whose job was to separate black coals from the red ones so that they would burn more evenly. Even after new machines, referred to as "self-feeding" furnaces, soon replaced these labor-intensive practices, there remained plenty of room for improvement. No matter how much scientists tried to perfect them, engines wasted most of the energy that went into them. Engine efficiency at midcentury stood at about 18 percent. Friction, noise, and overheating were all to blame.

Thomson's demon research represented the culmination of a lifetime of work. It was also the maturation of many childhood dreams he had shared with his brother. Perpetual motion was every child's fantasy, and the two Thomson boys were no exception. As the sons of a teacher of mathematics and engineering at the Royal Belfast Academical Institution, the Thomsons were some of the best-educated lads in northern Ireland. They grew up quickly, and William looked up to his older brother James, who learned about machines early on the hard way. In the 1840s, the younger William approached his brother with "a curious conclusion to which he had been led, by reasoning on principles similar to those developed by Carnot, with reference to the motive power of heat." He thought he had landed on a model for a perpetual motion machine. The scheme used ice's expanding force to push a piston inside a motor. "Because water expands while freezing," he recalled, "and therefore it seemed to follow, that if a quantity of it were merely enclosed in a vessel with a moveable piston and frozen, the motion of the piston, consequent on the expansion, being resisted by pressure, mechanical work would be given out without any corresponding expenditure; or, in other words, a perpetual source of mechanical work, commonly called a perpetual motion, would be possible." The scheme would violate the findings of the Frenchman Sadi Carnot, who earlier in the century had determined why engine efficiency was capped within unbreachable limits.

The older Thomson brother knew better. "This at first appeared to me to involve an impossibility," he remembered. James now had to figure out why. The freezing point of water, he surmised, must become lower with pressure. "To avoid the absurdity of supposing that mechanical work could be got out of nothing," he wrote, "it occurred to me that it is necessary farther to conclude, that the freezing point becomes lower as the pressure to which the water is subjected is increased."[31] William did not get his dream machine, but James got his paper published in the *Cambridge and Dublin Mathematical Journal* and in the *Transactions of the Royal Society of Edinburgh*.

James himself had been lectured on the topic of machine inventions by an older school friend who was a successful businessman and engineer. As a boy, he too thought he had found such a wonder, writing a letter to his friend describing a "scheme for propelling vessels against the stream." His interlocutor responded harshly and honestly. It "is very ingenious," but, sorry, it's been thought of already: "Altho' you have not been aware of it and have equal merit as an inventor, it was tried in France many years ago." This lesson was likely to apply to anything else that James could think of. His friend reminded him of "the great difficulty of hitting upon any useful scheme that has not been previously attempted, and the necessity of examining well before one expends time on such occupations." Basically, James was told to go back to his studies: "Indeed were I you, and I am sure you will excuse me as a senior apprentice, I would, for the present, avoid all attempts at inventing machinery of any description." Before talking so big, his friend told him, "it seems to me to be nearly as great a waste of time, making attempts at useful discovery without this previous knowledge, as for a person to labour at working out the highest problems in Astronomy without having first gone through the Calculus." He signed off curtly: "Believe me my dear James."[32]

From that moment on, the two Thomsons chased the dream of perpetual motion more maturely, studying the "numberless . . . schemes proposed and tried for procuring a perfect combustion of coals in furnaces" in detail before attempting to improve on them.[33] Still, they did not give up their attempts to develop machines that would come as close as possible to the dream. James would usually design them, while William would write the scientific papers that explained the science behind them, including his two famous articles on the demon.

As adults, the two brothers not only grew concerned about how fast locomotives ate up coal but also worried about the day when the sun would run out of energy. "As for the future, we may say, with equal certainty, that inhabitants of the earth can not continue to enjoy the light and heat essential to their life for many million years longer unless sources now unknown to us are prepared in the great storehouse of creation."[34] For Victorians like them, the most appropriate model for the universe was no longer the clockwork mechanism of the Newtonian era but rather that of a coal engine.[35]

What to do when its energy petered out? It was not too soon to start thinking about such an apocalypse.

MARVELOUS AGILITY

The science of Maxwell's time had another problem to contend with that also hinged on the possible existence of Maxwell's demon. The laws of mechanics were perfectly reversible, yet so much that took place in this world was patently not. The difficulty in reconciling reversible laws with irreversible processes stumped some of the greatest minds of the nineteenth century. Maxwell offered a possible solution. As Tait explained how Maxwell showed "the mode of escape from the difficulty," he noted that the only problem was that "it would require the intervention of beings, still finite, but infinitely more acute and able than any human beings (or even than the utmost ideal a human being can well conceive)" to effectively manipulate molecules against forces of equilibrium and dissipation.[36] Tait understood Maxwell's "imaginary beings" to be "small creatures without inertia, of extremely acute senses and intelligence, and marvellous agility," and explained that Thomson referred to these beings, "provisionally," as demons:

> He points out that if such imaginary beings, whom Sir W. Thomson provisionally calls demons—small creatures without inertia, of extremely acute senses and intelligence, and marvellous agility—were to watch the particles of a gas contained in a vessel with a partition full of trap-doors, also devoid of inertia; prepared to open and close these doors so as to let the quicker particles get out of the first compartment into the second, and to let an equal number of the slower ones escape from the second compartment into the first, they could, without doing any work themselves, give to the system the power of doing a certain amount of work without help from external bodies.

The agility of these demons notwithstanding, Tait warned that their powers were limited. "You must be careful," he warned his readers, "not to fancy that there is here any gain or creation of energy—not even a demon could effect that—there is a gain of transformability, a slight rise in the scale of availability—*voilà tout*."[37]

The most interesting cases involved samples with only a few molecules and showed that "restoration of energy is constantly going on, but on a very limited scale, in every mass of gas," Tait explained. Every once in a while, a fast molecule could appear where slow ones should dominate. Not all particles were perfectly distributed: "If there were only a few hundred particles in a small vessel of gas, the chances would be that if we suddenly cut off half the vessel there would be a sensible difference of temperature between the two parts."[38] These cases offered Tait hope that they could be scaled up, yet he was not sanguine about the prospect of building a nearly perfect, reversible engine that would consume almost no fuel. Not possible for us, according to Tait: "For, alas! we are not demons (in Maxwell's sense)."[39]

The reputation of Maxwell's demon was sealed in a description that appeared under the entry for "Energy" in the ninth edition of the *Encyclopedia Britannica*:

> Clerk Maxwell supposed two compartments, A and B, to be filled with gas at the same temperature, and to be separated by a partition containing a number of trapdoors, each of which could be opened or closed without any expenditure of energy. An intelligent creature, or "demon," possessed of unlimited powers of vision, is placed in charge of each door.

The entry concluded with the claim that "by the application of *intelligence alone* a portion of gas of uniform pressure and temperature may be divided into two parts, in which both the temperature and the pressure are different, and from which, therefore, work can be obtained at the expense of heat." Furthermore, "there seems no limit to the extent to which this operation may be carried." The author warned that these achievements had not yet been realized, as "the principle of the dissipation of energy has control over the actions of those agents only whose faculties are too gross to enable them to grapple with those portions of matter," but did not rule out that less "gross" agents would be so capable in the future.[40]

DÄMONEN

Maxwell's and Laplace's demons soon landed in Germany as Maxwell's *Dämonen* and Laplace's *Geist*. On August 14, 1872, the physiologist Emil du Bois-Reymond delivered an influential lecture at the second general meeting of the Assembly of German Naturalists and Physicians in Leipzig. What he said that day changed how scientists thought about Laplace's being, and about the limits of science, for the next one hundred years. As soon as it was published, the text was widely read, reprinted, and translated. In English it was titled "On the Limits of Our Understanding of Nature." Bois-Reymond referred to Laplace's being as "Geist." In French translation it was called a "prophet," and in English it was translated as a "mind" or "seer expert."[41]

The talents of this being were exalted by numerous readers who described it with colorful examples. The physicist Walther Nernst explained that "it was concluded, of course, not without irony, that with a precise knowledge of Goethe's constitution and of all external influences, that spirit [*Geist*] could literally dictate the Faust with the help of the relevant laws of nature; and of course not just the printed Faust, but all the earlier drafts as well." Yet "no one," he wrote, "has described with such graceful insistence the practical capacity of Laplace's formula as our great Berlin physiologist du Bois-Reymond."[42] The philosopher Ernst Cassirer was among many others who raised the same point. "Emil du Bois-Reymond," he wrote, "lifted the Laplacian formula [*Weltformel*] out of its long oblivion and placed it at the focal point of epistemological and scientific discussion."[43] To this day, Bois-Reymond's lecture remains one of the most powerful meditations on the limits of knowledge, and its success ensured his place as one of the most influential scientists of the nineteenth century. He would dominate the intellectual circles of Berlin, Potsdam, and beyond.

Bois-Reymond argued that scientists who understood the world in terms of moving atoms need not fear the uncertainties of chance. "If we were to suppose all changes in the physical world resolved into atomic motions," then "law and chance would be only different names for mechanical necessity." According to this view, chance was just a reflection of something currently not known, a practical nuisance that could be overcome. But just how difficult would it be to eliminate all our unknowns for all time? The physiologist wondered whether that would even be possible. It was an important "if," one that needed to be parsed out by diving deeper into Laplace's texts, his mathematics, and his philosophy. Bois-Reymond and others started to study in great detail the Frenchman's assertions about a mind that could calculate everything. What could it know and what could it not know?

While Laplace had been effusive about our ability to know and calculate everything, Bois-Reymond ended his talk by shouting "*Ignorabimus!*" Laplace's creation, according to the celebrated speaker, was "dumb" in one essential respect. It was nothing other than a mechanical expert, a dull calculator that did not feel and understand anything that it grasped.

Bois-Reymond nonetheless granted a lot to the Laplacian mind. By applying its "universal formula," it could see the past and the future. It could solve crimes and even historical mysteries. It could reveal the long-hidden identity of the "man in the iron mask" who had been imprisoned in the Bastille at the end of the seventeenth century and whose identity remains a mystery to this day. Could he really have been the illegitimate brother of Louis XIV, as rumors claimed? "Indeed, just as in lunar equations the astronomer need give but a negative value to time, in order to determine whether, when Pericles embarked for Epidaurus, the sun was eclipsed for the Piraeus, so could the mind imagined by Laplace, by suitable application of its universal formula, tell us who was the Man in the Iron Mask, or how the President was lost," wrote Bois-Reymond.[44]

Bois-Reymond based his views about predictability on the celebrated successes of astronomy, some of which had been pioneered by Laplace. "As the astronomer foretells the day whereon years hence a comet emerges again out of the depths of space into the heavens," he explained, "so could that mind by its equations determine the day whereon the Greek cross shall glitter from the mosque of St. Sophia, or when England shall have consumed the last of her coals."[45]

The German scientist lauded the mathematical techniques developed for the study of heat and energy distribution, admiring the clever technique of employing a t for time in equations. Ostensibly, these innovations would permit scientists to picture the beginning—and end—of all creation.

> If in his universal formula he set down $t=-\infty$, he could discover the mysterious primeval condition of all things. He would in the boundless space see matter already in motion, or unequally distributed, for, were the distribution equable, there could never be disturbance of equilibrium. Suppose he lets t grow *ad infinitum* in the positive sense, then he could tell whether Carnot's theorem threatens the universe with icy immobility in finite or only in infinite time.

The mind described by Laplace had been almost brought to life. "We resemble this mind, inasmuch as we conceive of it," Bois-Reymond insisted. The difference between it and actual minds was now "only a matter of degree." "We might even ask whether a mind like that of Newton does not differ less from the mind imagined by Laplace, than the mind of an Australian or of a Fuegian savage differs from the mind of Newton," he wondered.[46]

But even it could *never* know two things. "There are two positions where even the mind imagined by Laplace would strive in vain to press on farther, and where we have to stand stock-still." Laplace's mind could not ever unveil the mysteries of life and of consciousness. It could see "where and under what form life first appeared, whether at the bottom of the deep sea, as bathybius protoplasm, or whether with the cooperation of the still excessive ultra-violet solar rays, with still higher pressure of carbonic acid in the atmosphere." But what was the point of seeing these transformations? "Laplace's Mind *could* tell, with the aid of the universal formula," where and when life arose. "For, when inorganic matter coalesces to form organic matter, there is only a question of motion, of the arrangement of molecules into states of more or less stable equilibrium, and of an exchange of matter produced partly by the tension of the molecules, and partly by motion from without." All this would be known. Yet it would be a "dumb show," since such a "Mind" could never tell what suddenly made all these parts work as an organic whole. It would see atoms "crossing each other's course, but does not understand their pantomime." Hence, "the world of this Mind is still meaningless." Laplace's being could solve an "exceedingly difficult mechanical problem," but it would never solve the mystery of life and consciousness.[47]

By the first decades of the nineteenth century, Laplace's creation had become a veritable whipping boy for a number of prominent thinkers who argued that its knowledge was of an inferior kind. It was knowledge of the predictable, unchanging world, which could not be used to understand deeper transformations, and especially not those of kind rather than degree. The pompous pedant that arose from Laplace's imagination might be nothing more than a fachidiot, one that would always remain none the wiser when faced with complex challenges.

The renowned physicist and physiologist Hermann von Helmholtz was one of Bois-Reymond's closest friends and collaborators. The formulation of the law of energy conservation (in a paper that was later seen as key to thermodynamics) stands out among his many accomplishments, which also included work in optics, physiology, and even philosophy.

A few months after Bois-Reymond delivered his lecture, Helmholtz recalled not being able to get a catchy verse out of his head, one he learned as a child:

It rains when it rains,
And rains its course;
And when it has rained enough,
It thus stops again.[48]

Helmholtz was in his mid-fifties and a senior scientist when he confessed to having this rhyme stuck in his head. This "Verslein," he explained, had "been lodged in my memory since ancient times, evidently because it touches a sore spot in the conscience of a physicist, and makes a mockery out of him."[49] The author of the verse was none other than the writer and philosopher Johann Wolfgang von Goethe. Helmholtz decided to finally write it down in a perfect place for it: a scientific article on tornadoes and thunderstorms.

Using science to examine, let alone predict, the weather almost always ended in failure. Helmholtz explained that one would be very much the wiser to ask a local farmer about the portent of the clouds closing in than to consult a scientist isolated in his dark laboratory with his charts and instruments. Scientific progress on this topic was painfully slow "despite all the newly acquired insights into the context of natural phenomena, despite all the newly erected meteorological stations and endlessly long series of observations."[50]

Why was it so hard to predict the weather? Up in the sky, where the stormy clouds reigned, the German scientist could not help but see the hand of "the rebellious and absolutely unscientific Demon of Chance"—the *Dämon des Zufalls*. Forestalling the "rule of the eternal law" was this demon's specialty. Helmholtz wondered whether certain natural phenomena would remain forever resistant to the advances of science. If so, it was also worth asking, why? "Is it possible to show reasons why the rebellious and absolutely *unscientific demon of chance* is still defending this domain against the rule of the eternal law, which is at the same time the dominion of the understanding of thought?"[51]

For how much longer could the "Demon of Chance" keep up its reign of terror? Only "a spirit which had the exact knowledge of the facts, and whose operations of thought were carried out quickly and precisely enough to foresee the events" could beat him. This oppositional *Geist* would "see the harmonious power of eternal laws in the wildest capriciousness of the weather, no less than in the course of the stars."[52] As these two German scientists debated the strength of Laplace's creature in comparison to the "Demon of Chance," others started to look more carefully into the abilities of Maxwell's demon.

Maxwell's demons landed in Germany in the form of a short review in a specialized research journal. It described *Dämonen* who "have the ability to seize every single Molecule in every moment."[53] The demons had not traveled easily from England to the Continent, and neither German nor French scientists were much taken by them. Instead of leading scientists to espouse the statistical interpretation of thermodynamics advocated by Maxwell and his close colleagues, Continental scientists, for the most part, dismissed their significance to these explanatory frameworks. One of those scientists befuddled by demon-based British theories was Rudolf Clausius, one of the founders of thermodynamics.

Clausius's many accomplishments included having introduced the concept of entropy—which he coined from the Greek ἐν ("in") and τροπή ("transformation")—and clarified the relation of the two laws of thermodynamics to each other. He first encountered Maxwell's demon in the second edition of Tait's *Lectures on Some Recent Advances in Physical Science*. Clausius had cause for concern. The book criticized his own theory for not being able to explain how something could get colder than its warmer surroundings instead of reaching equilibrium. These cases appeared clearly when humans intervened, such as by closing windows on a hot day, and they could be sustained for a limited but significant amount of time. The German, Tait claimed, had not dealt successfully with these seemingly exceptional cases. Clausius, according to Tait, could not convincingly show "that it is possible, but possible only in a very curious way, and to an extremely limited extent, to get round this apparent difficulty,—to make a body colder than surrounding objects, and to get work from it in consequence." This difficulty, *"alone, is absolutely fatal to Clausius' reasoning*," he wrote, and thus invalidated his entire contribution. The problem, Tait claimed, "was first pointed out by Clerk-Maxwell not long ago." To solve it, "he showed that the mode of escape from the difficulty is, that it would require the intervention of beings, still finite, but infinitely more acute and able than any human beings (or even than the utmost ideal a human being can well conceive)."[54]

Clausius was appalled by Tait's comments. He criticized the Scottish scientist not only for distorting his contributions but for promoting some deeply flawed ones. Soon the two rivals would engage in a game of one-upmanship to determine who should get credit for the discovery of the laws of thermodynamics. Clausius struck back in a dedicated section titled "Another Objection from Tait," which he included in the second edition of his book. In it, he accused Maxwell and company of crafting a theory that worked "only when demons intervene." The advantages of his own work were clear in that he was "not concerned with what heat can do with the help of demons, but with what it can do by itself." Clausius went on to criticize Tait's explanation of nature for claiming "that what demons could do on a large scale, really goes on without the help of demons (though in a very small scale) in every mass of gas."[55]

The astrophysicist and expert in optical illusions Johann Carl Friedrich Zöllner soon entered the debate between Clausius and Tait. Zöllner, who held the prestigious chair of astrophysics at Leipzig, took up Clausius's objections and expanded on them in an article titled "Thomson's Demons and the Phantoms of Plato"—the latter referring to the famous allegory in *The Republic* of prisoners in a cave mistakenly interpreting the shadows reflected on their wall as reality itself.[56] Clausius's defender soon became Tait's enemy. Tait reviewed Zöllner's comments, with derision, in *Nature*. "Prof. Zöllner seems to think that Clerk-Maxwell, Thomson and myself *believe in the existence* of those imaginary beings (invented by Maxwell and called Demons by Thomson) who were introduced with the purpose solely of showing the true basis on which the Second Law of Thermodynamics has been received as a fact in physical science!"[57]

Zöllner was an easy target. In an article published in the *Quarterly Journal of Science* in 1878, the astrophysicist had inquired into the possible existence of four-dimensional spaces and of beings who might inhabit them. "If," he wrote, "there were beings among us who were able to produce by their will four-dimensional movements of material substances, they could tie and untie such knots in a much simpler manner." Mathematically speaking, the latter task was impossible for three-dimensional beings. Zöllner did not think that science had ruled out the possible existence of those beings, although he cautioned that "it is by no means necessary—nay, not even *probable*—that such beings should have a contemplative consciousness of the actions of their will."[58] In addition to these musings, Zöllner's reputation was fatally compromised when he fell for the tricks of the "medium" Henry Slade. In a spiritualist soiree with Slade, Zöllner became convinced that he had caught a glimpse of these beings. In his descriptions of the séances he attended, Zöllner offered his readers tangible details of the strange, sudden, and mysterious appearance of ghostly hands that came from nowhere and could tie knots on strings connected on both ends. Tait eagerly ridiculed Zöllner, effectively separating Maxwell's demons from Zöllner's and Slade's disreputable ones and distinguishing his research from that done in séances aimed at connecting with spirits and other inhabitants of the supernatural world. His attacks were so effective that the reputations of both the German scientist and the medium he had so trusted never recovered. As Zöllner's demons were effectively exorcised, the importance of Maxwell's grew.

Despite controversies about the role of imaginary beings in scientific explanations, Maxwell's *Theory of Heat* made important inroads into Germany, and the translator of Thomson's lectures soon took on the challenge of translating it. The first translation, appearing in 1877, was followed by another the next year.

A young physicist named Max Planck read both versions. He immediately expressed his dissatisfaction with that brand of thermodynamics by referencing the exact place in Maxwell's book where he had invoked his imaginary being. The problem motivated him to search for other explanations of heat not based on molecular thermodynamics. The alternative he developed, quantum mechanics, would eventually change the course of physics.[59]

ETERNAL RECURRENCE

Despite attempts by British scientists to get the upper hand with respect to priority claims around thermodynamics, resistance to their particular approach continued to be strong on the Continent. Soon Austrian scientists would hear about another molecular-sized demon, one whose discovery had allegedly preceded Maxwell's. Known as "Loschmidt's demon," he could also make the world run in reverse. Among his many accomplishments, the Austrian physicist Josef Loschmidt was known for having made some of the most exact determinations of the size and shape of molecules and for creating some of the most useful models for them.

Loschmidt also calculated the chances of eternal recurrence. Could a collection of molecules find themselves serendipitously in exactly the same place as they had once been? If they ended up in the exact state as before, their organization at the very next instant would also play out exactly as before, and so on. History would repeat itself. Following Loschmidt, more and more scientists throughout Europe calculated how long it might take for that to happen. If "after a time τ which is long enough to obtain the stationary state, one suddenly assumes that the velocities of all atoms are reversed, we would obtain an initial state that would appear to have the same character as the stationary state," he wrote. "Gradually the stationary state would deteriorate," he explained, "and after the passage of the time τ we would inevitably return to our original state." The conclusion was stunning:

"Obviously, in every arbitrary system the course of events must become retrograde when the velocities of all its elements are reversed." How could it be done? "The important problem of undoing what has happened has not been solved," Loschmidt added, "but it has received a simple formulation consisting in the simple instruction to suddenly reverse the instantaneous velocities of all the atoms of the Universe."[60]

Like so many others, the physicist was fascinated by "the terroristic nimbus of the second law" that "lets it appear as a destructive principle of all life in the universe," and he fantasized about how to circumvent it, becoming increasingly fascinated by the rare but real possibility that once molecules reached equilibrium a slight intervention might make the world run in reverse and everything would start again, only backward, before going forward again.[61] In years to come, more physics textbooks, including the influential *Elementary Principles of Statistical Mechanics* (1902) by Josiah Willard Gibbs, professor of mathematical physics at Yale University, would explain how the process would work and how long it might take for it to occur spontaneously.[62]

Scientists' speculations and calculations with respect to eternal recurrence had wide implications. The philosopher Friedrich Nietzsche worried about this very possibility, asking his readers:

> What, if some day or night a demon were to steal after you into your loneliest loneliness and say to you: "This life as you now live it and have lived it, you will have to live once more and innumerable times more; and there will be nothing new in it, but every pain and every joy and every thought and sigh and everything unutterably small or great in your life will have to return to you, all in the same succession and sequence—even this spider and this moonlight between the trees, and even this moment and I myself."

If all beings in the universe, including animate ones, were made up of the same material elements as those constituting sand and dust, this nightmare could seem closer to reality than we might otherwise think: "The eternal hourglass of existence is turned over again and again, and you with it, dust from the dust!"[63]

"God is dead. God remains dead. And we have killed him." In the same text where Nietzsche speculated about eternal recurrence, he wrote some of the most controversial lines of the century. These thoughts drew out the extreme implications of another widely discussed scientific speculation popular at that time: the possibility of traveling faster than the fastest messenger. "Lightning and thunder require time," wrote the philosopher. "The light of the stars requires time; deeds, though done, still require time to be heard."[64] The extreme velocity of light had fascinated philosophers and scientists alike for centuries, but the discovery that it was measurable, perhaps even breachable, had made it even more entrancing.

Nietzsche imagined someone far away in a remote corner of a universe who witnessed the death of God, lived to tell it, and came to earth to announce the news. In his estimation, this messenger would no doubt be considered a madman. But such a harsh judgment might simply be due to a mismatch between the relative speed of the messenger and that of the deed itself as both traveled through the universe. Nietzsche wrote down what the poor wretch who delivered the portentous news might come to realize upon noticing that no one believed him. "I have come too early," he would realize. "My time is not yet. This tremendous event is still on its way, still traveling; it has not yet reached the ears of men."[65]

At the time Nietzsche wrote these aphorisms, physicists were hard at work providing ever more precise measurements of the velocity of light.[66] Their research left a mark in Nietzsche's *Die fröhliche Wissenschaft*, usually translated as *The Gay Science*, so titled because of how it was inspired by the cutting-edge investigations of his era.

HEAT TRANSFER

Maxwell's demon might not have had the last answer when it came to heat. Some aspects of heat could be understood by the movement of particles, such as vibrating molecules bumping into each other. But others seemed more closely related to the properties of light. Radiant heat could be carried by invisible electromagnetic waves and propagated in a wavelike manner to raise temperatures, yet those mechanisms differed from those produced by convection and conduction. They could not yet be well understood through the application of statistical molecular thermodynamics.

How would a Maxwell demon deal with radiant heat *waves* instead of moving particles? This was the question that the astronomer Henry Turner Eddy, a professor at the University of Cincinnati, asked himself. "Since radiations are known to be moving in space apart from ponderable bodies and subject to reflections," they would slip from Maxwell's demons' fingers like waves overcoming all obstacles to reach shore. As he analyzed "Maxwell's 'sorting demon'" in detail, the American attacked Clausius and Thomson in equal measure. Eddy agreed that "if we suppose minute beings, endowed with senses sufficiently acute and having a corresponding agility, to guard minute openings in the diaphragm separating two portions of the same gas," it would be possible "to transfer heat from a colder to a hotter body."[67] Yet that could not be the entire story when it came to all the heat in the universe. The fact that "no one had as yet invented any machine, or discovered any principle on which it was possible to construct a machine," did not mean that one could not be found in the future. Studying the relation between radiant theories of heat and kinetic ones was urgent, argued Eddy, since "a large part of the exchange of heat in the universe takes place in the radiant form."[68]

Another American, the physicist Harold Whiting from Harvard, developed yet another objection to existing explanations of thermodynamics. In an article titled "Maxwell's Demon" and published in *Science*, Whiting used examples from astronomy to point out the relevance of possible Maxwell's demon–type violations of the second law. Circumstances as rare as those that permitted life to flourish on earth might also be creating other consequential effects in the universe. Something insignificant for us on earth might not be when the entire cosmos was taken in consideration. On this planet, those effects could seem minimal enough, but they were hardly so when considering all the masses of the universe. "Hence in the case of astronomical bodies, particularly masses of gas," wrote Whiting, "the molecules of greatest velocity may gradually be separated from the remainder as effectually as by the operation of Maxwell's small beings." Those particular molecules might take off from the rest. "When the motion of a molecule in the surface of a body happens to exceed a certain limit, it may be thrown off completely from that surface, as in ordinary evaporation." Its trajectory would be exceptional, yet comparable to the trajectories of rare mutants who managed to survive against the odds. "It seemed to me of interest to point out that what, as Maxwell has shown, could be done by the agency of these imaginary beings, can be and often is actually accomplished by the aid of a sort of natural selection," Whiting wrote.[69] This "natural selection," he stated, could create extraordinary feats with ordinary molecules. With these and other speculations, the problem of Maxwell's demon and the opportunity it posed moved from physics to biology.

During those years, Maxwell's demon started to be used to explain a wider set of effects that seemed to go against the usual order of things. Osmosis, the upward pulling of fluid in plants from root to crown, and the "spontaneous evaporation at the surface of a liquid into an unsaturated gas, such as the atmosphere," were added to a growing list in a "'sorting demon' class of molecular actions."[70]

Despite criticisms raised on the Continent and in America, the fame of Maxwell's demon quickly spread in the English-speaking world. The mathematician and geneticist Karl Pearson, a protégé of Darwin's cousin Francis Galton and one of the most prominent statisticians of his time, cited in full Maxwell's key paragraphs from *The Theory of Heat* in one of the most widely read science books of his era, *The Grammar of Science*. This "being," wrote Pearson, "has become known to fame as Clerk-Maxwell's demon." Pearson highlighted one phrase in Maxwell's rather lengthy description. "Clerk-Maxwell," he wrote in a footnote, "supposes the being's attributes 'essentially finite as our own.'" But since such finitude was "a peculiarity not usually associated with demons," Maxwell's creature might not need to be defined by such a controversial label. Clearly skeptical about the use of the term, Pearson described the demon as "having a perceptive faculty differing rather in degree than quality from our own." The gap between his perceptive abilities and ours was probably similar to those between us and other animals, or between "civilized man" and "savages" and children: "Clerk-Maxwell's demon would perceive nature as something totally different from our nature, and to a less extent this is in great probability true for the animal world, and even for man in different stages of growth and civilization." As Pearson noted in his conclusion: "The worlds of the child and of the savage differ widely from that of normal civilised man." The difference in the perceptual capacity of "normal civilised man" and that of Maxwell's demon could be similarly understood.[71]

The American writer Henry Adams never fully recovered from the impression that Pearson's *The Grammar of Science* left on him.[72] It motivated him to think about history in terms of entropy and thermodynamics and about the economy as a system "where only Clerk Maxwell's demon of Thought could create value."[73] Adams's perspective on this creature was marked by his experience as President Abraham Lincoln's ambassador to the United Kingdom, where he started to consider the actions of Maxwell's demon in geopolitical terms. "Germany," Adams wrote in his correspondence, "is and always has been a remarkably apt illustration of Maxwell's conception of 'sorting demons.' By bumping against all its neighbours and being bumped in turn, it gets and gives at last a common motion."[74] During those years, the memory of France's defeat at the hands of Germany during the Franco-Prussian War seemed as vibrant as ever. Both countries continued to arm themselves, thirsting for revenge rather than risk a humiliating defeat. During a trip to France, Adams "devoured" *Science and Hypothesis* (1902) by the French physicist Henri Poincaré, which gave him many more details about Maxwell's demon. He immediately started thinking about how that creature might serve his country. In 1903, a few years after President William McKinley was shot and Theodore Roosevelt assumed the presidency, Adams wrote with humor to his brother Brooks: "Clerk Maxwell's demon who runs the second law of Thermodynamics ought to be made President."[75]

WITH OR WITHOUT THEIR HELP

Demons became the focus of attention once again after Loschmidt died in the summer of 1895. Ludwig Boltzmann, who is best known today as the author of the famous entropy formula that carries his name ($S = k \log W$), was one of Loschmidt's closest friends and collaborators, and deeply invested in continuing Loschmidt's program. It is because of Boltzmann that statistical thermodynamics and Loschmidt's molecular view of nature associated with it are alive today. And it was Boltzmann who would stave off the near-death of Maxwell's demon at the hands of German researchers who did not accept such explanations.

Boltzmann gave a different history, interpretation, and parentage to the demon associated with Maxwell. The idea for the demon, according to him, originated in Loschmidt's Austria. During a memorial address commemorating his recently deceased colleague, Boltzmann claimed that Loschmidt had been the first to "invent a tiny intelligent being who would be able to see the individual gas molecules and, by some sort of contrivance, to separate the slow ones from the fast." This imaginary creature was "only hinted at in a few lines of an article," yet, Boltzmann claimed, it "was later introduced in Maxwell's heat theory and was widely discussed."[76] Little did it matter to Boltzmann that Loschmidt had not used that word. Loschmidt had helped Boltzmann improve on his work by imagining some of the most frightening conclusions that arose from the "reversibility paradox"; Boltzmann, in turn, had offered a solution to the paradox by developing his famous formula showing that molecular movements could be reversible at the level of atoms but rarely in the aggregate.

Is "almost never" an adequate scientific answer? Boltzmann was challenged by criticisms of statistical mechanics (such as Wilhelm Ostwald's "energetics" alternative and research on radiant heat waves, including Eddy's and later Planck's). He would not let a few rare exceptions ruin such a promising and comprehensive theory as statistical thermodynamics. Boltzmann responded to critics by offering a clear justification for writing off rare events, whatever their origin. The fact that they were rare was reason enough to discount them from the universe's general accounting sheet. Insurance companies survived and made a profit on that basis. Why should scientists fret? "One may recognize that this is practically equivalent to *never*," he wrote, "if one recalls that in this length of time, according to the laws of probability, there will have been many years in which every inhabitant of a large country committed suicide, purely by accident, on the same day, or every building burned down at the same time—yet the insurance companies get along quite well by ignoring the possibility of such events."[77] With Boltzmann, the demon of chance was weakened, but not finished.

Albert Einstein devoured Boltzmann's research while he was creating his own famous contributions to science. He too would strive mightily to contribute to the molecular theories on which Boltzmann's research was based. "The Boltzmann is magnificent," Einstein wrote to his girlfriend at the time. "I have almost finished it. He is a masterly expounder. I am firmly convinced that the principles of the theory are right," he concluded.[78]

Boltzmann committed suicide by hanging on September 5, 1906. Was his death just another sad statistic? A growing number of scientists continued to be absorbed in studying why, when, and under what circumstances those momentary violations to the second law would occur and why. The possibility of time travel, the fear and promise of eternal recurrence, the existence of other forms of heat transmission in the universe, and the relation of chance to fate were all at stake.

Rumors circulated that criticisms leveled by his colleague Ernst Mach drove Boltzmann to kill himself. The professional disagreement between Boltzmann, who defended statistical thermodynamics, and Mach, who attacked it, had quickly become personal, exacerbated by Boltzmann's position as chair of physics in Vienna and Mach's as chair of history and philosophy of science. The two became sworn enemies.

Mach was one of the most influential voices in science and philosophy during those years. Not ready to give in to explanations based on demons, and unwilling to espouse the molecular view of nature on which it was based, Mach foreswore statistical thermodynamics in its entirety owing to clear evidence of exceptions, no matter how rare, to the second law. The "interpretation of the second law in terms of mechanics seems to me rather artificial," wrote Mach. "If it were true that such a mechanical system forms the real basis of thermal processes," he continued, nothing would prevent these exceptions from becoming the norm. "One can hardly resist the conviction that a violation of the second law should be possible—even without the help of demons."[79] At stake in Mach's criticisms was the question of what should count as science. How much should science relay on actual sense impressions? According to Mach, these should be foundational.

By the end of the century, Mach's fights with some of his colleagues and their approaches developed into a full-blown war against "metaphysics." Younger, idealistic scientists and philosophers eagerly joined his antimetaphysical program and started to meet regularly as an informal group named Verein Ernst Mach, later known as the Vienna Circle. They would soon find a champion in Albert Einstein and his theory of relativity.

Einstein was a great admirer of Mach's parsimonious philosophy of science, arguing that when two hypotheses explained equally well the same phenomenon, the simpler and more economical one—that is, the one with fewer hypotheses—should prevail. The physicist believed that his theory of relativity met this criterion. But what appeared simple to Einstein was deemed complicated by most, especially by his critics. Coming to an agreement about what were the best and the simplest hypotheses that described the universe in its entirety would prove much harder. Yet if necessity was the mother of invention, disagreement became the parent of discovery.

4
Brownian Motion Demons

How can microparticles and microdroplets sometimes remain suspended in midair for prolonged periods of time, seemingly defying the laws of gravity, preventing dust from settling down and sometimes aiding the spread of communicable diseases? Maxwell's demon appeared to be more active than ever during the first decades of the twentieth century. When a beam of light enters a dark room, tiny dust particles can be seen held up in the air, dancing around in every direction. What makes them move so? Perhaps Maxwell's creature is pushing them hither and yon? When scientists looked at these particles through their microscopes, some thought they might be seeing the demon in action. At the beginning of the twentieth century, more and more scientists started considering the possibility that this demon was real in some sense, that he could be put to good use, and that there might be others like him.

What prevents these particles from descending? When a ball rolls down a path, friction slows it down until it stops completely. It loses momentum and speed progressively, before coming to a dead halt. But particles that measure around one-thousandth of a millimeter can float in the air and on the surface of liquids, defying such constraints. Their strange movement was eventually named "Brownian motion," after the Scottish botanist Robert Brown brought attention to them. Something, or someone, appeared to egg them on. Could their energy be harnessed to run engines or even to power computers?

Fascinated by these particles, Albert Einstein studied them in order to better understand the movement of molecules and the nature of heat. His research answered many questions—but it also opened up new riddles about the origin of life, about the behavior of chaotic systems, and even about the nature of intelligence and consciousness in our universe.

The young Albert Einstein was not one to ascribe the motion of strangely animated floating microparticles, or of anything else in the universe for that matter, to demons. He decided to look elsewhere for answers. Einstein's first research paper on this topic, "On the Motion of Small Particles Suspended in Liquids at Rest," was published in the prestigious *Annalen der Physik* in 1905. It was one of four Einstein articles that appeared that year (now known as his *annus mirabilis*) that would change the course of history. Every single one of these revolutionary papers dealt with topics that had been previously tackled by reference to demons. But in contrast to his colleagues, Einstein shunned such terminology. In doing so, he showed that what had been explained by reference to them could be explained solely by the new theories of physics he was developing.

The presenter of Einstein's Nobel Prize many years later recalled that "throughout the first decade of this century the so-called Brownian movement stimulated the keenest interest." Einstein's contributions resided, he said, in having "founded a kinetic theory to account for this movement by means of which he derived the chief properties of suspensions, i.e. liquids with solid particles suspended in them."[1] Einstein's colleague and friend Max Born summarized Einstein's work as having been key to the establishment of atomistic theories more generally. "These investigations of Einstein," he concluded, have "done more than any other work to convince physicists of the reality of atoms and molecules." Einstein himself was thrilled by this accomplishment. Born, now known for his contributions to quantum mechanics, also congratulated his friend for showing "the fundamental part of probability in the natural laws."[2] Einstein would *not* be so proud of that accomplishment, as it opened up an interpretation of nature long associated with the demon of chance.

Historians usually trace the earliest descriptions of these animated particles to Lucretius, who, in the first century BC, described in his poem *De Rerum Natura* the dust particles that could be seen floating in the air when sunbeams shone into a building. Other observers, including Robert Brown, looked at pollen floating on puddles of rainwater. These researchers thought that the particles might be moving because they were miniature organisms, living *animalculae* that shook, danced, and wiggled. To test this hypothesis, investigators tried an experiment: they changed the pollen to dust. The dust particles moved too. Next, they sealed them off in a jar to see how long the movement lasted. After a year, to their utter amazement, the particles were moving as they had been on day one.

The importance of Einstein's work on "Small Particles Suspended in Liquids at Rest" was at first difficult to recognize, in part because Einstein had arrived late. The French mathematician and polymath Henri Poincaré had obsessed about these particles in relation to Maxwell's demon for more than a decade. Historians have been fascinated by how "strikingly similar" Einstein's and Poincaré's work on the theory of relativity was.[3] Einstein's work on Brownian motion is another instance of a young researcher moving into an elder's research agenda. Einstein read Poincaré's research on the topic of Brownian motion as soon as it became available. It took him at least three years to publish something of his own.[4]

Throughout their lives, Poincaré and Einstein would have a distant and complicated relationship. The Frenchman's privileged pedigree loomed large over the only son of a German Jewish family that had fallen into deep financial straits, a rebel with an authority problem, a physics student who struggled with mathematics, a handsome womanizer who fathered a daughter out of wedlock before abandoning her, a second-class patent clerk and college graduate who failed to get a permanent job in academia for many years. Poincaré was bourgeois, Catholic, and an esteemed member of the political and scientific establishment of the Third Republic who counted as his cousin a prime minister and president of France and who, at an early age, had gained the reputation of being a "monster of mathematics." Poincaré was Einstein's senior by twenty-five years.

Poincaré was fascinated by Brownian motion. He was particularly intrigued by the claims of a scientist who believed that he had caught a glimpse of the wheeling and dealing of Maxwell's demon's jiggling, tiny, microscopic particles suspended in liquids. The physicist who witnessed such a spectacle was Louis Georges Gouy. In 1888, Gouy reported what he had seen to the scientific authorities. Reports of more and more witnesses started pouring in soon after. Scientists throughout Europe started scouring archives and rereading the annals of science looking for corroborating evidence.

Gouy recounted most of what was currently known about Brownian motion in the *Journal de physique, théorique, et appliquée*. He explained that what he had seen was widely known to microscopists but had not yet caught "the attention of physicists." It was "a most striking sight," he wrote, clearly captivated, "a sort of swarming or general trepidation." "Each particle seemed to move independently of its neighbors," he continued. They seemed to move "indifferently in all directions." Sometimes a particle "turned around on itself irregularly." The smaller they were the faster they seemed to move. When their size was down to about one-thousandth of a millimeter, they were so fast that their movement could barely be noticed.

Gouy saved the best for last. What if these particles could be harnessed to power miniature windmills? He speculated in the last paragraph of his article on whether these observations could lead to the creation of a perpetual motion machine: "we can recover work . . . in opposition to Carnot's principle," he argued. It was "certain that work is spent" and that "one could conceive of a mechanism through which a portion of that work can become available."[5] From that moment onward, these small particles acquired the dubious reputation of being minuscule violators of the second law of thermodynamics.

Gouy's investigations were particularly shocking for going against one of the sturdiest lessons of science. Since the early decades of the century, scientists had eloquently decried the madness and foolishness of chasing get-rich-quick schemes based on perpetual motion machines. Now Gouy claimed that such limitations might not apply in a world of "dimensions comparable to 1 micron." Studies on molecular theories of heat, Brownian motion, and Maxwell's demon could expand our understanding of the basic building blocks of our universe, but they could also be useful for exploring the possibility of harnessing their energy, combating entropy and decay, and stopping or reversing time. A frictionless mini-windmill that would never stop going would solve the world's energy problems and perhaps even those of the entire solar system. If only this Tom Thumb of energy could be scaled up! The dream of finding Brownian motors or computers still captivates scientists today.

Could Gouy's miniature windmill actually be built? "Imagine," he wrote, what would happen if we attached a very light ratchet wheel to the particle using a very narrow thread. It would "make the wheel turn and we could recover work." His speculations remained "evidently unrealizable" at present, he admitted humbly, but his reasoning was theoretically sound. Try as he might, he "could not see a theoretical reason that it would not work."[6] As if solving the dream of "producing" work "from ambient heat" in "opposition to Carnot's principle" was not exciting enough, Gouy finished his article off with another momentous speculation: "living tissues" might operate in the same way.

Gouy's investigations had been preceded by those of the British economist William Stanley Jevons. "It is natural to enquire how long the pedetic movement will go on," wrote Jevons. "Does it exhaust itself rapidly? The experiments which I have made lead to the opposite conclusion in a wonderful degree."[7] When he looked at particles in port wine, Jevons saw the long-sought philosopher's stone: "This experiment leaves no doubt in my mind that the sediment of port wine is in a state of perpetual motion, until it finally settles down and attaches itself to the glass." When he mixed china clay and water, the movement even seemed to increase, after years of being left untouched: "Thus, after a trial of eight or nine years' duration, we meet with the astonishing fact that the suspensory power and the pedetic motion apparently increase with time." The effect was so stark that Jevons concluded: "In fact this pedetic motion seems to be the best approach yet discovered to a perpetual motion."[8]

What made them move? Discovering their "long-continued supply of energy" was "required to explain the phenomena."[9] Researchers did not seem to be bored. A British microscopist studied them "at intervals for thirty years."[10] Jevons himself performed "nearly eight hundred" carefully recorded observations.[11] They tried pretty much any fine powder they could lay their hands on—from mineral to vegetable—and tested the effects of light and heat on it. Jevons described how to obtain this spectacle in detail. Take "a drop of old common ink which has been exposed to the air for some weeks" and examine "it under thin glass with a magnifying power of 500 or 1000 diameters."[12] Among many others substances, Jevons tried powder from clay of several kinds, earthenware, fire-brick, common brick, clay slate, silicified wood, glass of several kinds, talc, mica, pumice stone, asbestos, basalt, granite, and slag from iron and other furnaces; siliceous minerals such as topaz, tripoli powder, glacier sand, purest quartz crystal, agate, cornelian, chalcedony, or the finest white sand, silicates, and silica; oxides and other compounds devoid of silica; chalk, fluor spar, hematite iron ore, galena, oxide of tin, barium sulphate, calcium sulphate, titaniferous sand, bone-dust, charcoal of several kinds, coal, coke, emery powder, amorphous phosphorus, iron, steel, gold, and platinum, antimony, arsenic, and bismuth, steel, lead, bronze, alcohol, ether, amylic alcohol, chloroform, spirits of turpentine, wood spirit, tree sap; and finally, the yolk of an egg. Because all of these substances performed so well, Jevons came to the conclusion that the movement could not be due to any specific chemical. "Substances of the most widely different chemical characters will exhibit the phenomenon, and it is difficult to establish any clear differences in the activity of the motion as connected with the chemical nature," he concluded.[13]

THE "CUSTOMS OFFICER" OF THE UNIVERSE

Soon investigations of Brownian motion would intersect with those of Maxwell's demon. Poincaré wrote down his thoughts about Maxwell's demon in an article published in the first volume of the *Revue de métaphysique et de morale* in 1893. His earliest investigations on the topic already show him drawing conclusions from that creature that tempered his acceptance of the wholly "mechanistic conception of the universe" that, he argued, "has seduced so many good men."[14] He was not one of them.

Poincaré noted that the demon was born of attempts to explain irreversible effects from reversible laws: "To better understand them, Maxwell introduces the fiction of a 'demon' whose eyes are subtle enough to distinguish molecules, whose hands small and fast enough to grasp them." Poincaré explained what this creature could do: "For such a demon, if one believes the defenders of a mechanical approach [*les mécanistes*], there would be no difficulty in passing the heat of a cold body to a hot one." He then discussed the difference between reversible and irreversible effects by asking readers to consider what happened when one mixed a grain of barley with one hundred liters of wheat. The grain got lost in the wheat, and "it would be almost impossible to find it again." Such an action, he explained, was irreversible in practice, but not in theory. Exponents of Maxwell's demon used it to explain why at the molecular level actions were reversible but at our level they were not. "Maxwell made an ingenious effort to triumph. But I am not so sure he succeeded," wrote Poincaré, concluding with an adverse assessment of the contributions of the Scotsman.[15]

Poincaré, like the increasing number of researchers who focused on these topics, worried that statistical explanations were being used to write off certain effects.[16] Some scientists were satisfied with the simple explanation that the chances of effects being reversed at macroscopic scales were low, while at microscopic scales they were high. Poincaré reminded readers and colleagues, however, that even when the chances of something happening were low, given enough time, they were likely to happen. What was more, given an *infinite* amount of time, they *had* to happen. This was true when it came to heat transmission and dissipation. "According to this theory," wrote Poincaré, "to see heat pass from a cold body to a warm one, it will not be necessary to have the acute vision, the intelligence, and the dexterity of Maxwell's demon; it will suffice to have a little patience." "A little patience," in Poincaré's tongue-in-cheek formulation, could be as long as many orders of magnitude more than the presumed age of the universe.[17] While some scientists felt justified in writing off low-probability effects when thinking about the universe as a whole—from its lowly beginnings to its momentous end—others, such as Poincaré, believed that they were compelled to take low-probability effects into consideration or risk missing the full story of the cosmos.

In Poincaré's estimation, British kinetic theories had bizarre implications, which the Frenchman enjoyed describing to a wider public. One of them was that in a bounded universe, at some moment in time, given the tumultuous movement of the molecules it contained, all molecules would necessarily end up serendipitously just like they were at an earlier moment. At such a crossroads, everything would start to happen once again as before: every previous state of the universe would repeat itself identically, creating a state of endless recurrence. Yes, it would most likely take "an enormous amount of time" for this strange occurrence to happen, according to the mathematician, but if one followed the current molecular theories of science to their logical conclusions, it would surely happen at some point. "Such a state," he explained, "would therefore not be the definitive death of the universe, but a sort of sleep, from which he will wake up after millions and millions of centuries." Poincaré hoped that astronomers might soon witness such a spectacle. "One would like to be able to stop at this point and hope that some day the telescope will show us a world in the process of waking up, where the laws of thermodynamics are reversed," he wrote.[18] The strange possibility of eternal recurrence detailed by Poincaré would later be related to two demons: Loschmidt's demon and Zermelo's demon, the latter referring to the work of the German logician Ernst Zermelo.[19]

In the year 1900, Poincaré turned the fate of Maxwell's demon around completely. He found it hiding in the most unlikely of places—right under his nose. "One would think he is seeing Maxwell's demon at work," he wrote in amazement, after lifting up his head from a microscope where he witnessed moving Brownian particles.[20] After him, more and more scientists started seeing demonic forces right in front of them. What they saw was extremely shocking.

The Frenchman presented his ideas about Maxwell's demon and Brownian motion to a wider public at the International Congress of Physics of 1900.

William Thomson (already knighted Lord Kelvin) and Gouy were both present. "Consider here the original ideas of Monsieur Gouy about Brownian movement," Poincaré told his audience. "According to this scientist, this singular motion should violate Carnot's principle." His presentation immediately raised the profile of Gouy's work and of Brownian motion in general. These investigations were about much more than just the movement of dust. Poincaré explained that his interest in the demon had first been kindled by the "reversibility paradox" that arose from using perfectly reversible laws of mechanics to explain a macroscopic world that was evidently not reversible. "Everything seems to advance in one direction, without hope of return," he noted.[21] Molecular physicists had no adequate explanation for "such rebellious" irreversible phenomena, which characterized most of human experience and which had led them to "suppose that irreversibility is nothing other than apparent." At a macroscopic scale involving too many particles, our "coarse eyes" saw the world tending toward uniformity. But Maxwell's demon, living comfortably in the midst of atoms and hyperfine particles, saw the world differently with his ultra-fine senses. The French scientist added yet another ability to this unlikely creature, in addition to those already pointed out by Gouy: it could reverse the direction of time. "Only a being of infinitely subtle senses," speculated Poincaré, "such as the imaginary demon of Maxwell, could unravel this tangled skein and make the universe run in reverse."[22]

As thermodynamics became more widely accepted, more and more listeners flocked to listen to scientists lecture on the topic. Demons started traveling from international scientific conferences to world's fairs. At the Universal Exposition in St. Louis in 1904, the public gathered to hear Poincaré's keynote lecture stressing the importance of Maxwell's demon and Brownian motion for physics. His widely celebrated talk was translated, published, and republished many times. According to Poincaré, Maxwell's original speculations, in light of Brownian motion, had left the realm of pure theory and become newly urgent. Until then, all objections to the second law of thermodynamics "remained theoretic and were not very disquieting." But once Brownian motion entered the picture, "the scene changes." "To see the world return backward, we no longer have need of the infinitely subtle eye of Maxwell's demon; our microscope suffices us," explained Poincaré.[23]

Sublime black swans. Four-leaf clovers. We are right to disregard most rare events as inconsequential because we know how uncommon they are, but a few are not so easily dismissed. Poincaré was among those who never really did write off the tiny possibility that the actions of Maxwell's demon were consequential. At the right time and place, minute changes could act or combine in such ways as to cause exponential effects. The evolution of the universe, the nature of nebulae, and avoiding the heat-death of the universe were contingent on what Maxwell's demons could or could not do. If the universe was finite, they could achieve only so much. If it was infinite, then they could do much more.

"Our little demons," he wrote, "will accumulate to the right all fast molecules and to the left all slow ones." They could separate a gas that was at one temperature into "two parts with different temperatures." The result could be stunning: "They will have *turned around* Carnot's principle." Poincaré recounted the work of astronomers who had already found evidence of these "petits demons" in the universe and who "thought to have found an analogous mechanism that is produced naturally" at the outermost fringes of planetary atmospheres and in nebulae. Having seen the demon's actions under his microscope, and surmising that he must also be working away in other places in the universe, Poincaré elevated Maxwell's demon to the official status of "customs officer" of the universe.[24] It was this being who ultimately determined where molecules could go to meet, mix, and accumulate throughout space.

THE CATALYST, THE TRIGGER, AND THE SCALE-TIPPER

The scientist and engineer Alan Archibald Campbell-Swinton, now known for sketching out the first theoretical model for television, was among many other scientists who became fascinated with Brownian motion's entropy-retarding qualities. In his presidential address to the Röntgen Society in 1911, titled "On Scientific Progress and Prospects," he explained that, with Brownian movement, "we should have immediately to hand the means of producing the perpetual motion dreamed of by mediaeval philosophers." Intensely optimistic about the prospects raised by more research on these questions, he proposed Brownian motion as one of the most interesting research areas of the new decade: "The unordered molecular motions of which the Brownian movements give us an indication—motions which constitute heat—would merely be directed for a time in the particular manner needful to give us the power that we require." "Here at last," he concluded, "we should have the perpetually burning lamp of the story books, which consumed no oil; the perpetual fire of the burning bush, which requires no fuel."[25]

How did Laplace's demon fare in the context of these new investigations? The philosopher Wildon Carr cited the "well-known passage" where

"Laplace imagined an ideal calculator to whom the state of the universe at any future moment would be fully known." For these thinkers, prediction was possible only if the world was predictable, so that Laplace's being was reduced to a mere calculator—something like an automaton, and no better than it. Yet in the eyes of most, this was clearly not the case. "That prediction may be possible the world must be viewed as uniform, as subject to natural law." Carr compared such an "imaginary calculator," who was bereft of any real understanding, to someone who would remain "stone deaf" in a "world of sound."[26]

In contrast to the predictability of Laplace's rule-following actions, Maxwell's on-the-fly operations seemed potentially more pertinent owing to their sheer shock value. Little did it matter that these operations were little. Carr described the effects of Maxwell's demon in terms of a spark used to detonate an explosive pile, or a hair trigger that, when broken, could set off an ultrafast shutter, catching everything and everyone by surprise. "The often-quoted illustrations of the operation of the hair trigger, or the firing of the electric spark to explode the mixture in the cylinder," were similar to "the idea of a Maxwell's demon who times the opening and closing of a frictionless shutter," he explained.[27]

Consider triple aces and double zeros when they appear in the croupier's hand or on the roulette table. At the right place, next to the right catalyst, rare events, though rightfully ignored in most contexts, can become explosive. With the perfect lubricant, almost everything is possible: the slightest push suffices for something to go on and on. With minimum effort, the point of perfect equilibrium can easily tip, setting off a pandemic, an avalanche, an earthquake, or a tsunami. For demons that can only act rarely, context matters tremendously. In certain cases, they can trigger extreme reversals of fate, causing unpredictable boons or disasters.

Scientists had determined that Maxwell's demon could function only in a really tiny world or in very rare cases. But they had not killed him off entirely. The punctual effects of a single molecule moving in an unexpected direction could be more consequential than the entire regular movement of the rest of the crowd.

RANDOM WALKS

Brownian motion was a crowded field by the time Einstein entered it. In his first paper, he repeated some of the same conclusions reached by previous scientists. "If it is really possible to observe the motion to be discussed here, along with the laws it is expected to obey, then classical thermodynamics can no longer be viewed as strictly valid even for microscopically distinguishable spaces," he explained in 1905.[28] But in contrast to the Maxwellians and to Poincaré, Einstein tried to explain this anomaly without resorting to demons. In his view, underlying molecular forces were solely responsible for pushing these particles hither and yon.

The year Einstein published his studies on Brownian motion, the statistician Karl Pearson published a question in the journal *Nature*. His intention was to crowdsource a solution to a difficult mathematical problem: the problem of "the Random Walk." If "a man starts from a point O and walks l yards in a straight line, he then turns through any angle whatever and walks another l yards in a second straight line," and "he repeats the process n times," where would he most likely end up? Where should a search team go to fetch such an aimless wanderer? When n was large in comparison to l, the answer was clear: "In open country the most probable place to find a drunken man who is at all capable of keeping on his feet is somewhere near his starting point!"[29] When n was small in comparison to l, the inebriated man was much harder to find. Pearson's example of the "Random Walk," or the "Drunkard's Walk," was almost immediately identified with Brownian motion and soon became famous.[30]

The French scientist Jean Perrin joined these investigations, eventually earning a Nobel Prize for his contributions. He was most fascinated by how Brownian motion could be used as evidence supporting a molecular view of nature. Perrin came to the conclusion that the particles' movements were not driven by anything that was even close to demonic; their movement was simply due to the underlying movement of the molecules that made up the liquid on which they floated.[31] "The only irrefutable explanation for this phenomenon," explained the presenter of Perrin's Nobel Prize in Physics, "ascribes the movements of the particles to shocks produced on them by the molecules of the liquid themselves." Perrin's research had brought "a definite end to the long struggle regarding the real existence of molecules."[32]

Perrin at first liked to stress that the second law was only an approximation. "The demon, imagined by Maxwell, which, being sufficiently quick to discern the molecules individually, made heat pass at will from a cold to a hot region without work," could "be recalled" to show how the second law was subject to certain exceptions. Brownian motion, according to Perrin, showed the *ascertainable reality* of the actions of this imaginary demon.

"But this would no longer be admissible, for this rigidity is now in opposition to a *palpable reality* [*réalité sensible*]," he wrote in 1909.[33] "Brownian motion never ceases," he explained a few years later. "It is eternal and spontaneous."[34] Could one take advantage of the particles' inexhaustible energy? No, according to Perrin. "If we were no bigger than bacteria," he explained, we could use Brownian motion as a perpetual machine "to build a house, for instance, without having to pay for the raising of the materials." But in fact, humans were much larger, and Brownian motion particles, alas, were too small to produce any significant work at human scale. "Common sense tells us, of course, that it would be foolish to rely upon the Brownian movement to raise the bricks necessary to build a house."[35] But running the risk of seeming foolish has rarely prevented humans from trying. Consider the chances of producing useful work from Brownian motion. Statistically, it would take "considerably more than 10 to the 10 to the 10 years" to see "a brick rise to a second level by virtue of Brownian movement," but a brick would eventually go up.[36]

Perrin, basing his work on Einstein's, immunized science from some of the most alarming consequences of Maxwell's demon. That being was too small to produce useful work, and too slow for these effects to emerge in any useful way. The chances of such were too slim to cause any significant harm or produce any meaningful quantities of work. Laplace's demon might be safe from Maxwell's demon after all.

A HUMANITARIAN CASINO

The Polish physicist Marion von Smoluchowski had also been studying Brownian particles years before and concurrently with Einstein. "It is quite remarkable how often Smoluchowski and Einstein simultaneously and independently pursued similar if not identical problems," noted one of Einstein's biographers.[37] Smoluchowski, in contrast to Einstein, wrote explicitly about the relation between Brownian motion and Maxwell's demon. "It does not seem totally impossible to chase molecules with the aid of a Maxwellian demon," he asserted excitedly in his research.[38] While the *results* of these three Brownian motion researchers (Einstein, Perrin, and Smoluchowski) were nearly identical, the *conclusions* drawn by them were not. As Brownian demons received more and more scrutiny, one idea in particular—an idea for harnessing their energy—would fascinate scientists for the next century.

Smoluchowski was not willing to simply write off speculations about using these motions to circumvent the second law. He ventured his final conclusion on the topic in a Göttingen lecture, one sentence of which would be cited over and over again. He began by stating, unsurprisingly, that "there is no automatic, permanently effective perpetual motion machine, in spite of molecular fluctuations," then added, "but such a device might perhaps function regularly if it were operated by intelligent beings."[39] Scientists would be obsessed with that intriguing *if* for the next one hundred years. With this assertion, Smoluchowski left open a possibility that microscopic "chance" fluctuations could be harnessed in a productive way. Einstein's work on Brownian motion won him numerous accolades, but it was Smoluchowski's speculations about it in relation to "intelligent beings" that fueled the imagination of scientists to come. Einstein was satisfied by the statistical interpretation of thermodynamics; Smoluchowski was concerned about how the power of our minds and intelligence could be used to harness molecular energy. When considering the "almost never" possible, Einstein and Perrin emphasized the "never." Smoluchowski stressed the "almost."

Statistically, the to-and-fro movement exhibited by Brownian particles resembled that of the back-and-forth hand movements of a skilled croupier taking the chips from a roulette player on the betting carpet. But the movement of the particles was different in one respect: the odds that they would be in one place versus another were not stacked against the player and for the house. They resembled a "humanitarian Casino that allowed a bankrupt player to continue playing, and which would eventually pay him for hitting the jackpot."[40] Their perfectly random motion presented opportunities. The good news was that, because that motion delivered its boons randomly, people were rarely completely unlucky. The bad news was that they could not be habitually lucky either. One of the consolations of ill luck was the assurance that others would be getting a shot at the jackpot. People could delight in momentary bouts of schadenfreude over the misfortunes of others because they knew full well that next time the joke might be on them. Anyone with enough "time and capital," argued Smoluchowski, could win in a "fair" game of chance under one condition: they only needed to *stay in the game and wait for a win* and not be forced to leave the game abruptly when down on chips.

The physicist did not recommend gambling as a "permanent source of income" given the amount of "time" and "capital" needed to show a "profit."

Time, Smoluchowski claimed, stood in a power of two relation to net gains. With "absolute certainty," he stressed that science proved that any player would *eventually* be confronted with a winner-take-all outcome. Smoluchowski was intrigued with the possibility of capitalizing on these outcomes, however rare they might be.[41]

"We no longer need a Maxwell Demon," wrote Smoluchowski, since we can now see "with our bare eyes" how an "automatic device" could work like him. A single-acting valve or a spool with delicate hairs like eyelashes could let certain particles pass through the hole of a vessel and prevent others from entering.[42] Constructing such mechanisms was no doubt difficult, but was it impossible? Smoluchowski reminded readers that "difficulties in technical design are not an objection if the matter is possible in principle." But there was another problem that did not seem to be merely practical: it was a "fundamental impossibility."[43] Any regular valve, he argued, that could do the work of the demon would eventually fluctuate and stop working. We would need a different kind of mechanism, an entirely different kind of valve. Imagining what that valve might look like seemed impossible "today," but Smoluchowski kept thinking about it.

Smoluchowski's published lecture was provocatively titled "Limits on the Validity of the Laws of Thermodynamics." Therein the physicist continued to compare Brownian motion to games of chance. He also reconsidered the quest to build a *perpetuum mobile*. He had given up on any purely "automatic" machine, but left open the possibility that an "intelligent creature or thing" (*intelligente Wesen*) might produce those entropy-free effects.[44] What kind of creature or thing could do that? The answer was clear: a *deus ex machina*.

> A perpetual motion is possible, all you need is an experimenting human perceived as a "Deus ex machina" by the usual standards of physics, one who is continually accurately informed of the current state of nature and who may set in motion or interrupt at any moments the macroscopic processes of nature without doing any work. He therefore needs by no means to be a Maxwell demon with the ability to trap single molecules, but would nevertheless be quite different from real living beings in the above-stated points.[45]

Despite advances in thermodynamics and statistical mechanics, Smoluchowski concluded that because of these questions, science in "modern" times was just as "speculative" as it had ever been.[46]

5

Einstein's Ghosts

Karl Pearson's The Grammar of Science, *one of the most widely read books of the late nineteenth and early twentieth centuries, contained pithy descriptions of two demons. One was Maxwell's. The other was "the colleague of Maxwell's demon." If Maxwell's demon was much like an ideal porter, his mate was more of an ideal teleporter. This colleague could travel at faster speeds, having the power of instantaneous transmission. He could act at enormous distances simultaneously and instantaneously.*

Descriptions of this being were common in the popular science books of the nineteenth century, where he went by different names. When he traveled at the same speed as light, he could freeze any scene and analyze it at leisure. No detail would go unnoticed. Children, including the young Einstein, were fascinated by stories featuring this creature. Criminals feared him as much as they feared Laplace's demon's perfect record-keeping. When he traveled at speeds faster than light, he could see and act in most unusual ways, and he could even witness the development of world history unraveling—undeveloping in reverse.

Einstein chose Pearson's book to read with a group of his closest friends. In revolutionary work he produced shortly thereafter, he succeeded in disproving the possible existence of this being. The character was so important for the theory of relativity as to become immortalized in a famous limerick:

> *There was a young lady named Bright,*
> *Whose speed was far faster than light;*
> *She set out one day*
> *In a relative way,*
> *And returned on the previous night.*[1]

"She" was not the only demon eliminated by Einstein. Arthur Eddington, a British astronomer who proved the theory of relativity during a famous

1919 eclipse expedition, hailed Einstein as an "exorcist" who had banished another demon who could act instantaneously at a distance: the "demon of gravitation."

As more and more scientists rallied around him, Einstein described his own contributions as eliminating the "ghosts" of absolute time and space.

"The colleague of Maxwell's demon" was a being endowed with great speed as well as great visual powers. The British scientist Karl Pearson described him as someone "gifted with an immensely intensified acuteness of sight so that he could watch from enormous distances the events of our earth." How would the world look to him as he hopped from star to star? "Now suppose him to travel away from our earth with a velocity greater than that of light," since, Pearson concluded, "clearly all natural processes and all history would for him be reversed." Pearson drew a stunning conclusion from the possible existence of this demon—that this being was "of much interest from the standpoint of the pure relativity of all phenomena."[2] In other words, *everything was relative*, including time.

Descriptions of faster-than-light travelers circulated widely in popular science circles of the nineteenth and early twentieth centuries. Pearson first heard about them from a friend who was an astronomer.[3] Einstein, who was more than two decades younger than Pearson, encountered them as a child. He first learned about these travelers around the age of twelve, when a close family friend gave him a popular science book as a gift. The young Einstein devoured its time travel adventure stories, and the volume soon became one of his favorites. Years later, in his autobiography, he would describe reading them "with breathless attention." These stories were more than just amusing and entertaining; according to Einstein, those early childhood imaginings provided the "kernel" or "germ" of the discovery "of a paradox upon which I had already hit at the age of sixteen." The "result" of that discovery, in turn, was the theory of relativity.[4]

"Through the reading of popular scientific books," Einstein recounted, "I soon reached the conviction that much in the stories of the Bible could not be true." As a child, he had been profoundly religious, but his "deep religiosity" ended suddenly. It "found an abrupt ending at the age of twelve." Science took the place that religion had once occupied in his life. "The religious paradise of youth," which he had once worshiped, "was thus lost." The young Einstein dropped the Bible and began reading—"fanatically," he said—all the science books he could lay his hands on. These science books, he recalled, became for him "holier" than the Talmud.[5]

This demon's ability to see the world at past times depended on light's finite velocity. Since the nineteenth century, a growing science-literate public learned that the image of the stars in the heavens arrived at their retinas only after being delayed by light's finite speed. All that they could see of the stars were images of the past. The sun was about eight minutes away, Uranus about two and a half hours. For an observer looking at us from those places, the situation was the inverse: they would see us in the past. Depending on how far away an observer traveled, a delay of a few hours could become days, years, centuries, or longer. Curious consequences followed.

Einstein famously pictured himself propelled through space chasing after a light beam: "If I pursue a beam of light with the velocity c (velocity of light in a vacuum), I should observe such a beam of light as an electromagnetic field." As he grew older, he came to the stunning conclusion that he could find no evidence anywhere in the universe of an electromagnetic field at rest. "There seems to be no such thing, however, neither on the basis of experience nor according to Maxwell's equations," he concluded.[6] From a very early age, Einstein grew intent on following through all the consequences of those childish thoughts.

TIME TRAVEL STORIES

There was a lot for Einstein to draw from. Faster-than-light travelers galivanted across the universe of his youth, hopping on light beams to travel swiftly through the air.

Einstein's favorite author was Aaron Bernstein, a successful popular science writer who enthralled innumerable children of his age. Bernstein might have fascinated Einstein for another reason as well. He moved freely between writing about science and religion, comparing biblical claims against scientific ones and figuring out which ones to trust. All of his tales were based on the most up-to-date knowledge of science, including research on the speed of light. Bernstein disclosed that the original source that led him to craft his narratives was a "little book" titled *The Stars and World History*, originally published anonymously in 1846 and written by a "sharp thinker" who led adventurous readers along this particular "thought game [*Gedankenspiele*]."[7]

The original author who had first published under pseudonymous initials turned out to be Felix Eberty, a German jurist and author who described an "observer, who had the power of following the reflection of a transitory event upon the wings of light."[8] This fast traveler competed against an even faster one who could jump to a faraway star and leisurely wait for light waves, much slower than he, to reach him. Eberty described this being as a "higher or highest spirit" and as "a being simply endowed with a higher degree of human power," comparable to an "inhabitant" of a distant star "mounting away with the images and rays of light."[9] This observer could move rapidly to alter the time taken by world events to reach him.

Eberty stuck as close as possible to the science of his day. Placed at a twelfth-magnitude distant star, he "would see the earth at this moment as it existed at the time of Abraham." By jumping from one distant star to the next, "before the eye of this observer the entire history of the world, from the time of Abraham to the present day, passes by in the space of an hour." By moving quickly, "he will be able to represent to himself, as rapidly as he pleases, that moment in the world's history which he wishes to observe at leisure" and "comprehend with his eye the whirling procession of these consecutive images." He would possess a "microscope for time," with which he could analyze biological processes such as the blooming of a flower or the butterfly in flight.[10]

Like many others of his generation, Eberty had witnessed great gains in the speed of transportation and communication technologies during his own lifetime. These advances had led thinkers of his day to extrapolate to future ones. "For as with a steam-carriage we can travel a geographical mile in ten minutes, and with the electric telegraph can ring a bell at a distance of ten miles in a second," he wrote, "so the supposition that we may be enabled to move from one place to another with a speed surpassing the rapidity of light, rests upon possibility."[11]

These beings' great powers were ideal forensic tools—perfect for solving crimes and misdemeanors. "Not only upon the floor of the chamber is the blood-spot of murder indelibly fixed, but the deed glances further and further into the spacious heaven," wrote Eberty.[12] When Eberty's book was translated into English, Reverend Thomas Hill, president of Harvard University, wrote the preface. In it, he marveled at the possibility of total surveillance—not only of the present but of the past. "The circulation of this book would be, I am convinced, of benefit both to science and religion," he concluded.¹³ Additionally, Hill cited approvingly the claim that everything, including every misdeed, could eventually be found out.

The stories that reached Einstein via Bernstein asked readers to picture themselves traveling progressively faster, until they reached the speed of an electric telegraph. "Are we traveling on water? On horse? With a train? None of those! We travel with the help of the electric telegraph!"¹⁴ When thinking about fast communication, one no longer needed to imagine "winged horses" tied to "chariots in the clouds."¹⁵ The electric telegraph, in Bernstein's retelling, had now taken on the role of these creatures. Another volume speculated about breaching the speed of electricity, and how this could lead to witnessing the world in reverse.

> In one point in space, the light of the scenes of the French Revolution is just coming into view. And even farther away, the invasion of the barbarians has just become the order of the day, Alexander the Great is still conquering the World. . . . And even farther away in space, the representation of earth's past by way of light will just be advancing into the future, historical events that have long been dead for us will just be coming to life.¹⁶

Bernstein told his readers that these possible scenarios were "no longer a mere fantasy." They were not "ghostly [*geistreiche*] invasions, but rather real conclusions grounded on natural truths."¹⁷

FLAMMARION'S "LUMEN"

While Einstein gleaned these stories from reading Bernstein in German, Henri Poincaré read them in French as told by the astronomer and popularizer Camille Flammarion. "Flammarion," Poincaré noted, "once imagined an observer moving away from the earth at a velocity greater than that of light," without offering any scientific objections.¹⁸ What would he see? "For him time would have its sign changed, history would be reversed, and Waterloo would come before Austerlitz."¹⁹ Poincaré added another element to previous speculations: our understanding of causality and chance would flip. Not only would we see neat images succeeding each other in reverse, but we would slowly descend into chaos as the universe moved away from equilibrium. "Well, for this observer, effects and causes would be reversed; unstable equilibrium would no longer be the exception; because of universal irreversibility, everything would seem to him to come out of a sort of chaos in unstable equilibrium; nature in its entirety would appear to him as subject to chance."[20]

In Flammarion's speculations about ultrafast beings, the main character's name was Lumen, from the Latin term for "light beam." Lumen had once been human but was now living in the afterlife. He had flown out of an open window at the moment his body died, but diligently and periodically returned to earth to converse with his dearly beloved. Still shackled in some ways to the world of mortals, he was unable to completely let go of his past. Lumen was a restless soul. "I had no body; yet I was not incorporeal," he explained. "I can tell you neither by what law, nor by what power, souls can transport themselves so rapidly from one point to another," but he could do it. After going away to a distant corner of the universe, he would be able to return "to earth in less than one day."[21] This creature was "a bird of the upper regions," who could surreptitiously descend where he willed, making only a slight "rustling" sound, "as if some new-comer had walked over dead leaves."[22] Sensitive mortals could sense his presence ever so subtly: "As if a breath touched my forehead."[23]

Oh, what powers did Lumen have! To zoom into time and expand it! He could see "a flash of lightning magnified in duration 60,000 times." He could hold still a waterfall, an avalanche, and even an earthquake. He could maintain in its vision even a flash of light. "What a pandemonium!" Lumen exclaimed upon realizing his new abilities. "A world now hidden by its very fugacity from the imperfect vision of mortals" was all his.[24]

Flammarion noted in detail all the paradoxes that ensued from the simple fact that "light is transmitted far more rapidly, but not instantaneously, as the ancients believed."[25] When looking back at the earth in past times, Lumen was able to see himself, even at the moment of his birth and death. Interlocutors were stunned by his descriptions. "You cannot be two persons," a friend protested.[26] But in one sense, that was exactly what he was. "Can you find in all creation a paradox more formidable than this?" Lumen replied.[27] Lumen and his friend discussed how messages, including those carried by light, could arrive long after senders had died. Due to the possible existence of faster-than-light travelers, the universe was a place of constant "conversation between the living and the dead."[28]

Flammarion, much like Eberty and Bernstein, portrayed the creature as having the ability to obtain a perfect record of history. Criminals would not be able to hide their evil deeds from future investigators, unless they had taken the precaution of preventing all light leaks. "A crime is perpetuated," explained Flammarion in one of the classic retellings, he "escapes undetected," "he has washed his hands," he thinks "his deed is *past* forever," but he is wrong. The crime will "transmit itself eternally into infinity," carried by a wave. For a demon traveling at exactly the speed of light and looking at an image of a slice of history, that image would be frozen, enabling the demon to study it at leisure. The world, Flammarion noted, would appear "like a photograph that changes not, while its original grows old in the lapse of time."[29]

A RIGID BLOCK UNIVERSE

Laplace's creature was also a time traveler of sorts, but one who encountered no surprises along the way. Being omniscient, he had access to the universe's full picture. The science fiction writer H. G. Wells chose this model for *The Time Machine* (1895) to describe how this creature must experience history: "If 'past' meant anything, it would mean looking in a certain direction; while 'future' meant looking the opposite way."[30]

Time travel was not so much fun for a being like Laplace's, but it could certainly be fun for a lesser being who had been taken along for the joyride. In Wells's *Time Machine*, a group of incredulous guests gathered around the main character as he explained "the gist" of his invention. If in theory one could travel in time like fingers running swiftly across the piano keys, why not in practice? That is, if Laplace was correct and our universe was as he supposed it to be, then could we not avail ourselves of our knowledge of things to go forward and backward in time? The professor who built the machine in Wells's story demonstrated how it worked by invoking the example of Laplace's omniscient observer: "Suppose you knew fully the position and the properties of every particle of matter, of everything existing in the universe at any particular moment of time: suppose, that is, that you were omniscient." The consequences would be stunning: "If you understood all natural laws the present would be a complete and vivid record of the past. Similarly, if you grasped the whole of the present, knew all its tendencies and laws, you would see clearly all the future." For this being, nothing would be lost, nothing forgotten. "To an omniscient observer there would be no forgotten past—no piece of time as it were that had dropped out of existence—and no blank future of things yet to be revealed." Time would basically disappear for him: "Indeed, present and past and future would be without meaning to such an observer: he would always perceive exactly the same thing. He would see, as it were, a Rigid Universe filling space and time—a Universe in which things were always the same. He would see one sole unchanging series of cause and effect to-day and to-morrow and always."[31] Scientists would soon refer to this rigid universe as the "block universe."

Wells did not delve into elaborate technical details. Laplace's omniscient observer was sufficiently credible to convince readers to go on with his adventure.

FASTER-THAN-LIGHT TRAVELERS

A celebratory but critical eye was cast on the time travel stories of these years by Richard Proctor, an astronomer and a fellow of the Royal Astronomical Society, in his *Other Worlds than Ours* (1896). Proctor considered in detail "a little treatise called 'The Stars and the Earth,' published anonymously several years since." In it, "some results of modern discoveries respecting light were dealt with in a very interesting manner."[32] But the astronomer inferred that something was not quite right about them. He protested that the author seemed to have forgotten that the earth was spinning around, so that all light waves sent from earth would reach a faraway observer as an almost impossibly tangled braided pretzel.

In his own account of the universe, Proctor needed no demon to accomplish such traveling feats, but merely a man endowed with "supervision" and perched somewhere faraway in the universe: "To a being placed on some far-distant orb, whence light would occupy thousands of years to wing its flight to us, there would be presented, if he turned his gaze upon our earth, and if his vision were adequate to tell him of her aspect, the picture of events which thousands of years since really occurred upon her surface."[33] If one increased little by little the abilities of this "being," readers could learn other lessons about the universe. What if his "powers of locomotion" were such that he acquired the power of "instantaneous flight"? Added to the powers of vision, the results could be stunning: "But now conceive that powers of locomotion commensurate with his wonderful powers of vision were given to this being," he commanded. "In an instant of time he could sweep through the enormous interval separating him from our earth."[34] For Proctor, the new telegraph-based news delivery services were the obvious point of comparison: "Precisely as a daily newspaper gives us a later account of what is going on in London than of events happening in the provinces, of these than of events on the Continent, and of these again than of occurrences taking place in America, Asia, Africa, or Australasia, so the intelligence brought by light respecting the various members of the solar system belongs to different epochs."[35] An "intelligent Neptunian" was just farther away. It would see the earth during a moment that had long since passed for us.[36]

In another account written by the Jesuit priest and theologian Joseph Pohle, the faster-than-light traveler appeared completely deprived of a body and was referred to only as a superfast eye. It could observe world history as a set of "swiftly changing *tableaux vivants*" that could also be displayed in reverse: if an observer could move faster than the speed of light, then "the history of man and earth would be turned upside down. People would first be seen on the death bed, then the sick bed, then in the prime of life and finally as an infant in the cradle."[37] Its author reiterated the usual descriptions of this creature's world, where men could die before being born. "Perceptual relationships would be downright paradoxical if the eye moved at a faster pace than the light: then the imaginary case would have to happen that the events were reversed."[38] Like Bernstein, the author of this text was interested in separating good science from fraudulent religious claims and criticized the "all too ready credence which in our own time thousands of well-meaning Catholics" gave to demonic occurrences. Many of these, he argued, were simply the result of "bogus revelations."[39] Yet these fraudulent examples could *not* completely disprove other *daemoniaci* possibly acting on this world.

A SPIRIT CALLED LUMEN

These stories had a serious side to them—the scientific premises on which they were based came from the most advanced research in physics, astronomy, and cosmology. Poincaré discussed faster-than-light signals, in addition to describing demons and Brownian motion, in his presentation at the St. Louis Universal Exposition. "What would happen if one could communicate by non-luminous signals whose velocity of propagation differed from that of light?" he asked. For Poincaré, the future discovery of such signals was not "inconceivable." Something or someone might one day be found traveling faster than light. Gravitational waves might prove to hold the key to this mystery, suggesting that such signals could exist: "And are such signals inconceivable, if we admit with Laplace that universal gravitation is transmitted a million times more rapidly than light?"[40]

Einstein had other ideas about superfast travelers. For him, the fact that the speed of light could not be surpassed was not just a technological limitation. His theory of relativity forbade the existence of superfast travelers. He eventually came to the conclusion that this impossibility was a property of the universe itself.

In his special relativity paper of 1905—arguably the most famous text in the history of science—Einstein noted that if the speed of light was independent of the source of its motion (as various experiments had already shown), the time marked by clocks and the lengths of rulers would dilate in comparison to each other, depending on their velocity. To reach this conclusion, Einstein once again considered what a source of illumination would look like for "an observer approaching a light source with velocity V."[41] When the velocity of this observer traveling through space approached that of light, things started to look very strange. The classical understanding of time and space, which had relied on action-at-a-distance effects, risked bursting at the seams. Einstein ventured that, in consequence, there might not be such a thing as "empty space," in the traditional sense, or universal time.[42] Faster-than-light effects or travelers need not fret. "For superluminal velocities," he wrote, "our considerations become meaningless."[43] Einstein was not yet ready to completely rule them out from the universe and from science.

THE GHOSTS OF TIME AND SPACE

At the beginning of 1914, Einstein completed the manuscript for an article that would be published that spring. In it, he mentioned ghosts, using the German word *Gespenster*. In the French translation of the article, Einstein's characters were called *revenants*. The physicist therein described his work as doing away with these creatures. The main topic of the article was the concept of absolute space. Einstein no longer considered it a stable framework, or a kind of stage on which everything was supposed to happen. What most of his contemporaries considered to be real, he noted, was not. His first target was nothing less than the traditional understanding of space.

"I could never be made to believe in ghosts, so I cannot believe in the gigantic thing of which you speak to me and which you call space," he wrote. "I can neither see such a thing nor think about it."[44] The French translation was even eerier, as it compared the classical notion of space to a belief in something or someone that had returned from beyond the grave.[45] With those words, Einstein equated the traditional notions of space he was intent on dispelling with a belief in these creatures.

A few months after Einstein's article appeared in *Scientia*, Austrian Archduke Franz Ferdinand and his wife were shot and killed by a lone gunman. Soon thereafter, Russia, Germany, France, and Britain were drawn into World War I. How could the action of a single bullet create such great repercussions?

The years that followed were hard for Einstein. He had just moved to Berlin when the war broke out, and he was increasingly critical of German militarism. By the end of 1915, he had finally managed to expand his theory of relativity into a more general one, incorporating gravitation and acceleration. In a letter to a friend, he admitted to having "had one of the most stimulating, exhausting times of my life," but "also one of the most successful." To another friend he described himself as "contented, but kaput."[46] Einstein had worked in parallel with the mathematician David Hilbert, at times collaboratively and at times in competition with him. Together they developed a new way of understanding the universe as the warping of spacetime around lumps of masses. Time slowed down the closer observers got to these masses. Space became warped around these masses too. The entire structure of this universe was non-Euclidean. In it, parallel lines could cross, the shortest line between two points was not straight, and time was simply another dimension next to the three dimensions of space.

Einstein's theory was so complicated that almost no one saw the need for it. Hilbert himself did not seem to have been as excited about it as his colleague. How could Einstein reframe it in a way that would convince others of its importance? He needed to work harder to flesh out its full implications. Einstein decided to change gears, momentarily leaving aside technical and complex scientific work to focus on popularizing the theory. With the intention of bringing "someone a few happy hours of suggestive thought," he hoped to wordsmith his work into greater acceptance. By the end of the year, he had completed his first popular science book.

Einstein enticed readers into envisioning the repercussions of relativity theory by asking them what the universe would look like for someone "hastening towards the beam of light" or "riding on ahead of the beam of light."[47] In many respects, the book was typical of the genre, but in contrast to most other popular science time travel stories of that era, Einstein's observers could no longer see the magical reverse-effects because they were unable to breach speed-of-light limits. At most, they could travel along at that speed, and what they saw would be unexceptional in some ways: they would see nothing different from what they saw when they were at rest. The past was no longer accessible to faster-than-light travelers in the wonderful way that Eberty, Bernstein, and Flammarion had described.

Those stories were not the only ones that could be brushed aside as farcical. German audiences had started to be captivated by new films that rivaled traditional operatic and theatrical productions and inaugurated the horror genre. Einstein explained that some of the concepts he believed in elicited comparable fears. "A feeling not dissimilar to that generated by those ghosts of the theater" sent "a mystical shiver" down spines. It "seizes non-mathematicians when they hear the term 'four-dimensional.'"[48] He urged calm, noting that the frightening "ghosts of the theater [*Theatergespenst*]" could be dispensed with by those armed with a proper understanding of how "commonplace" the idea of a four-dimensional universe could be. Further along in the narrative, Einstein described a being who could mess with a person trapped inside a windowless room. This being could attach a "hook" on the roof outside, knot it with a "rope," and start to pull. Einstein asked readers to imagine themselves feeling pressure on their legs as if they were being held captive inside an elevator that was being pulled upward from the outside by such a creature until it reached "fantastical" speeds. If the tug was delivered at a constant speed, the push from the acceleration would be practically indistinguishable from that of gravitation, argued Einstein, offering an argument in favor of their equivalency. Who was this "being," asked the physicist? The answer "was immaterial to us," he responded, since the entire purpose of introducing the creature was to see that the notion of it could be dismissed once a curvature in four-dimensional space-time was accepted.[49] The astronomer Arthur Eddington, who planned a famous eclipse expedition to test Einstein's theory, would soon refer to "the agent which plays these tricks" with a lift as a "demon."[50]

Einstein wrote again about "ghosts" in an unpublished manuscript summarizing the main points of his theory. The main topic of the piece was no longer space, but time. "Only ghosts [*Geister*]," Einstein argued, would be able to perceive an enormous clock stretched across all of space. Evidently, the sounds of "an eternally uniformly occurring tick-tock" could not be captured by mortals. Still, almost everyone, argued Einstein, believed uncritically in these ghosts. If one "[asked] an intelligent man who is not a scholar" for the time, Einstein continued, his reply would show that he took time to be this ghostly "tick-tock."[51] The physicist quickly chased away these ghosts. "There is no audible tick-tock everywhere in the world that could be considered as time," he concluded. In several places in his published work, Einstein portrayed respectable members of his community, as well as most of the lay public, as haunted by ghostly notions of time and space. "The ghost [*Gespenst*] of absolute space," he explained to a colleague, "haunts" [*spukt*] physics.[52]

The following year Einstein had a breakdown. His health deteriorated rapidly, and he was not sure he would make it. His marriage had fallen apart, and he had just written an obituary for Ernst Mach; he had to write another one for Smoluchowski. The last years had been a *"Weltkatastrophe,"* he wrote on a cold December day.[53]

EINSTEIN THE EXORCIST

Just when he was about to lose all hope, Einstein found a champion in Eddington, a British Quaker and fellow pacifist who became intent on testing the physicist's hypotheses. Eddington would make Einstein world-famous. With Eddington's help, he was finally able to convince the general public that believing in faster-than-light speeds was unjustified: no one and nothing could travel faster than that.

Eddington obtained the necessary funding and managed to avoid the draft by organizing an expedition of astronomers to a remote island off the coast of Africa to observe and photograph an eclipse that would take place in 1919. The trip was hard work, but far better than being in the deadly muddy trenches of World War I.

The astronomer and his team came back from Africa with great news. Einstein himself was not too surprised. He had anticipated the result when he first completed his theory at the end of 1915. For him, Eddington's expedition was a large-scale confirmation of results that had already been observed qualitatively (the spectral-line shift, for instance) and others that had been known since the mid-nineteenth century (such as the perihelion of Mercury). Einstein had some regrets about missing a previous opportunity to test his theory, but not much. To his colleague Arnold Sommerfeld he explained: "This is not so painful for me anyway, because the theory seems to me to be adequately secure, especially also with regard to the qualitative verification of the spectral-line shift."[54] Although Einstein himself had been convinced of the merits of his theory for years, it would take much more to convince his colleagues and the public at large of its importance. Eddington was key to his success.

Eddington presented the results of his fact-finding mission in a well-orchestrated meeting filled with scientists and journalists at Burlington House in central London. In the days that followed, newspapers across the world celebrated Einstein's accomplishments with bombastic headlines, describing a revolution in science. Suddenly everybody cared. Einstein was transformed from a successful university professor into a worldwide sensation. A recent historian argued that the "modern world began on 29 May 1919 when photographs of a solar eclipse confirmed the truth of a new theory of the universe."[55]

During the years following the eclipse expedition, Eddington promoted Einstein as more than a great scientist—he was to be an exorcist. "Einstein has exorcized the demon," he explained in the Romanes lectures delivered at the Sheldonian Theater in Oxford in 1922, using a phrase that he would repeat in various venues for more than a decade.[56] According to the astronomer, the most important consequence of Einstein's relativity theory was its elimination of the "demon of gravity"—that is, its success in disproving the actions of "an intangible agency or demon called gravitation."[57]

Who exactly was this demon? According to Eddington, Newton's theory was based on the actions of a "mysterious agent" (considered godly in Newton's time, but now typically seen more as a demonic figure).[58] Perhaps Eddington was right to blame Newton. Newton did indeed posit "some agent" at work as the cause of gravitation. In correspondence with a prominent theologian, a scholar of his era, Newton had claimed that "gravity must be caused by some agent acting constantly according to certain laws." Newton himself was not sure what type of agent it was: "whether this agent be material or immaterial I have left to the consideration of my readers."[59] Eddington explained Newton's reasoning. "This new phenomenon must be accounted for," wrote Eddington, "so he invents a *deus ex machina* which he calls gravitation to whose activities the disturbance is attributed."[60] Such inadequate explanation had survived for centuries, until Einstein came and did away with this mysterious "tugging agent."[61] Eddington himself, he stated unapologetically, had "disrespectfully" called Newtonian explanations of gravitation a "demon."[62]

The characterization of gravity as an agent became even more prominent in the early decades of the nineteenth century. The astronomer John Herschel explained the action of gravity by reference to a consciousness and a will in his *Treatise on Astronomy* (1834). "All bodies with which we are acquainted, when raised into the air and quietly abandoned, descend to the earth's surface in line perpendicular to it," he wrote. "They are therefore urged thereto by a force or effort, which it is but reasonable to regard as the direct or indirect result of *a consciousness and a will existing somewhere*, though beyond our power to trace, which force we term gravity."[63] Was there really such an agency responsible for forcing things to fall down each and every second of the day and everywhere?

Eddington described the process that had led Einstein and his followers to disprove Newton's demon of gravity. One of the consequences of Einstein's theory pertained to the bending of light by gravitational forces as it travels through space. The idea that light can be bent by gravity was hardly new. Astronomers had considered the possibility for more than a century. As electromagnetic waves were sent as light and radio signals across longer and longer distances, the question became more relevant technically. But Eddington saw gravity's effect on light as a competition between two radically different ways of understanding the universe. "You cannot deflect waves by tugging on them," he wrote.[64] If astronomers used Einstein's theory, the deflection of light rays during the solar eclipse would be twice as large as they would be using Newton's theory. Why would a demon work only half as hard in that location? The difference, according to Eddington and Einstein, could prove that effects traditionally ascribed to the demon of gravitation were really due to "a curvature of the world" in the region surrounding massive bodies. "The larger deflection was quantitatively confirmed by the eclipse observations," Eddington concluded.[65]

What had been explained as the work of a demon could now be explained by warps in space-time. Convincing readers of the merits of Einstein's theory, however, was not easy. Eddington proceeded by elaborating on how demons were usually invoked when reality did not seem to match expectations. Eddington used a hypothetical example involving mapmakers and map users. Because the scale in Mercator projection maps shrank as one moved away from the center, Greeks who used their maps to travel to Greenland would be likely to think that Greenland was unusually large compared to what their maps had told them. For Greenlanders who traveled to Greece, the country would seem much smaller than what they had expected from looking at their maps. Not understanding the challenges of mapping a three-dimensional sphere onto a four-dimensional surface, they might simply blame the altered dimensions on the actions of a demon. "They would, I suppose, invent a theory that a demon resided in that country who helped travellers on their way, making the journeys appear much shorter than they 'really' were."

It was natural for people to normalize their experiences and to see others' as anomalous. It was also natural for them to invent demons when their perceptions did not match their expectations. But another solution, one that did not involve demons, was also possible, Eddington noted. Greeks in Greenland could change their maps so that the size of the territory was the same as the one they were used to interpreting. "We might equally well start our flat map with its centre in Greenland; then it would be found that journeys there were quite normal, and that the activities of the demon were disturbing travellers in Europe," he explained. Such a move, in Eddington's understanding, would be identical to what Einstein proposed. Instead of inventing ad hoc hypotheses to fit a four-dimensional space-time universe into a three-dimensional one of space plus gravitational tugging agent, scientists could simply adopt the new model. "We now recognize that the true explanation is that the earth's surface is curved; and the demoniacal complications appeared because we were forcing the earth's surface into an inappropriate flat frame which distorts the simplicity of things," he concluded.[66]

Eddington surmised that scientists might disapprove of his choice of the word "demon" and would probably propose another, rather fancier term. "No doubt the scientists would preserve their self-respect by using some Graeco-Latin polysyllable instead of the word 'demon,' but that must not disguise from us the fact that they were appealing to a *deus ex machina*," he explained. For him, the word was perfect: "The name demon is rather suitable, however, because he has the impish characteristic that we cannot pin him down to any particular locality."[67]

For years to come, Eddington continued to portray Einstein as an exorcist. "Einstein," he told his audience in lectures given in 1927, "pitched out the respectable causal demon who called himself Gravitation."[68] Before Einstein's theory of relativity, "falling apples" were "accounted for by an intangible agency or demon called gravitation which persuades the apples to deviate from their proper uniform motion." The astronomer discussed the example of the distortions that result when mapping a three-dimensional surface onto a two-dimensional surface over and over again. "As we change from one observer to another—from one flat space-time frame to another—

the scene of activity of the demon shifts."[69] These shifts revealed a particular characteristic of demons: they were hard to see when close by and easier to detect from a distance. The scene of the demon's activity "is never where our observer is, but always away yonder." In other words: "The demon is never where you are; it is always the other fellow who is haunted," he explained.[70]

The astronomer next discussed Einstein's example of "the man in the lift," referring to the illustration used by Einstein to explain how forces due to acceleration were the same as gravitational forces. A person inside an elevator would explain the push against his feet as due to gravity, but a person outside it would explain the same push of the ground against his feet as due to the lift's acceleration. The Newtonian framework required one explanation for the person in the elevator and another for the person outside, and Einstein had shown that the two experiences were equivalent. "It is necessary to postulate the activity of a demon urging unsupported bodies upwards" in one case, but not in the other. A curved space-time universe, and one that explained acceleration and gravitation at the same time, did away with the need to invoke two theories. "Is not the solution now apparent?" asked Eddington, answering: "The demon is simply the complication which arises when we try to fit a curved world into a flat frame." The solution was simple: "Admit a curvature of the world and the mysterious agency disappears," he explained.[71]

How should scientists refer to "the agent which plays these tricks," especially the one mentioned earlier by Einstein when he discussed a "being" who trapped someone in a lift and pulled it up at an immense velocity?[72] When discussing his use of the word "demon," Eddington cautioned his audience in the same way he had done before: "Of course no scientist would use so crude a word . . . to denote the mysterious agent." But be what it may, the fancier terminology, he claimed, "is only camouflage" that justified him in continuing to compare "the invisible agent invented to account for the tug of gravitation to a 'demon.'" As long as such demons were unexamined, science would remain stunted. He then proceeded to ask a more difficult question. "Is a view of the world which admits such an agent any more scientific than that of a savage who attributes all that he finds mysterious in Nature to the work of invisible demons?"[73]

Were the demons of modern science identical to those of "the savage"? Eddington explained the slight, but important, difference between the two. It hinged on causality—in other words, on the degree of mischievousness and irresponsibility adjudicated to those demons. The Newtonian "could point out that his demon Gravitation was supposed to act according to fixed causal laws and was therefore not to be compared with the irresponsible demons of the savage." True enough, but "the savage" could respond by saying that his demons would not be prone to getting totally out of line either. "I suppose that the savage would admit that his demon was to some extent a creature of habit and that it would be possible to make a fair guess as to what he would do in the future; but that sometimes he would show a will of his own." Eddington concluded that the demon of "the savages" and the demon of the Newtonians were, if not nearly identical, more like relatives—in fact like brothers. "It is that imperfect consistency," he explained, "which formerly disqualified [the savage's demon] from admission as an entity of physics along with his brother Gravitation."[74]

After years of garnering only moderate attention, Einstein's popular science book with light-beam-chasing examples and "beings" pulling elevators with a person trapped inside suddenly became a commercial success. Over the course of Einstein's lifetime, it was translated into more than twelve languages, and fifteen editions appeared in the original German. Innumerable imitators embellished his accounts, crafting ever more colorful versions. His colleague Paul Langevin was most imaginative, picturing a voyager on Jules Verne's spaceship.[75] Eventually outer-space-traveling "twins" were introduced into these narratives, as well as flies landing in stagnant pools to illustrate the curvature of space, dishes in the dining room of trains that looked oval to those outside, fireworks seen from hot-air balloons, and aviators smoking cigars and looking at wristwatches with hands that slowed down or came to a halt the faster they flew. *The Einstein Theory of Relativity*, one of the most widely watched documentaries of that time, used innovative animation techniques to depict a rocket ship launched from earth traveling through the years and overcoming them, before stopping in 1492 to catch Columbus "in the act of discovering America." The companion book to the film explained the value of Einstein's theory by comparing its merits to the demystification by science of phenomena previously explained only by reference to ghosts and spirits. "As people on the earth have done with regard to the fourth dimension theory; they would affirm that such things couldn't be, no matter what the appearance might show,—*or else it was the work of ghosts and disembodied spirits!*"[76]

Einstein continued to fight vigorously against previous explanations of the universe that referred to near-instantaneous travel. His final coup against the faster-than-light traveler known as "the colleague of Maxwell's demon" was delivered in the form of a short, succinct preface he wrote for the republication of Felix Eberty's *The Stars and World History*. He praised the book "because it shows . . . [the author's] critical spirit against the traditional concept of time" of his era, but he also pointed out that the author's conclusions were now outdated and proven wrong by his own work. "The theory of relativity, itself often accused of leading to bizarre conclusions, can in fact save us from some even stranger ones."[77]

Einstein's preface served as a short obituary for an unworthy parvenu, a miserable upstart, who had introduced unnecessary paradoxes into science. It finally put an end to the colleague of a demon that had captivated the nineteenth century and given rise to the ghosts of absolute time and space.

6
Quantum Demons

The discovery of radioactivity disrupted much of what had been previously known about demons. In 1911, at the Solvay Congress in Brussels, Marie Curie listened to her colleague Max Planck present his revolutionary research on energy "quanta." She immediately started to think of a new demon, analogous to Maxwell's, who could manipulate these packets of radiating energy.

Quantum demons were the main suspects in investigations into the strange goings-on of subatomic particles, such as electrons and photons. These particles could be found to move within spaces smaller than 0.000000000000 00000000000000000000162 meters. One scientist recently compared them to the UFOs "in bad science fiction movies" that "can sit still for hours ... and suddenly jump up and run away."[1] Because these particles could not be pinned down, researchers started wondering if quantum-sized demons might be fiddling with their position whenever scientists determined their momentum, and vice versa. Because they seemed to act randomly, the suspects were considered a threat to Laplace's demon. Scientists soon started imagining yet other creatures who could predict their whereabouts. After years of research, they determined that in the aggregate, the actions of quantum demons were practically undetectable. They camouflaged themselves perfectly against statistical regularities, as masters of the poker face gifted with the perfect alibi. Yet the scientific community became split into those who thought the behavior of these particles was essentially unpredictable at the quantum level and those, including Albert Einstein, who thought that their unpredictability was just apparent and due to our meager knowledge of their mysterious modus operandi. Soon Einstein would get into a vicious debate with a growing number of colleagues who started to consider nature as something that might have pockets of uncertainty ready to stump anyone.

The bizarre behavior of these particles threw into doubt the long-held belief that all things in nature could be localized and found in a certain place in space at a certain moment of time. The additional possibility that they might be connected to each other in strange ways, that is, of quantum entanglement, bothered Einstein profoundly. He considered these connections as "spooky" and "ghostly." Yet despite Einstein, some research pointed to the possible existence of connections in this universe that were faster-than-light, bringing the colleague of Maxwell's demon back to life. A younger generation of scientists started to consider that Einstein's approach might be wrong.

Scientists would soon thereafter discover that quantum demons were capable of devastating feats.

The atom started disintegrating as soon as statistical molecular thermodynamics became more firmly established. Its stability started to be amply questioned. New elements, such as uranium, were found to give off energy without, apparently, losing mass. The first law of thermodynamics was at risk. The French scientist Henri Becquerel, one of the first researchers to discover radioactivity, was enthralled by how uranium salts seemed to give off energy continuously without requiring an external energy source and without depleting themselves in any measurable way. Pierre and Marie Curie were also fascinated by these strange elements. Radioactivity appeared to them to be "a source of light" that "functions without a source of energy," they marveled, and "in contradiction, at least apparently, to the principle of Carnot."[2] Research on these new intriguing substances—radium, uranium, and later plutonium and others—the rays they emitted, and other subatomic particles, such as electrons and photons, would mark the rest of the century.

The explanations of science's demons that had been developed during the previous centuries started to be amply questioned. Laplace's and Maxwell's demons might not be capable of all that was demanded of them. Some started to wonder if Maxwell's demon actually got tired because the information he required to act was thermodynamically costly. Perhaps his memory function was limited, his hands were simply too coarse, or his eyesight was too weak to handle particles that were much smaller than atoms. As scientists worked hard to figure out all the limitations of Laplace's and Maxwell's demons and how to circumvent them, they started imagining new demons who could explain the quantum world. They started to think differently not only about the relation of mass to energy in the universe but about knowledge itself. The ideal of objective knowledge that they represented started to falter.

Einstein was intrigued by the promises of these new radioactive substances. In the last of his four *annus mirabilis* articles, he had discussed radium salts as evidence for the need to rethink the relation between mass and energy.[3] Scholars would later see in this work his earliest formulation of the famous $E=mc^2$ equation; along with another paper published that year on light "quanta," it would later be recognized as having started the quantum revolution in science—a revolution that Einstein himself contested.

Quantum mechanics led scientists to reconsider demons that had already been under intense scrutiny. In the context of this research, Laplace's being was designated a demon. Maxwell's being, in turn, was reinterpreted as a creature with a limited ability to manipulate subatomic particles. Unlike the two elder creatures, demons that operated at the quantum scale seemed to have a strong penchant for indeterminism and chaos. If Laplace's creature was law-abiding and Maxwell's demons were law breakers, quantum demons were law *benders*.

What scientists found out went on inside atoms shocked them. They had to develop new concepts—uncertainty, indeterminacy, complementarity, nonlocality, entanglement—to explain the bewildering subatomic world. Classical physics had worked wonders in explaining many of the regular conditions of everyday life. Relativity had taken it to the next level, successfully exploring the universe at scales where objects moved nearly at the speed of light and masses were as heavy as those of the planets and stars. The insights of classical physics and relativity could not, however, be carried over to the realm of the really, really small.

Quantum mechanics was developed by about a dozen brilliant scientists working closely with each other. Among those who referred to demons in their work were the Hungarian mathematician John von Neumann, the Germans Max Planck, Max Born, Werner Heisenberg, and Grete Hermann (the only woman in the group), and the American brothers Karl and Arthur Compton; president of MIT and Nobel Prize laurate, respectively, the Comptons were both key members of the Manhattan Project for building an atomic bomb. All of these scientists agreed about the phenomena they saw before their eyes, yet they remained radically divided on how to interpret them. They agreed that when the world was considered at a human scale or bigger, these effects dampened until they became statistically insignificant, yet they disagreed about how fundamental quantum physics should be to our understanding of reality. Einstein believed that the interpretation given to some of these strange effects by the Danish physicist Niels Bohr and his allies was incomplete in that it did not account for the theory's limitations. He led the crusade against Bohr's interpretation of quantum mechanics, arguing that it offered only an incomplete description of reality. These two camps, one led by Einstein and the other by Bohr, were also divided about demons.

RADIOACTIVITY

At the 1911 Solvay Congress in Brussels, scientists faced the daunting challenge of having to investigate nature at the subatomic level. In the opening speech, the physicist Hendrick Lorentz reflected on the current state of science. "We feel today that we are caught in an impasse," he explained, "as the old theories are showing themselves to be more and more impotent at dissipating the great obscurity that surrounds us from all sides."[4] But Einstein, the youngest physicist invited, was excited by the new developments. In a letter to his best friend, he referred to this yearly professional gathering as the "'witches' sabbath" (*Hexensabbat*).[5]

The German physicist Max Planck stunned those in attendance by delivering a lecture focused not on atoms but on much smaller "quanta" of energy. Born would later describe Planck's insight as "the most revolutionary idea which ever has shaken physics."[6] Marie Curie, the only woman at the conference, intervened during the discussion. She had started to think of a new demon, one that was analogous to Maxwell's but could manipulate the packets of radiating energy described by Planck. Curie opened with a curious remark: she asked Planck about the velocity of the emission of the energy quanta. "Does M. Planck assume that the emission of an energy element is made instantaneously?" If so, the speed of the processes leading to radioactive decay would have to be faster than light. She then started to wonder about the possibility of a "mechanism that would permit to interrupt such emission." "It is quite probable," she said, "that such mechanism will not exist at our level, but rather may be comparable to Maxwell's *demons*." Just as Maxwell's demons could do thermodynamic work for us, the demon first imagined by Marie Curie might work at the quantum level. "They would permit us to obtain differences due to radiation laws described statistically, just like Maxwell's demons permit us to obtain differences due to the consequences of Carnot's principles."[7]

A new world of possibilities started opening up. A few years later, in 1913, the *New York Times* reported on new investigations into radioactivity under the dramatic headline "Science on Road to Revolutionize All Existence."[8] The possibility of finding new means to disrupt the general degradation and dissipation of energy started to fascinate more and more researchers. Radioactivity "seems to claim lineage with the worlds beyond us," explained the British radiochemist Frederick Soddy, who took on the work started by the Curies and Becquerel. "Radioactive substances," he noted, "apparently were performing the scientifically impossible feat of evolving a store of energy presumably out of nothing." They seemed to be powered by "a perennial supply of energy."[9] "Radium," he explained, seemed to be capable of giving "out light and heat like Aladdin's lamp, apparently defying the law of the conservation of energy, and raising questions in physical science which seemed unanswerable." Radioactivity had revived the ancient dream of creating a perpetual motion machine. "The driving power of the machinery of the modern world is often mysterious," he observed, "but the laws of energy state that nothing goes by itself, and our experience, in spite of all the perpetual motion machines which inventors have claimed to have constructed, bore this doctrine out, until we came face to face with radium." That magical element "is no longer the radium we know," he cautioned, but with it the attainability of such an impossible dream now seemed closer: "The 'physical impossibility' of one era becomes the commonplace of the next," he concluded optimistically.[10]

What did that subatomic world look like? What would a mini-demon, one much smaller than Maxwell's, see? What would it teach us about the universe? Soddy collaborated closely with the Nobel Prize–winning chemist Ernest Rutherford, determining with him that radioactivity did *not* violate the law of energy conservation.[11] Rutherford was the world's foremost expert on the inner workings of the atom. When he successfully probed the inside of the atom by aiming subatomic particles from radioactive substances at a thin foil of gold, he came to the surprising conclusion that atoms were made up of a heavy central nucleus surrounded by much lighter electrons. The philosopher C. D. Broad, who was closely following Rutherford's recent work on "the experimental splitting of atoms," grew increasingly curious about what happened inside. In the Tarner Lectures delivered in 1923 at Cambridge University, Broad asked readers to follow him in a curious thought experiment by journeying deeper and deeper into the netherworld of the interior of an atom and getting close to its very nucleus. "Let us replace Sir Ernest Rutherford by a mathematical archangel," he urged. What would "a mathematical archangel, gifted with the further power of perceiving the microscopic structure of atoms as easily as we can perceive haystacks," discover? If "the archangel could deduce from his knowledge of the microscopic structure of atoms all these facts," as he gained this coveted information he would likely lose other knowledge. A being who would know the internal world of atoms would not be able to know other essential aspects of the world available at human scale, such as color, smell, and temperature. Acknowledging that some of his conclusions must appear somewhat "unscientific and superstitious" to his audience, Broad insisted that the subatomic realm was likely to bring in big surprises: in that context, he said, "I know no reason whatever why new and theoretically unpredictable modes of behavior should not appear."[12] And appear they did.

Since the beginning of modern science, explanations of nature based on causal principles had been successfully used to eliminate a widespread belief in supernatural forces at play in the universe. But some very important natural phenomena—including historical development—seemed to a budding generation of early twentieth-century German intellectuals to be escaping causality's grip. The German writer Oswald Spengler eloquently articulated the contradictions pertaining to causality in his best-selling *The Decline of the West* (1918). "The principle of causality," he wrote, "banishes the demonic."[13] Yet his own experience had taught him that it was largely inoperative when it came to the general unfolding of world history. "Anyone who has absorbed these ideas," he wrote, "will have no difficulty in understanding how the causality principle is bound to have a fatal effect upon the capacity for genuinely experiencing History."[14] Spengler followed recent developments in quantum mechanics closely, arguing that the discipline showed that the principle of causality was faltering beyond repair. He was ready to let it go.

Throngs of readers avidly identified themselves with Spengler's criticism of causality, but Einstein resisted his wide appeal. In a letter discussing Spengler with Born, Einstein admitted that the "business of causality plagues me a great deal too." Yet he would never "forgo complete causality."[15] He explained to Born that the whole point of science, the entire point of being a scientist, was to fight against indeterminism. "I would rather be a cobbler or even a casino employee than a physicist," he wrote, indicating that he would rather quit the profession that made him famous than to capitulate to the new acausal ideas emerging around him. "I won't be driven into abandoning strict causality before entirely different defenses have been tried against it than hitherto," he concluded.[16]

One particular experiment led scientists to consider new kinds of demons who might be interfering in the subatomic world. Known to history as the "double-slit experiment," it has since fascinated scientists, philosophers, and the general public. The setup was extremely simple. A beam of light or electrons was aimed at a screen with two openings (slits) through which the rays could pass. Scientists could run two tests with this single setup. In one, they let the rays pass through the two slits; in the other, they closed one of the two slits half the time and the other slit the other half of the time. When the two slits were open, the results showed an interference pattern on the screen. When only one of the two was available, the interference pattern disappeared. The experiment appeared to show that the rays of photons or electrons behaved one way if they were detected as passing through one of the two slits, and another way if they were not. In every case, particles would choose evenly, going through the right-hand slit half the time and through the left-hand slit the other half. Yet in one case they would act as a wave and create interference patterns on the screen, and in the other, when they were given only one slit, they would act as particles and the interference pattern would disappear. They seemed to behave differently depending on which future options were available to them and whether, strangely enough, they were going to pass through one or two slits. How could a photon or an electron know what it would encounter at a later moment—whether it was to face a single slit or two in its path—in order to alter its behavior? How would it know it should act like a particle if given one option and as a wave if given two?

The most popular explanation of the strange behavior of quantum particles was developed by Bohr and his allies at his institute in Copenhagen. They claimed that photons and electrons—along with everything else in nature—were particulate at the same time as they were wavelike, an idea known as wave-particle duality. According to wave-particle duality, the act of measurement altered the results of any experiment; moreover, not everything about a particle, including things like position and momentum, could be known with precision beyond a certain limit. Bohr embraced explanations based on the principle of nonlocality, which no longer considered electrons and photons as localized in time and space but rather as also spread out, and he defended the principles of indeterminacy and uncertainty, maintaining that certain properties of subatomic particles could not be known with precision. Laplace's demon could have his finger in the pie almost everywhere, except in this portion of the quantum realm.

The double-slit experiment baffled Einstein and bothered him profoundly. He told his good friend Born about his discomfort. "The thought that an electron subjected to a ray chooses of its free volition the instant and direction in which it wants to bounce away," told Born, "is intolerable to me."[17] To Einstein, the way some of his colleagues were thinking about the experiment was loopy. Their explanations embraced ideas of nature acting in strangely unpredictable ways—or even worse, in a strangely animate way in relation to us, describing nature as intimately connected to living consciousnesses. Einstein believed that there had to be a better explanation. For the rest of his life, he searched for determinate laws—even if none had been found so far—that would explain where all things (including photons and electrons) go and where they were to be found in space at any given moment in time. Others disagreed, arguing that Einstein should simply embrace the new reality and stop looking for hidden causes or explanations.

Born was about to change science and our understanding of the universe by making a key contribution to quantum mechanics. He determined that quantum wave equations were statistical in nature. Unlike the calculations used to describe classical waves, such as water waves and string undulations, calculations in quantum mechanics contained only information about the probability of where a certain particle could be found. The inspiration for his breakthrough, he explained, came from "a remark by Einstein on the relation between a wavefield and light quanta." At the time he came up with his Nobel Prize–winning idea, Born, like many others around him, was trying to understand why the waves of quantum particles differed from classical ones. In contrast to the latter, quantum waves seemed to spread out everywhere, affecting other objects even though they had no energy or momentum. In an article published in the prestigious *Zeitschrift für Physik* (1926), Born recounted how Einstein's "remark" led him to his statistical interpretation of quantum waves. Einstein, according to Born, "said that the waves are there only to guide the corpuscular light quantum, and spoke in this sense of a 'ghost field' [*Gespenterfeld*]" or, literally, a "ghost wave." Born described the seemingly magical grasp of these waves on things, and how they determined "the likelihood that a quantum of light, the carrier of energy and momentum, takes a certain path, but the field itself has no energy and no momentum."[18] For Born, investigating the *Gespensterfeld*, "or better 'pilot wave' [*Führungsfeld*]," as he would rebaptize it, led him to interpret quantum waves statistically. In a letter to Einstein, he explained that the "idea to consider Schrödinger's wavefield as a 'Gespenterfeld' in your sense of the word proves to be more useful all the time."[19] Einstein's problem had presented an opportunity for Born. A statistical interpretation explained the waves' "ghostly" behavior to a tee.

The next big breakthrough for the field arrived the following year. During the fifth Solvay Congress (1927), Arthur Compton and the aristocratic Frenchman Louis de Broglie, known for his work on the wavelike qualities of quanta that would soon earn him a Nobel Prize, also considered quantum mechanics by reference to the "pilot waves" related to Einstein's "ghost waves." Some scientists hoped that the behavior of subatomic particles and waves could thus be understood dynamically and not just statistically.

Einstein remained silent during most of the day of Compton and de Broglie's presentation. But that night he and Bohr locked horns in a heated debate on which the success of the new field of science hinged. The physicist Paul Ehrenfest was among the handful of physicists who witnessed the debate with awe. At one point, even he was driven to tears by Einstein's insistence on the need to find causal laws hiding behind quantum effects.[20] Ehrenfest told a friend that Einstein's stubborn performance reminded him of a tireless and trapped demon: "Einstein [is] like the little devil in the box [*Teuferln in der Box*]," who, like a toy jack-in-the box, would "jump up each morning fresh" to work away like a "perpetual motion machine trying to disprove the uncertainty relation."[21]

Einstein's cause was not helped by the publication that year of Heisenberg's groundbreaking paper articulating the uncertainty principle claim that the position and momentum of a subatomic particle could *never* be known beyond certain limits.[22] Bohr, in contrast to Einstein, just loved it. He developed its full consequences at the International Physics Congress, held on the border of Lake Como. When faced with the possibility that some things in the world were intrinsically unpredictable, Einstein continued to protest. In discussions with Bohr later that fall, Einstein uttered the famous phrase: "God does not play dice with the universe."[23] Bohr, in turn, replied with a less famous, but perhaps more sensible request: he asked Einstein to please stop telling God what to do.[24] Bohr, despite all his disagreements, would always remember Einstein fondly, especially for "his favoured use of such picturesque phrases as 'ghost waves' (Gespensterfelder) guiding photons."[25]

As long as quantum uncertainty was based on outcomes that could change statistically, Einstein refused to consider it complete. In his view, if we could process all the data, nothing would be left up to chance. "His conviction seems always to have been, and still is to-day, that the ultimate laws of nature are causal and deterministic," explained Born, describing how for Einstein "probability is used to cover our ignorance . . . only the vastness of this ignorance pushes statistics into the fore-front."[26]

Progress in analyzing the behavior of atoms and molecules in gases increased with the development of improved vacuum technologies. The physicists Karl Compton and Henry De Wolf Smyth tried a new experiment at the Palmer Laboratory at Princeton University: building a "Mechanical Maxwell Demon" that separated molecules in gases using mechanical vanes. The two Princeton physicists joined forces with a third researcher from an industrial laboratory and published their results in the prestigious *Physical Review*. "Long ago," they explained, "Maxwell suggested that molecules of differing velocities might be separated by the intervention of his infinitesimal but highly intelligent 'demons.'" Inspired by this research, "it occurred to the writers that vacuum technique had progressed to the point where a mechanical Maxwell demon might be possible."[27] The rest of the text described an instrument that separated heavier from lighter molecules by pushing molecules of a particular gas through two radial slits rotating in parallel. With their mechanical Maxwell demon, Compton and his coauthors succeeded in measuring the kinetic energy of different molecules.[28]

A promising strategy for analyzing the behavior of atoms and molecules involved manipulating them via their electrical properties. "It is a curious consequence of the electrical nature of matter that we can study atoms and molecules more easily when they are ionized than when they are in the normal electrically neutral state." This provided a completely different avenue than that of statistical mechanics: "If we are dealing with ions we can control their paths and speed, measure the ratio of their charges to their masses and in general study their behavior as individual particles but when dealing with normal molecules we still have to depend on statistical effects such as pressure, density, and temperature."[29] Soon the method would be used for isotope separation.

After building their early "mechanical Maxwell demon," Compton and Smyth continued to think about demons throughout their lives. Their first investigations of Maxwell's demon would be a short prelude to what was to come. Karl and his brother Arthur spent the years 1926 to 1927 at Born's theoretical physics mecca in Göttingen, where they crossed paths with Robert Oppenheimer. Upon returning to the United States, their careers took off. Arthur would receive a Nobel Prize later that year for his work studying photons (light particles) shot at and scattered off electrons, an effect now known as "Compton scattering." Karl became president of MIT and played a key role in the development of the atomic bomb. He would be joined in that endeavor by Arthur, who vigorously supported building and using the atomic bomb and who led the construction of the first self-sustaining nuclear chain reaction at the University of Chicago.

Most histories of the atomic bomb focus on a select group of scientists around Einstein and Oppenheimer. Yet Oppenheimer, although widely known as "the father of the atomic bomb," was a little-known theoretical physicist hired well after some of the most important decisions about the project had already been made. Scientists who made decisions behind the scenes were often not the ones credited by the press. When Neumann later reflected on how everyone seemed to focus on Oppie's role in the bomb's creation, he was simply incredulous.[30] While publicity-friendly scientists took center stage, the Compton brothers, who were ultimately responsible for mobilizing a team to build a new atomic weapon, shunned media attention. Smyth would join the brilliant Compton brothers in these endeavors, first as a researcher of dirty radioactive weapons and later as the person in charge of curating the narrative about the bomb's creation presented to the public. As the author of the official report about the creation of the atomic bomb, Smyth would consider it in terms of a larger effort to create a perpetual motion machine that could expend more energy than it consumed.

In contrast to some of the Jewish scientists who had recently emigrated from Europe and whose politics often veered left, such as Einstein and Oppenheimer, the WASP-ish Compton brothers and Smyth, their colleague at Princeton, were trusted members of a close circle of well-heeled and mostly like-minded collaborators that included James Conant, the president of Harvard University, and Vannevar Bush; the vice president of MIT, Bush was the first presidential science adviser and the head of the Uranium Committee charged with analyzing the possibility of building a bomb. Smyth and the Comptons had easy access to the highest echelons of government circles reaching all the way to the top—access that was closed off to potentially unpatriotic foreigners. President Franklin D. Roosevelt would listen carefully to what the Compton brothers had to say about the possibility of building a bomb, about the chances that the Nazis might too, and more. Karl Compton would spend the rest of his career advocating for "intelligent planning," and he served nearly two decades as a top scientific adviser to Roosevelt and later Harry S. Truman. At MIT, Karl would go on to think of Maxwell's demon as a model for organizing society and Arthur would focus on quantum mechanical "daemons" who could alter the paths of subatomic particles.

Quantum mechanics soon had an impact on research into the nature of logic and mathematics. The mathematician Richard von Mises, director of and professor at the Institute of Applied Mathematics at the University of Berlin and later Harvard University, was convinced by the new studies on radioactivity, quantum theory, and Brownian motion that a proper understanding of statistics required leaving behind outmoded concepts of mathematical logic. Statistics, in his view, was no longer just a useful shortcut for manipulating large numbers and studying effects whose entire story was not yet properly understood. Rather, these new discoveries showed that statistics itself reflected the actual state of the world in a way that was more accurate and fitting. As mathematicians and logicians started to incorporate the idea of quantum indeterminism into their fields, they increasingly described the traditional mathematical techniques associated with Laplace as faulty and as having been haunted by a demon.

To Mises, the quest to know all of nature more and more precisely through measurement and calculation was driven by the idea of a "mathematical demon" that had emerged from Laplace's speculations: "The most extreme formulation of Newton's determinism is to be found in Laplace's idea of a 'mathematical demon,' a spirit endowed with an unlimited ability for mathematical deduction who would be able to predict all future events in the world if at a certain moment he would know all the magnitudes characterizing its present state," Mises wrote, noting in a footnote that "Der Laplacesche 'Dämon'" had appeared first in Laplace's *Essai philosophique*.[31] He argued that this demon, alongside the idea of causality that he sustained, was nothing but a two-century-long prejudice. There might be no point in trying to get at the truth by working on obtaining more precise measurements and calculations. Good math and good logic, in his view, should reflect this underlying uncertain reality.

If the concept of causality was in trouble, traditional ideals of exactitude, precision, and finality seemed broken beyond repair. After a certain point, attempts to get more precise knowledge simply backfired on scientists and logicians. "Instead of finding something which, from the point of view of classical physics, is simpler than the original phenomenon," explained Mises, "we arrive at phenomena that are more and more complicated (from a deterministic point of view)."[32] In a later publication, Mises speculated on why Laplace's "Dämon" had been born in tight connection with the discipline of astronomy. It was much more difficult to think of him as an effective operator on planet earth, with all its terrestrial complexities. "In the creation of his 'demon,'" argued Mises, "Laplace, too, probably thought that he would encounter quite some difficulties in solving his task on earth, as compared to celestial bodies."[33]

The earthly troubles besetting Laplace's demon were getting worse. What would happen if more and more scientists and mathematicians stopped believing in strict causality? Other demons might take advantage of his weakness. "We can scarcely deny the charge that in abolishing the criterion of causality we are opening the door to the savage's demons," warned Eddington.[34] Yet most of his colleagues were ready to welcome these demons, holding the door wide open for them. New studies on radioactivity only helped their cause. During the 1928 National Radium Conference, the US surgeon-general told listeners that radium "reminds one of a mythological super-being."[35] This being was still weaker than Laplace's predictable creature—and weaker than the minute Maxwell's demon, who could tire so easily and could hardly ever impinge on the world at human scale—but it would not remain so for much longer.

SZILARD'S EXORCISM

Meanwhile in Berlin, Leó Szilard, a young Hungarian student who had moved there to finish his engineering studies, which had been interrupted during the war, pursued his *idée fixe* of building a perpetual motion machine. After meeting Einstein, he switched to physics and became one of his closest collaborators.

Szilard would soon reach fame for creating one of the first nuclear chain reactions (he would hold the patent for it); for writing the letter sent and signed by Einstein to President Roosevelt warning him of the consequences of building an atomic bomb; for being one of the most important scientists to work on the Manhattan Project; and for infuriating, on one too many occasions, the general in charge of the project. But first, Szilard dedicated himself fully to the quest to understand why a machine could never produce more work than it expended. He spelled out the reasons for this in a highly original solution to the problem; published in the prestigious journal *Zeitschrift für Physik*, it was widely known as "Szilard's exorcism." Therein he described a theoretical model for a highly efficient engine that would be known as "Szilard's engine" or "Szilard's demon."[36] It would come closer than anything known before to almost subverting the second law of thermodynamics. All it needed was a tiny bit of knowledge that he could obtain by performing a simple measurement.

When Einstein first met Szilard, the professor was impressed by the younger student. As experience that comes with seniority often dictates, Einstein cautioned his protégé against relying too much on fanciful, overambitious ideas. "It is not a good thing for a scientist to be dependent on laying golden eggs," he told the young Hungarian.[37] Einstein insisted that instead of chasing after seemingly impossible goals, Szilard should get a steady income and a job; he could even find work in a patent office, Einstein suggested, as he himself had done when he was younger. Despite the sensible advice Szilard received from his teacher and mentor, the young upstart continued to pursue pie in the sky.

Szilard's so-called exorcism was published in 1929 with a long title translated as "On the Decrease of Entropy in a Thermodynamic System by the Intervention of Intelligent Beings."[38] It changed how scientists thought about Maxwell's demon for the rest of the century. The main question it tackled was similar to the one that had fascinated Maxwell and his colleagues: could intelligent beings decrease entropy? Szilard reread Smoluchowski's work on Brownian motion and cited its central insight: "As long as we know today, there is no automatic, permanently effective perpetual motion machine, in spite of the molecular fluctuations, but such a device might perhaps, function regularly if it were appropriately operated by intelligent beings." But why? What could be so special about "intelligent beings" that only they possessed such abilities? Szilard zeroed in on the conclusions of the Polish scientist. Such a machine could be built if the "intelligent beings" operating the machine and taking advantage of the fluctuations were not considered part of the system. If "we view the experimenting man as a sort of *deus ex machina*," then "a perpetual motion machine," Szilard elaborated, "is possible." He stressed that current limitations on building one had nothing to do with humans not being small enough or not "possess[ing] the ability to catch single molecules like Maxwell's demon." So what exactly seemed to give intelligent beings the power of acting like this *deus ex machina*? The requisite smart operator, rather than needing to be tiny and nimble, only needed to be someone who was "continuously and exactly informed of the existing course of nature and who is able to start or interrupt the macroscopic course of nature without any expenditure of work." Szilard noted that it was this quality, "a sort of memory faculty," that was necessary for taking a measurement. That measurement-taking activity did not come for free. The physicist determined that if a semipermeable wall was installed between two partitions in order to select molecules from which work could be extracted, this "simple inanimate device can achieve the same result as would be achieved by the intervention of intelligent beings." But such "intelligent" intervention would not be able to create an endless supply of energy because it would expend as much energy in "measuring" the molecules as it would gain in extracting work from them.

Szilard's article reviewed attempts to build perpetual motion machines from random, fluctuating motions. He determined that, in these cases, our ability to harness work was no better than what chance afforded us. "We are in the same position as playing a game of chance, in which we may win certain amounts now and then, although the expectation value of winning is zero or negative."[39] Were the chances of winning better when intelligence was introduced? Could intelligence be used to beat chance? If so, how well could we escape the tentacles of entropic fate by using our wits?

Szilard pictured an intelligent being starting and stopping his engine at exactly the most favorable moments. The results of acting in this way looked so good that they had to be too good to be true. Szilard admitted that, "as long as we allow intelligent beings to perform the intervention, a direct test is not possible."[40] Szilard came to the pessimistic conclusion that, by replacing the intelligent being with a mechanical device, such as a semipermeable wall that acted as such, "exactly that quantity of entropy which is required by thermodynamics" could be generated.[41] Was the proposed substitute *really* intelligent? Other scientists would soon think of ways of bypassing the limitations of Szilard's engine by arguing that real intelligence could do more than what a mere filter (the semipermeable wall) could accomplish.

Another way of obtaining potentially endless sources of energy was being explored in California, on the other US coast. Working in the College of Chemistry at the University of California at Berkeley, the eminent scientist

Gilbert Newton Lewis started to think more carefully about Maxwell's demon. As he wrote in his book *The Anatomy of Science* (1926), he was particularly intrigued by a demon who could choose between molecules and "through the exercise of this *conscious choice* would ultimately restore the original condition" of different kinds of gases that had been mixed together. "That such a reversal of a so-called irreversible process may occur, even without a demoniacal agency," he reminded readers, "was first recognized by Willard Gibbs, who announced that such a reversal is not an impossibility but only an improbability." According to Lewis, this idea "has grown to be an important, and almost the most important, guiding principle of modern physics."[42] As if the possibility of reversing the regular course of nature was not titillating enough, Lewis explored whether these results might also reveal the secret of life.

A reviewer of Lewis's book for the *Mathematical Gazette* noted that "the author suggests that living things are cheats in the game of physics and chemistry and, like Maxwell's demon, may take advantage of the local fluctuations from the average state, and that it will never be possible to explain vital phenomena by physics and chemistry alone."[43] Soon after its publication, Lewis became even more involved in the new physics, coining the term "photon" for light particles.[44] In another groundbreaking article, he considered the complex connections between Maxwell's demon, entropy, living beings, and information.

One sentence in Lewis's paper would be cited over and over again by later researchers: "Gain in entropy always means loss of information, and nothing more."[45] Lewis first started exploring these weighty questions in "The Symmetry of Time in Physics," published in 1930 in *Science*. There he came to conclusions similar to those reached by Smoluchowski and Szilard: "Without the aid of demoniacal devices," any method "requires at least as much work as the old-fashioned thermodynamical method of forcing the system into the particular distribution."[46] Lewis considered the case of single molecules. "In the simplest case, if we have one molecule which must be in one of two flasks, the entropy becomes less than $k \ln 2$, if we know which is the flask in which the molecule is trapped." According to him, this unit symbolized the minimum amount of energy required to *know* something. The quantity was no doubt small, but no longer negligible given how much science had advanced in its study of the atomic and subatomic realms. "In dealing with the individual molecules," he noted, "we are perhaps arrogating to ourselves the privileges of Maxwell's demon; but in recent years, if I may say so without offense, physicists have become demons."[47]

Lewis's paper, appearing one year after Szilard's, started by envisioning a filtering device in the shape of a "cylinder closed at each end, and with a middle wall provided with a shutter."[48] Could such a device be constructed so that it could unmix mixtures? If so, separated substances that had once been mixed could be reused as renewable fuel. During World War I, Lewis had worked at the Gas Service and the Chemical Warfare Service, where cylinders divided by filters (also known as semipermeable membranes) to separate different elements were standard equipment. But separating isotopes of the same element was much harder. Soon Lewis and his colleagues would be looking for ways to separate the small percentages of the highly reactive Uranium 235 isotope mixed in with the more stable Uranium 238 in order to build a bomb.

Lewis took the demon thought experiment up a notch. His investigations of Maxwell's demon increasingly impinged not only on cosmology but also on larger philosophical questions about the essence of life and its relation to physics and consciousness. Lewis thought about what would happen if Maxwell's demon manipulated living organisms. He started by comparing the behavior of healthy versus sensory-deprived mice. Did they use "information" to act differently from the way mere physical systems did? "I have learned," Lewis explained, "that it is possible to perform an operation upon the brains of mice so that they respond to no external stimuli, but can still run aimlessly about." Their random movement would lead to equilibrium. "If a large number of these mice are placed in one end of a box," Lewis explained, "that end is now heavier than the other." The mice would soon wander around, and about half of them would reach the other side of the box. "This distinction rapidly disappears as the mice, in their random movements, cover with great uniformity the bottom of the box, so that we may no longer discern any tendency of the box in one direction or the other."[49] Sensory-deprived mice acted like random molecules, he found, but most sentient beings were not like these operated-upon lab rats. Where exactly lay the difference? Lewis circled once again into the old Cartesian riddle of determining what exactly was the difference between purely mechanical and living systems. The answer had to hinge on entropy, information, and intelligence, he concluded. "Unless there is in sentient beings the power to defy the second law of thermodynamics," entropy would be leaked into the system, rendering it unable to produce work by itself. His use of the term "unless" proved intriguing for decades to come.

AMPLIFYING QUANTUM EFFECTS

In the fall of 1930, a few days after the dramatic gains by the Nazis in Germany's federal elections, Szilard wrote to Einstein complaining about the worsening political situation in Germany. The US stock market had crashed a year before, the economy was getting worse, and day had turned into night in America's "Dust Bowl." As the Great Depression spread quickly around the world, the global economic and political situation became so dire in the early 1930s that Mephistopheles's tempting offers could seem worth taking. Europe seemed on the brink of another world war.

How could one understand such unanticipated historical events? Szilard responded by pairing up with Einstein to literally "beat the heat." At the same time that he was publishing theoretical physics papers, Szilard sought to develop a practical circumvention of the laws of thermodynamics by devising and patenting a novel technology for building a more efficient refrigerator. While entropy naturally led to temperature equilibrium, a fridge created the opposite effect: temperatures became colder within a well-insulated enclosure. By the end of that tumultuous year, Einstein and Szilard had submitted thirty-seven patent applications and had twenty-eight of them approved. When they sold some to Electrolux in Stockholm and to other companies in Europe and America, they were able to pocket some desperately needed profits.

"From week to week I detect new symptoms," Szilard wrote to Einstein. "If my nose doesn't deceive me, that peaceful development in Europe in the next ten years is not to be counted on." Einstein and Szilard might have to relocate or go into exile. "Indeed, I don't know if it will be possible to build our refrigerator in Europe," Szilard concluded.[50]

A refrigerator was not going to save the world, but radioactivity and atomic transmutation might. Could quantum effects be made consequential at human scale? At a time when some of science's core principles were being debated, scientists forged ahead with new research that was likely to better the world and less likely to destroy it. The scientific community was divided along ideological and even metaphysical lines. The philosophical foundations of quantum mechanics and relativity theory seemed increasingly incompatible. Could an experiment be constructed that might help scientists decide whether they should back Einstein's program (of continuing to look for deterministic laws) or be satisfied with Bohr's so-called Copenhagen interpretation? Could such an experiment shed light on what was going on in the world more generally during a complex historical crossroad?

Einstein could use the help of allies in America, but Arthur Compton was firmly against Einstein's program and in favor of Bohr's. Compton and Bohr's alliance was fitting. Bohr's coup against Einstein was based on the "Compton scattering" research that had earned him the Nobel Prize. When considering the effect named after him, Compton imagined quantum "daemons" who could manipulate these tiny particles. Rigged with sticks of dynamite that could create much bigger explosions, Compton's quantum daemons could either let them trigger the payload or intervene to "save the day." "Free will," surmised Compton, might work in a similar way, by giving us the power to create great effects on the outside world.

In a discussion about the uncertainty principle and free will published in a 1931 issue of *Science*, Compton described an ingenious setup to amplify quantum effects by using a photoelectric cell triggered by a photon to explode a stick of dynamite. Scientists knew that if photons were aimed toward a double slit, they would pass half the time to the right and half the time to the left. If only one of the slits was rigged to an explosive, then it would go off 50 percent of the time. "By means of suitable amplifiers it may be arranged that if the first photon enters cell A, a stick of dynamite will be exploded (or any other large-scale event performed); if the first photon enters cell B a switch will be opened which will prevent the dynamite from being exploded." What could scientists learn from this arrangement about nature at human scale? "What then will be the effect of passing the ray of light through the slits?" he asked. Most physicists would agree that in this experiment, "the chances are even whether or not the explosion will occur." But Compton insisted that the real lesson of the experiment was that it showed that the chances of the explosive going off *could never be known* beyond a 50 percent limit. The moment scientists intervened to know exactly where a photon was going, its weird wavelike interference effects disappeared, yet when given a choice, photons would continue to choose evenly between the two slits. "That is, the result is unpredictable from the physical conditions," concluded Compton.[51] Like other quantum physicists, Compton argued that statistical uncertainty did not arise from our ignorance of underlying causal effects, but rather that it was a fundamental fact of nature and, more importantly, could be consequential at human scale. A single subatomic particle could act like the proverbial butterfly wing—or the pebble that caused an avalanche, acting as the domino that set the others off on their path of destruction.

Compton argued that a mechanism similar to the one that made photons act in the strange ways they did was comparable to that which led to the emergence of life from physical systems. Living systems, in his view, were much like quantum amplifiers. To reach these conclusions, he gained inspiration from the work of the Canadian biologist and physiologist Ralph Lillie, who, after carefully studying Maxwell's demon, concluded that Heisenberg's uncertainty principle provided wiggle room for a statistically low-chance effect to have repercussions at human scale, leading to the emergence of life. Lillie, according to Compton, "has pointed out that the nervous system of a living organism likewise acts as an amplifier, such that the actions of the organism depend upon events on so small a scale that they are appreciably subject to Heisenberg uncertainty," he explained. "This implies that the actions of a living organism can not be predicted definitely on the basis of its physical conditions."[52] Lillie, who had at first been critical of extending the lessons of Maxwell's demon to the realm of biology, started to consider the possibility of some sort of "directive influence" that resulted in the creation and perpetuation of life. A minor selection in the right direction (one so small that it could be hiding within the Planck scale) could explain the origin of life and the secret to its perpetuity.[53] Something similar, speculated Compton, might be behind free will.

In this view, it could well be that humans and other living creatures were the means through which demons operated in the macroscopic world. Compton developed his thoughts about quantum mechanics and free will further for the prestigious Terry Lectures delivered at Yale (1931–1932), then reworked them into the book *The Freedom of Man*, published three years later. By then, he would be thinking about the possible intervention of a new "daemon" in the quantum world hiding behind the cover of statistical uncertainty. A few years later, he was researching the possibility of building an atomic bomb. Soon afterward, he would be supervising the first atomic pile as a key member of the Manhattan Project.

That spring at Bohr's institute in Copenhagen, some of the most respected physicists of the day wrote and put on a show parodying Goethe's *Faust*. What would they do with the new knowledge coming out of the laboratories? Many of Compton's colleagues were present. The physicist Wolfgang Pauli was tapped to play Mephistopheles, who tempted Faust deep into the quantum world of new subatomic particles; Arthur Eddington took on the role of the archangel Raphael, and Robert Oppenheimer formed part of the chorus alongside Heisenberg. As Mephistopheles tempted Faust with the promise of *"unendlicher Selbstenergie,"* he chanted:

> Despise reason and science, if you will
> Mankind's highest power!
> Let dazzling tricks and magic spells
> Strengthen you in the quantum spirit![54]

Demon-talk was not all fun and games. That summer Max Planck, arguably the highest authority on these matters, was busy traveling across Europe to lecture about quantum mechanics and Laplace's demon. His contributions to quantum mechanics had made him world-famous, but he was not ready to embrace Bohr's Copenhagen interpretation in its entirety. It could be tempting to think of Planck as an enemy of Laplace's creature, but his views about this creature were far more complicated, and far more positive. As Planck was delivering these lectures across Europe, the Nazis were gaining seats in the Reichstag.

Planck considered the limits of the *idealen Geist* who could predict the future and know the entire past. Recent contributions had proven that "there is a point, one single point in the immeasurable world of mind and matter, where science and therefore every causal method of research will always remain inapplicable." That space of inapplicability would remain so "even in the case of the super-intelligence postulated by Laplace." What would happen if a lowly mortal tried to converse or consort with Laplace's super-intelligence? "The inquisitive human being who would do so, is quite likely to hear this answer to his question: 'You resemble the intellect which you comprehend, not *me!*'"[55]

The urge to abandon a belief in Laplace's being merely because of the current evidence against him was strong, Planck acknowledged, but should be resisted. "And if after this reprimand," wrote Planck, "he persists in his obstinacy and declares the concept of an ideal intellect to be meaningless and unnecessary, if not illogical, let him be reminded that not all statements which lack a logical foundation are scientifically worthless." To declare him so nugatory and to readily erase the concept of causality from science with a "short-sighted formalism" would be a grave mistake that might hamper great discoveries in the future. Such an abandonment would "stop up the very fountain at which Galileo, a Kepler, a Newton, and many other great physicists quenched their thirst for scientific knowledge and insight."[56]

In Planck's account, the search for Laplace's being was still valuable, even if at present he proved to be an inscrutable and powerful force reminding us of the limits of our knowledge vis-à-vis his own. He might not be at all as we imagined him to be or found where we thought he might be. Science might not even be the best discipline to get to know him. "We must be on our guard against the temptation to make the ideal intellect the object of a scientific analysis, to regard it as something analogous to ourselves, and to ask of it how it obtains the knowledge which enables it to make precise predictions." To know what it knew, Laplace's being could be acting in ways that contemporary scientists might not even begin to suspect.

MAXWELL'S DEMON IN THE QUANTUM WORLD

Einstein's friends in Berlin included a twenty-nine-year-old prodigy from a wealthy Jewish Hungarian family who cotaught physics with Szilard. John von Neumann would soon be recognized as one of the world's foremost experts on quantum mechanics with the publication of his *Mathematical Foundations of Quantum Mechanics*. The German edition appeared in 1932 and immediately became a classic. For decades to come, it was considered to have disproven the possibility of "hidden variables"—that is, the idea that as-yet-unknown mechanisms could give scientists the means to reinterpret quantum effects in a fully deterministic fashion.

Neumann's contributions to science were based on a startlingly different understanding of the incapacities of Maxwell's demon. The demon could not work at the quantum level, he explained, because he could not measure the position and momentum of a particle with complete accuracy: a measurement of one led inevitably to an uncertainty in the determination of the other. The two could not be separated—not with deft fingers, a door, a cricket bat, a semipermeable membrane, a filter, or anything else—*without changing the system* in an essential and irreversible way. Heisenberg's uncertainty relations, in Neumann's formulation, resulted from the limitations of Maxwell's little guy.

Laplacian and Maxwellian demons could work only if full knowledge of the position and momentum of the particles they handled was available to them. "If we knew all the properties of the molecule before diffusion (position and momentum), we could calculate for each moment after the diffusion whether it is on the right or left side."[57] Neumann argued that this information was simply not available in nature. A moving molecule should be thought of as being simultaneously on one side of a partition and also on the other. Classical thermodynamics assumed that "states whose differences are arbitrarily small are always 100%, separable," yet this assumption was unwarranted in the microworld. "The separation is completely impossible," he stated.[58] In other words, the limitation was not merely practical and *de facto*, but *de jure*.

Neumann expanded on Szilard's research, giving a new explanation for the conclusions reached by his compatriot. The book was centrally preoccupied with Laplacian determinism, Maxwell's demon, and Szilard's research; these topics played key roles in the celebrated chapters on measurement and reversibility. His challenge was clear: "to solve a well-known paradox of classical thermodynamics" long associated with Maxwell's demon: how could mechanical processes be reversible in principle but rarely in practice?[59] According to Neumann, quantum mechanics could provide the answer to this nearly century-old question. His quantum-based explanation of entropy was "based on the fact that the observer does not know everything, that he cannot find out (measure) everything which is measurable in principle."[60] Having learned about Maxwell's demon's limitations from reading Szilard's exorcism, Neumann was most eloquent about why Maxwell's demon could not accomplish much at our human scale: the reasons lay in the quantum realm.

"A thorough going analysis of this question is made possible by the researches of L. Szilard, which clarified the nature of the semi-permeable wall, 'Maxwell's demon,' and the general role of the 'intervention of an intelligent being in thermodynamical systems,'" wrote Neumann, elaborating on Szilard's interpretation of what exactly happened when the demon tried to manipulate individual molecules.[61] Proof of why two states (such as position and momentum) should be understood as superimposed at the quantum level resided in the fact that they could *not* be separated without introducing an increase in the entropy of the system and thus changing it irreversibly. From then on, the possibility of knowing nature without changing it became amply disputed. Even the most carelessly cast glance could change the world forever.

In Szilard's exorcism, the intervention of an intelligent being in thermodynamical systems led inevitably to an increase in the entropy of the system. In Neumann's elaboration of his ideas, this action led to an increase of entropy as well and to the introduction of irreversibility into the system due to *a change in the state of nature*. Szilard had blamed difficulties in determining the exact position of a certain particle as due to the cost of obtaining that knowledge. But in the quantum mechanical view advocated by Neumann, observers could not know the position and momentum of a particle, not because they could not have all of the measurements for free, but rather because this information did not yet exist in a recoverable form. Obtaining it involved separating and thus changing the system.

The consequences of this new understanding of the limitations of Maxwell's demon were nothing short of revolutionary. Its implications reached far beyond quantum mechanics and affected science and philosophy more generally, potentially leading to the espousal of nonlocality, indeterminacy, and uncertainty, and to the view that the universe might be a place affected by consciousnesses. Neumann was trenchant in his critique of causality. "There is at present no occasion and no reason to speak of causality in nature," he insisted, "because no experiment indicates its presence, since the macroscopic are unsuitable in principle, and the only known theory which is compatible with our experiences relative to elementary processes, quantum mechanics, contradicts it." Neumann lamented that old habits die hard. "To be sure, we are dealing with an age old way of thinking of all mankind, but not with a logical necessity (this follows from the fact that it was at all possible to build a statistical theory), and anyone who enters upon the subject without preconceived notions has no reason to adhere to it." Quantum mechanics "has opened up for us a qualitatively new side of the world," and although "one can never say of the theory that it has been proved by experience," it is "the best known summarization of experience." Should it be abandoned merely because it offends our view of causality? Neumann concluded with a final rhetorical question, "Under such circumstances, is it sensible to sacrifice a reasonable physical theory for its sake?"[62]

To convince his readers, Neumann asked them to follow him on a *Gedankenexperiment* similar to Maxwell's. While in Maxwell's example scientists compared the demon to a semipermeable wall or valve used to separate molecules, in Neumann's case "our semi-permeable wall is essentially different from this thermodynamically unacceptable one." Semipermeable walls or valves, he argued, did not work perfectly in the quantum world because the position and momentum of the molecules could not be determined beyond certain limits. Maxwell's examples dealt with the "exterior (i.e., whether it comes from the right or left, or something similar)" of molecules, but Neumann's dealt with characteristics of molecules that were unobservable not in practice but in principle.[63] There were simply no elements in a perfectly determinate state, Neumann concluded, for a demon to play with.

By exploring exactly what Maxwell's demon could or could not do, Neumann and his followers developed the concept of the "classical observer" or "macroscopic observer." For such observers, measurement worked as we were used to in classical terms where "everything which is measurable at all, is also simultaneously measurable, i.e., that all questions which can be answered separately can also be answered simultaneously." But in the quantum world, a measurement of something led to our inability to measure something else. "The reason that the non-simultaneous measurability of quantum mechanical quantities has made such a paradoxical impression," Neumann explained, "is just that this concept is so alien to the macroscopic method of observation."[64] Scientists until then had thought that when Maxwell's demon looked around him, the world certainly appeared much larger, but they had imagined it to seem similar in qualitative terms, except for scale. Quantum physicists argued that it was really not—it was a different universe that functioned according to different laws and that retreated just as we peeked into it.

PHILOSOPHY, LOGIC, AND COMMUNIST POLITICS

At the Ivy League campus of Yale University, the philosopher Henry Margenau described Laplace's demon as an entire approach to science and philosophy that was now widely questioned. Its historical importance was such that it still "pervades all of classical physics and has been eminently fruitful in the development of that science" by enabling scientists to "predict the future."[65] Margenau used the term "Laplace's demon" interchangeably with that of "omniscient demon."[66] The philosopher had initially been unwilling to give up such a demon—and a belief in causality—even in light of the new discoveries that came from the quantum realm. A few years later, he would no longer be as certain, and his faith in this demon started wavering. "The fact that the postulation of a universal formula, as phrased by Laplace, involves the hypothetical existence of an omniscient demon," explained Margenau, "has been considered unsatisfactory by numerous investigators."[67] It was not reason enough to dismiss Laplace's idea entirely. "Hence it is not legitimate to say that the demon does not exist or is impossible and therefore reject the proposition." "A real criticism," argued the philosopher, "should attack its meat and not its form."[68] In later work, Margenau reflected on the use of the word "demon." "Whether or not this word is appropriate, we shall use it because of its brevity," he wrote, while acknowledging that "the demon has been a bone of contention."[69]

"How Laplace would turn in his grave if he knew what happened to it!" wrote Robert Millikan, known for his work on the electron and for following Einstein and Bohr in winning the Nobel Prize.[70] Determinism seemed to be under siege. Demons could be master logicians; perhaps they could even access higher forms of logic than those available to us. For philosophically minded mathematics, the noxious effects of quantum mechanics on the health of Laplace's demon started to show that what seemed logical from one perspective might not be so from another. If it was hard for us to see a logical connection between two things, say A and B, that did not mean that such a connection could not exist. The mathematician Léon Lichtenstein (a distant relative of Norbert Wiener) paired up with André Metz (an ardent critic of Henri Bergson) to rethink the notion of mathematical inference by reference to Laplace's demon. They suggested that steps that seemed complicated or impossible to us might be simple and logical for him. "For a being of the genre of Laplace's demon, whose inner visual field could be regarded as unlimited," Lichtenstein and Metz explained, "the path between the A-statement and the B-statement would be shortened in such a way that we would see the starting point and the arrival point coincide: B would be recognized as presenting, as it were, the same meaning as A and, therefore, as obvious."[71]

Philosophers had to adapt. In Vienna, the philosopher and political economist Otto Neurath led a group of scientists and philosophers known as the Vienna Circle who were committed to fighting for science. He was among many others who saw Laplace's demon at the center of an entire physical-philosophical-political program that had characterized the previous century. The creature no longer appeared as credible as it once had. Neither did the hope of creating a rational structure for science based on combining sense data with analytical mathematical principles, a main hope of Neurath and his allies. "The fiction of an ideal language constructed from pure atomic sentences," he wrote, "is no less metaphysical than the fiction of Laplace's demon" (*Laplaceßchen Geißtes* in the original).[72] If that foundation for science faltered, what would take its place? For some scientists and philosophers, the demon's apparent weakness, and possibly even his demise, threw a wrench into their hopes that a unified structure for science and society could be built up logically from sense data, a goal pursued by optimistic logical empiricists and logical positivists.[73]

The failure of a causal understanding of nature was deeply felt far beyond scientific laboratories. The Russian revolutionary Leon Trotsky, who broke rank and became one of Stalin's strongest opponents during those years, mocked the naive determinism of his political adversaries by reference to Laplace's fancy. Poking fun at humans' ability to determine the future, let alone the future of the Soviet state's attempts to control history through five-year plans, Trotsky published a short pamphlet titled *Soviet Economy in Danger* (1932). "If there existed the universal mind that projected itself into the scientific fancy of Laplace," he wrote, "such a mind could, of course, draw up a priori a faultless and an exhaustive economic plan, beginning with the number of hectares of wheat and down to the last button for a vest."[74] The problem with Stalinist communism, according to Trotsky, was that it assumed that such a Big Brother could in fact exist.

The adherence by prominent Soviet socialists to a cult of Laplace's being underpinned the techno-scientific optimism of mainstream communism and was often posited as the raison d'être of collectivization. During those years, the utopian transhumanist thinker Nikolai Fyodorov (often anglicized as Fedorov) became an inspiration to many young socialists. Trotsky and Fyodor Dostoevsky, among others, drew attention to his legacy. Fyodorov's posthumously published *Philosophy of Common Task* claimed that Laplace's creation could be brought to life if everyone worked together, literally putting their minds to it to collectivize intellectual work. Two brains might think better than one, and three better than two, and so on. A "collective mind," in this view, would sum up to Laplace's ideal. "For the vast intellect able to encompass in one formula the motions both of the largest celestial bodies in the Universe and of the tiniest atoms," wrote Fyodorov, "nothing would remain unknown; the future as well as the past would be accessible to him." He confidently concluded, "The collective mind of all humans working for many generations together would of course be vast enough—all that is needed is concord, multi-unity."[75]

Easier said than done. The optimism of the prerevolutionary days started to wane in light of Stalin's economic and cultural programs and the prosecution of his opponents in what came to be known as the Moscow Trials. Trotsky himself had started to lose hope in the new system. "In truth," he wrote, "the bureaucracy often conceives that just such a mind is at its disposal; that is why it so easily frees itself from the control of the market and of Soviet democracy. But in reality the bureaucracy errs frightfully."[76] Such mistakes were used to argue against democracy and to justify the complete control of the economy by the state. Armed with a critique of Laplace's mind, Trotsky started to contest the basic principles of communism and to argue for the need for a partial restoration of capitalism, pleading mercifully for deregulation. Current Laplacean-inspired "blueprints" for the future of communism, he claimed, were off. It was hard, if not impossible, to plan for the future, and the culprit, according to Trotsky, had been Laplace's dangerous idea. Trotsky would soon be murdered while in exile in Mexico.

For more than a century, Laplace's being had embodied the ideal of a certain type of causal knowledge. He motivated hopes that one day we might know it all by *collectivizing* thought in a way that inspired revolutionaries and conservatives alike. But quantum indeterminism and uncertainty posed challenges to the vision he represented. His weakness in the context of new quantum-level experiments worried scientists and philosophers alike. For half a century, Maxwell's creature had represented a new way of understanding intelligence and a kind of progress that was compatible with Laplace's, but even better. This demon showed scientists the possibility of escaping the fatalism of a state of increasing loss. But as scientists analyzed him in detail, they noticed evident limitations that were as disappointing as they were disquieting. As fascism spread throughout Europe, scientists continued to hope that learning more about these two demons might show scientists how to build a better world, while new research into the quantum world led scientists to consider the possible existence of other demons that acted in previously unthought-of ways.

ATOMIC ENERGY

On January 29, 1933, the *New York Times* ran a dramatic headline for a story on new research on atomic energy: "PREDICTS UTILIZING OF ATOMIC ENERGY; President Compton of MIT Declares Science Now Ready to Harness Vast Force; HIGH VOLTAGE ION AS TOOL; He Cites Generator Experiments Delivering as High as 10,000,000 Volts; BUT RELIES ON CHEMISTRY; He Looks to Laboratory Process to Liberate Atom for Increased Power Output."

The following day, on January 30, 1933, a follow-up article authored by Karl Compton was published in the *Technology Review* and Adolf Hitler became chancellor of Germany.

Einstein would soon emigrate, resettling at Princeton University with a position at the Institute for Advanced Studies. His two Hungarian friends also left. Neumann had left for Princeton earlier that year, and Szilard moved to England before emigrating to America. Throughout the next months, the *New York Times* ran stunning headlines about transmutation and radioactivity, claiming that the cyclotron used to smash atoms and particles into each other to create these effects was the "new miracle worker of science, the most powerful cannon yet found for liberating relatively enormous stores of energy locked up in the inner core of the atom."[77] The clever mechanical engines and explosives of the nineteenth century proved to be no match for radioactive substances. The popular press enthusiastically covered articles in *Nature* and the *Physical Review* that described the exciting possibilities for creating potentially unlimited sources of energy and stores of precious materials, such as gold, by forcing atoms and subatomic particles to crash into each other.

During those years the biologist J. B. S. Haldane, a militant atheist known for his pioneering work in genetics, described scientific experimentation as a method safe from the actions of creatures associated with religion or myth: "When I set up an experiment I assume that no god, angel, or devil is going to interfere with its course," he wrote.[78] Similar thoughts were voiced by Arthur Eddington during his spring Messenger Lectures to students at Cornell University. "If tomorrow I find ice instead of boiling water in the kettle," he told his audience, "probably I shall exclaim 'The devil's in it,'" even though, he said, "I do not believe that devils interfere with cooking arrangements or other experimental proceedings."[79] Eddington believed that in most cases, especially if one considered an expanding universe, "improbable" simply meant "impossible." Chance effects averaged themselves out so that the regular course of nature ruled the world. While boiling kettles were mostly safe from demonic intervention, Eddington found other more vulnerable access points. Particularly worrisome for Eddington was the possibility of eternal recurrence: "If we wait long enough," he speculated, "a number of atoms will, just by chance, arrange themselves as the atoms are now arranged in this room; and, just by chance, the same sound waves will come from one of the systems of atoms as are now emerging from my lips; they will strike other systems of atoms arranged, just by chance, to resemble you, and in the same stages of attention or somnolence."[80]

Eddington imagined himself giving his lecture over and over again, caught in a recurrent cosmic nightmare in 1934. His own "delivery of the present course of Messenger Lectures will repeat itself many times over an infinite number of times in fact before t reaches ∞," he warned his listeners.[81] Additionally, Maxwell's demon could play a selective role in key cases. More importantly, natural equilibrium did have an adversary in man. "The mind of man," explained Eddington, "in virtue of its conscious purpose, must play to some extent the part of Maxwell's sorting demon." What could be done by this creature, whose role man frequently took on? "Being a *sorting* agent, he is the embodiment of anti-chance," Eddington concluded.[82] Like many other scientists of that period, he was concerned about how Maxwell's demon abilities were modified by quantum effects. "I have sometimes wondered," he wrote, "whether it would not be possible to baffle Maxwell's sorting demon by one of the modern developments of atomic physics, viz. Heisenberg's Uncertainty Principle." Try as hard as he might, the astronomer was unable to find a way to strengthen the current limitations of his micro-sorting abilities. "I am afraid the demon is too clever for me," he concluded.[83]

ROOSEVELT AS THE "MAXWELL DEMON" OF AMERICA

Newspaper headlines about the world-changing potential of radioactivity were followed by headlines covering Karl Compton's most recent thoughts about demons. When he took the podium at the American Academy for the Advancement of Science (AAAS), the *New York Times* covered the scientist's presentation.[84] Compton compared the president of the United States, the paper reported, to Maxwell's demon: "We may call President Roosevelt the 'Maxwell Demon' of America."[85] Compton was not the first to think of someone holding high office in terms of the demon. The imaginative Henry Adams had considered such a possibility much earlier. Yet what had once been highly speculative was now science policy.

Compton's "Science and Prosperity" speech was published as an article in the journal *Science*, in a version that differed slightly from the one reported by the *New York Times*. Instead of referring to President Roosevelt directly, Compton focused on the powers of the federal government more generally: "We may call the Federal Government the 'Maxwell demon' of America," he wrote.[86] The elder Compton brother aimed to achieve results at the level of humans in society similar to those accomplished by machines at the atomic level. "We must realize that society," he urged, "like a gas, is chaotic and powerless if its human molecules, like gas molecules, have complete independence."[87]

Compton's concerns in those years centered on planning for the use and exploitation of "natural resources in land minerals, timber and water."[88] He aimed his message at the Agricultural Adjustment Administration, the Tennessee Valley Authority, the Federal Aviation Commission, and a score of other federal institutions. He knew well the catastrophe facing the pine belt east of the Cascade Mountains, the problems caused by draining the Mississippi River Valley for arable farmland, the water shortages in California, and the nationwide crisis caused by soil erosion.

The president of MIT sought to employ the lessons of Maxwell's demon to become "a little more civilized" and fight against the forces of decay. "Human affairs, if left to themselves," he argued, "tend toward chaos." For this reason, he advocated for strong interventionist policies, since "a business organization, if left without a guiding hand, becomes a disorganized business. A farm, if left to itself, becomes a wilderness. An economic policy of 'let nature take her course' leads inevitably to economic chaos." Physics—and the physics of Maxwell's demon in particular—provided him with ideas for how to better society:

> It is a significant fact that, in physical science, only one way has ever been suggested by which the tendency toward chaos can be circumvented. . . . There is only one way to "beat" the second law of thermodynamics. . . . It was Maxwell who showed how this might be done in the case of molecules of a gas through the agency of a hypothetical intelligent being, who has been dubbed "Maxwell's demon" and who . . . separate[s] the fast from the slow molecules of a gas, which in nature remain mixed together.[89]

The same mechanisms that could lead us to create a sound environment could also lead to the development of a sound population, he argued. The problem of "hereditary weaknesses," in his accounting, was the second-largest problem next to environmental problems. The physicist told his listeners that the number of "defectives . . . runs into the hundreds of thousands." These "defectives," he asserted, "constitute one of the greatest drains on our economic resources." "The welfare of the race requires their elimination," argued Compton. But how? His answer was clear: "by such means as have been found successful in repressing undesirable or developing desirable physical and mental traits in domesticated animals."[90]

THE FREEDOM OF MAN

As Karl Compton took on ever more prominent public roles in higher education, science administration, and government, his brother Arthur was busy creating new laboratory experiments to understand science's demons even better. The two brothers had been raised in a devoutly Christian family with a father who was a Presbyterian minister and a mother, born a Mennonite, who worked for the church's Missionary Society.

While Karl lectured publicly about demons, the younger Arthur started to frame the debate between quantum mechanics and relativity as depending on the possible existence of photon-fiddling or electron-fiddling demons. In a 1935 coauthored article (known as EPR after the initials of its authors Einstein, Boris Podolsky, and Nathan Rosen), Einstein published his strongest statement against prevailing interpretations of quantum mechanics. He and his colleagues argued that current explanations positing an essential indeterminism in nature based on the nonlocality of particles and their entanglement when separated at a distance were "incomplete." A year later, he would describe action-at-a-distance effects—entangled particles separated from each other affecting each other—as "spooky" (*spukhafte*) in a letter to Born.[91] If nothing else explained this strange behavior, scientists might have to accept it as a violation of the limits Einstein had placed on the speed of light. The old demon of chance that Laplace had chased away appeared strong, and the superfast "colleague of Maxwell's demon" might come back to science.

Compton directly confronted Einstein's arguments in *The Freedom of Man* (1935). He criticized the physicist's accusations of incompleteness head on: "Einstein, while admitting the correctness of our present quantum mechanics, has recently called attention to an apparent incompleteness in this theory of the world."[92] Compton insisted that "the strong support of Einstein's position" was due to the intellectual inertia of those "long schooled in the classical tradition."[93] Finding himself "partly responsible for this dramatic reversal of the physicist's point of view," he decided to argue for the opposite side. He would declare himself in favor of Bohr's

"Copenhagen interpretation" that espoused a belief in an essential indeterminism in nature.[94]

Einstein, according to Compton, was simply wrong. To buttress his claims, Compton devised an ingenious "while perhaps somewhat fantastic" thought experiment based on the double-slit experiment that bothered Einstein so much. He compared the subatomic particles going through the slits to "rifle bullets" shot at a distant target.[95] The comparison was fitting. It not only permitted scientists to think of subatomic particles by reference to classical physics, but also illustrated the new ways in which subatomic particles were being used to bombard atoms into new radioactive elements. His example included the possible actions of an intervening *daemon* in the system.[96]

One of the most baffling results of the double-slit experiment was that scientists could not determine which slit a particular electron or photon would go through without affecting the system. When they ran one experiment with only a single slit open, they could find individual particles going through it. But when they ran it again with two slits open, they found these particles going through both slits evenly and acting in a wavelike fashion. "We might predict the average result of many trials," explained Compton, "but no conceivable physical experiment could tell us where a particular electron would be at a definite time in the future."[97] In his view, the result was "truly a matter of chance. That is what we mean by saying that the law of causality does not hold: knowledge of the initial conditions does not enable us to predict what will happen, for with the same initial conditions we cannot consistently produce the same effect."[98] To believe otherwise was tantamount to believing that a demon could control them and that the "daemon controlling the shutter might be conscious of their qualities." Yet unless this demon could be found, scientists had no justification for speaking about strict causality in nature. "Yet if these [demon's] qualities cannot be revealed by experiment, it makes no experimental (that is, physical) difference whether they exist or not." His conclusion was trenchant: "Thus it is a physical principle that the law of causality must be abandoned."[99] Einstein could simply not let go of the feeling that they "must be overlooking something" that would make "prediction possible" in the case of the two slits.[100] Compton was eager to dismiss his position by claiming that they seemed to hold on to a belief in "daemons" "which our experiments cannot reveal."[101]

What was most alarming about Compton's conclusions was that although he did not believe in "daemons" who could bring determinism to the world of subatomic particles, he did not rule out the possibility of other kinds of demons who were intervening in the double-slit experiment. Compton considered them smart enough to operate in a way that would make these imps impossible to detect. They were not working behind the scenes to maintain a causal universe. Rather, they used the appearance of causality in nature as a cover for their covert actions. All they needed to do was to keep the overall probability of their actions interfering with each of the two slits at a fifty-fifty ratio. In his view, the way things occurred on average should not lead scientists to neglect the peaks and the troughs. Fluctuations, no matter how small, could be rigged to the macroscopic world, making the effects of these demons noticeable. Compton's thought experiment consisted in envisioning one of the two slits connected to an explosive device ready to go off the moment a particle passed through it. "Undetectable" demons, he explained, might know where a single photon was heading and intervene. "The daemon controlling the shutter might consider a photon which would enter the photoelectric cell that would result in exploding the dynamite a 'bad' photon, and one which would enter the cell where it would prevent the catastrophe a 'good' photon," he explained. "Being directly conscious, of these nonphysical characteristics which will determine their direction, the daemon may then close the shutter to all approaching 'bad' photons until a 'good' photon has passed through and saved the day." Compton continued by asking, "Has the daemon in this way contravened any physical laws?" The answer was no. "It is not necessary that the daemon should produce any physical action . . . a choice of a single event does not disturb the statistical probability of emission," he wrote in a footnote.[102] Nothing in the science of his day, he argued, could demonstrate the inexistence of such a stealthily "undetectable" demon:

> Under the conditions as described, however, only an individual event is determined by a daemon, and the statistical equilibrium is unaffected. Thus, the intervention of the daemon is physically undetectable. That is, the action has occurred according to physical laws, in spite of the fact that the course of the event was not governed by chance but was determined by the whim of a demon.[103]

Chance, according to Compton, was the perfect invisibility cloak for quantum demons' whims—a perfect alibi. These demons did not have powers typically ascribed to Laplace's. They would not lead back to Laplacian determinism at a fundamental level. They were in a sense more masterful, as they could sometimes act unpredictably as long as they acted predictably most of the time.

Compton cited the famous passage by Laplace about "an intelligent being who at a given instant knew all the forces animating Nature" and for whom "nothing would be uncertain," as "the future as well as the past would be present to his eyes."[104] Because of quantum mechanics "no physicist could now subscribe to La Place's thesis."[105] What they should be focusing on, he insisted, was other "daemons" who could act surreptitiously at key moments to amplify the uncertainties of the quantum universe.

To make a further point about the common mistake of considering the universe, including the decisions made by humans, deterministic, Compton explained how his thought experiment could be easily implemented in practice. He set the scene by taking his readers to a typical laboratory running the double-slit experiment. A scientist could decide when to go out for lunch depending on the path chosen by the photon. "He decides that when the next photon enters photocell A he will go home"; when it does, "he stops his experiment, picks up his hat, and departs."[106] The timing of the lunch hour would be at the mercy of a completely statistical event. But who or what was really making the decision? Compton did not rule out the possibility that a physically undetectable demon might be in charge. It was not only lunchtime that could be at stake. The possibility that humans were reacting in prescribed ways in anticipation of events whose exact underlying operating processes remained hidden from them was terrifying—more so to some than to others during those years, when Hitler was in charge of the mighty German army. Compton, ever the devout Christian, understood the possible actions of demons in the context of quantum mechanics as proof that nothing in science prevented us from considering man an agent who could act, within statistical limits, according to his free will and conscience.[107]

SCHRÖDINGER'S DEMON BOX

By the summer of 1935, Einstein had dived deep into debates about explosive-rigged quantum experiments in conversations with the Austrian Erwin Schrödinger, quantum mechanics' most controversial rising star. Almost a decade earlier, he had developed the famous Schrödinger equation to describe the behavior of nature at quantum levels, using the Greek symbol Ψ for it. At the time of his correspondence with Einstein, Schrödinger started wondering about what went on inside a box considered to be a "diabolical device" or "Hell machine" (*Höllenmaschine*).[108] In it a living being, such as the now-famous cat who carries his name, could be trapped with a poison whose release would be triggered by one of these particles. In one interpretation of his "Hell machine," scientists believed that the cat might lie in a new indeterminate half-dead and half-alive state of being tied to the particle's ambiguous existence. In another interpretation, they thought that it might simply be that not enough was known about the particle responsible for the cat's life to ascertain whether it was dead or alive.

Einstein's and Schrödinger's experiments featuring rigged atomic particles were as violent as Compton's, although crueler. In a letter to Schrödinger, Einstein made it patently clear that even in light of his failure to combat the prevailing interpretations of quantum mechanics, he would continue to look for still-hidden or overlooked causes that would return scientific explanations to the track of determinism. He disagreed with those who interpreted experimental results as evidence for nonlocality and who made the claim—ridiculous in Einstein's opinion—that something could be simultaneously "exploded and not-exploded." "Through no art of interpretation," he concluded, could this "be turned into an adequate description of a real state of affairs."[109]

Schrödinger was at the right time and place to consider these questions.[110] He had secured funding for a sabbatical from Imperial Chemical Industries, the British industrial powerhouse that produced poison gas, explosives, and radioactive weapons. The company had given him a two-year sabbatical position, which afforded him "the leisure to write this article."[111]

Schrödinger used his time to imagine how a Geiger counter, an instrument used to measure radioactive decay, could be adapted to release a poison into a closed chamber. A signal of radioactive decay could make a shutter go off, setting off a hammer in motion to break a flask containing cyanide. The chances of the poison spreading thus depended on the chances of a particle decaying. "There is a tiny bit of radioactive substance, so small, that perhaps in the course of one hour one of the atoms decays, but also, with equal probability, perhaps none," he wrote. "If it happens, the counter tube discharges and through a relay releases a hammer which shatters a small flask of cyanide." He then imagined a feline caged in the box. The chances of the flask shattering and the cat dying were thus exactly the same chances of radioactive decay. The chances of the atom's decay could only be known probabilistically, and the cat's fate could only be known to that same uncertain degree. The "function of the entire system would express this by having in it the living and the dead cat (pardon the expression) mixed or smeared out in equal parts."[112] The physically spread-out, half-living, half-dead cat represented, in scientists' lingo, nonlocality as well as uncertainty. In another letter to Einstein, Schrödinger described his diabolical device again and the "smeared-out" existence of the feline. This time, the example he used was that of decaying uranium. "Confined in a steel chamber is a Geiger counter prepared with a tiny amount of uranium," Schrödinger wrote, "so small that in the next hour it is just as probable to expect one atomic decay as none. An amplified relay provides that the first atomic decay shatters a small bottle of prussic acid. This and—cruelly—a cat is also trapped in the steel chamber."[113]

LAPLACE'S DEMON ASSISTANTS

In the same year when Compton's account of the possibility of new demons who acted on the quantum level was published and Einstein and Schrödinger were corresponding about hellish machines with caged felines, the mathematician Grete Hermann published a groundbreaking essay about the bizarre implications of the orthodox interpretation of quantum mechanics. Although largely overlooked at that time, years later the essay would be remembered as one of the earliest and smartest challenges to Neumann's proof against the possibility of hidden variables. If Hermann was right, the universe might function according to deterministic laws after all.

Hermann was a rarity. She had completed her PhD in Göttingen with the mathematician Emmy Noether, earning sterling credentials in male-dominated fields. Werner Heisenberg, whose seminars she attended, wrote admiringly of how "the young philosopher Grete Hermann came to Leipzig for the express purpose of challenging the philosophical basis of atomic physics." Heisenberg considered her "a fabulously cleverwoman."[114]

Hermann's work opened with a strong description of "the idea of the Laplacian demon" as a being "who has complete knowledge of the present state of nature, who oversees all laws of nature, and who can predict the future course of events on the basis of this knowledge."[115] She elaborated on her thoughts in a section titled "Das Ende des Laplaceschen Dämons."[116] Hermann was not convinced that the formalism of quantum mechanics required ruling out the possibility that yet-unknown or hidden properties would one day be found, and that these would resurrect the old ideal of determinism associated with Laplace. The future, she insisted, remained open on this very question. "But with what right does it anticipate future research and declare the attempt to try to find such features moot from the start?" she asked.[117] Hermann fought against the interpretations of her colleagues who willingly celebrated the death of Laplace's demon. "The old ideal of physics—tied to research aiming to constantly work one's way beyond any limit to reach the clairvoyance of Laplace's demon—has not been abandoned without a struggle," she asserted.[118] Most of Hermann's colleagues had tried to make that demon into a phantom that did not exist. "It is no longer the unattained ideal that is nevertheless to be aspired at beyond all limits," she wrote, "but proves to be a phantom that has been rejected through the progress of experience as a construct not corresponding to nature."[119] She disagreed. In her conclusion, she argued that this "phantom" was not contradicted at all.[120]

Endangered by her socialist sympathies, Hermann would soon have to flee Nazi Germany. She moved to Denmark before going to Paris and then London. She was not done, however, with writing about demons in science. She would return to the subject after the war.

During the years leading up to World War II, scientists continued in many other publications to express their understanding of quantum mechanics by reference to demons. In a letter to the editor of *Nature*, one writer explained that uncertainty relations could be thought of as the work of a new type of demon who moved particles faster than any of the others. Planck's constant and Heisenberg's uncertainty relations might result from the handiwork of a mischievous new imp that did not let you measure a particle's momentum when you tried to measure its position. Like a deft magician moving dice inside two upside-down cups, you just could never know how many dice were in the second cup after peeking into the first. The writer reported that a prominent physicist "recently invoked Maxwell's demon" in "illustration of the uncertainty principle" by describing a little creature playing with the momentum and position of a particle as if on a magician's mat. "Let us call forth the demon again and place him in the rectangle, where he may displace the point to another position." The play area of the demon was none other than $\hbar > \Delta p \Delta x$, where $\hbar$ was a factor of Planck's constant. Free to act fiendishly in that range, this "act of devilment on the part of the imp" corresponded "to no physical change," as none could ever be measured or detected at human scale.[121]

In the context of these investigations, philosophers continued to adjust their understanding of the universe. The philosopher Ernst Cassirer opened his *Determinismus und Indeterminismus in der modernen Physik* with a chapter arguing that scientific objectivity no longer needed to be based on the old ideals of absolute certainty derived from *Laplace'sche Geist*. "I begin with the image of Laplace's Geist," he wrote, "not because I regard this connection as logically appropriate or even as particularly apt psychologically, but for exactly the opposite reason." By then Laplace's demon was an obligatory starting point. "In all the discussions of the general problem of causality raised by the present state of atomic physics, the image coined by Laplace has played an important, indeed decisive role," he explained.[122] Indeed, with Laplace's and Maxwell's creatures in mind, scientists from all over Europe and North America pored over the vast stores of accumulated scientific knowledge, hoping to understand this new land of opportunity, solve the world's energy problems, or at the very least win the war. Even lowly librarians could be enlisted in an entropy-reducing mission, as "clearly the classifying librarian fulfills the function of a sorter at the post-office—he is a generalized Maxwell-demon, defying the disorderly tendencies of the second law of thermodynamics."[123]

"A MYTHOLOGICAL SUPER-BEING"

By the late 1930s, prominent physicists and chemists had turned their attention fully to the study of radioactive substances, mainly radium and uranium. The physicist Ernest Lawrence, working at UC Berkeley with Gilbert Newton Lewis, found a way to accelerate particles and bombard them against all kinds of new targets. Soon enough, "radioactive elements fell in our laps as though we were shaking apples off a tree," recalled a physicist employed at Lawrence's cyclotron.[124] More and more scientists, most prominently among them Otto Hahn, Lisa Meitner, and Enrico Fermi, started to investigate what happened when atomic and subatomic particles collide with each other. When physicists used neutrons as bullets, they noticed that two neutrons were released from inside atoms for every single neutron shot at it, along with stores of energy. If one of the released neutrons was reutilized, it could in turn be used to release another two. The result could be exponential.

Perhaps no one was better prepared to pursue this train of thought than Leó Szilard. The physicist quickly realized the possibilities of a two-for-the-price-of-one equation. From then on, he would think obsessively about the possibility of creating a chain reaction. How could he do it? Neutrons behaved so differently from ordinary particles. "You could produce neutrons," he explained during a public lecture, "but you could not collect them in any bottle because, having no charge, the neutrons would pass through the walls of the bottle."[125]

In January 1938, Szilard emigrated to America. He immediately set to work on nuclear fission, collaborating with Fermi, who had fled Benito Mussolini's Italy. A few months later, on March 12, 1938, the Nazis marched into Vienna. That same month Fermi and Szilard proved that it was possible to create a fission nuclear chain reaction using uranium. The following fall, Compton once again lectured about demons in quantum mechanics and about God in science. In prestigious lectures delivered at the University of North Carolina, he introduced his favorite cast of characters: demons who could go faster than atomic particles to prevent or cause an explosion and daemons who could never be detected by experiments. While he insisted that "the intervention of the daemon is physically undetectable," he once again explained that their actions hid behind statistical regularities: "That is, the action has occurred strictly according to physical laws, in spite of the fact that the course of the event was not governed by chance but was determined by the whim of a daemon."[126] According to Compton, these undetectable quantum-level effects could be immensely consequential when amplified. "Living organisms," he argued, acted "like an amplifier of very great power which may be set in operation by events on a scale comparable with the elementary events which we know to be indeterminate."[127] He turned as well to Laplace's "absurd" idea of an intelligent being who was capacious enough to know everything. After reviewing the contributions of Heisenberg and the principle of uncertainty, Compton forcefully insisted that "no physicist could now subscribe to Laplace's thesis."[128]

The summer after the invasion of Poland and the declaration of war against Germany by France and England, Szilard drafted a letter together with Eugene Wigner, his colleague and childhood friend from Hungary who had studied with Neumann. They urged Einstein to sign the letter as well, and to find a way to get it to the president. Warning President Roosevelt about the possibility that a bomb could be built with radioactive material, the letter described what could be the production of "large quantities of new radium-like elements" with "vast amounts of power." "A single bomb of this type," they wrote, "might very well destroy the whole port together with some of the surrounding territory." Moreover, "some of the American work on uranium," they claimed, "is now being repeated" in Germany.[129] The president of the United States was briefed on the contents of the Einstein-Szilard-Wigner letter on October 11, 1939. In his polite response, Roosevelt said that he would convene a committee to explore the military uses of uranium further.

The Harvard professor Percy Bridgman had an idea. Decades earlier, he had become famous for developing a no-nonsense philosophy known as "operationalism," which argued that complex physical concepts could be purged from some of the most mysterious interpretations attached to them if only physicists would focus simply on the results of measurement operations. But things had changed since those halcyon days, and quantum mechanics posed new challenges to Bridgman's approach. In his book *The Nature of Thermodynamics* (1941), Bridgman explained that scientists were "unsatisfactorily met by the pious hope that for some inscrutable reason no demon would ever be able to crash the gate of our laboratories." He gave one evocative illustration of the changing state of affairs: "By the use of a Geiger counter and amplifier, a battleship may be destroyed by a single radioactive disintegration."[130] Radioactivity had revived the old dream of circumventing the second law. "Today, when it is so easy to conjure the capricious happenings of the atomic world," it was again necessary to consider the question of whether "a sufficiently ingenious combination of means now in our control might violate the second law on a commercially profitable scale."[131] One place where the demon was clearly visible, Bridgeman noted, was in Brownian motion effects, which made it "possible to actually see fluctuation phenomena which are the basis of operations by the demon." He too considered how these random effects could be put to profitable use when combined with the new forces coming from atoms. By then, the experimental practice of detecting decaying radioactive particles using Geiger tubes and amplifiers was commonplace, and nobody doubted the real-world consequences of quantum phenomena. "It is possible," wrote Bridgman, "to bring up these small-scale things to the control of the events of ordinary life."[132]

In his physics text, Bridgman discussed the use of the "mental experiment" by scientists, which he considered an "important operation which is a kind of hybrid between paper and pencil operations" and actual experiments. The one featuring "Maxwell and his demon," according to Bridgman, was not entirely resolved.[133]

THE DEMON TEAM

In light of the continuing press coverage about the potential uses of new radioactive sources of energy, Karl Compton decided that the National Defense Research Council should take a second look at the possibility of building a new type of weapon. The council hired his brother Arthur to undertake the research. Within a few months of looking into the matter, the younger Compton concluded that a "full effort toward making atomic bombs is essential to the safety of the nation and the free world."[134] By the following fall, fully convinced of its urgency, he communicated his concerns to his brother and Karl's well-placed coterie. Arthur Compton's report was handed over to President Roosevelt on November 27, 1941. A few days later, on December 6, the president responded with the news that he had procured several million dollars for the initial six months of work on building an atomic bomb, and he promised practically unlimited funds for continuing work.[135] The Manhattan Project was set in motion.

The next day the Japanese attacked Pearl Harbor. Karl Compton joined both the National Defense Research Council and the Secretary of War's Scientific Advisory Committee. Arthur would head the laboratory in Chicago, becoming Oppenheimer's boss before the latter was appointed head of the Los Alamos National Laboratory in New Mexico. When the time came to start work on the weapon, Arthur hired Szilard as chief physicist, and Neumann was put to work calculating implosion mechanisms for the bomb that would maximize its damage. Both Compton brothers were usually present at the interim committee meetings that led to the use of the weapon. Meetings in which Karl participated as a member were often attended by Arthur as an invited scientist.

Four years later, in the spring of 1945, chemical engineers from the Hanford reactor in Washington State handed off to the assembly team in New Mexico enough material to start building a bomb. Although working at full speed, scientists and engineers had been unable to complete the weapon before Germany surrendered. Did that mean that the Americans should stop working on the bomb? One lone wolf decided to quit and leave the ranch. The rest continued to work after armistice day, with the same urgency as before. War was still raging in the Far East. Why not use the bomb on Japan?

Perhaps the target for this expensive packet of radioactive material could be shifted some degrees eastward.

The atomic bomb "might be a Frankenstein which would eat us up," worried Secretary of War Henry Stimson in his private diary a few days after Germany capitulated.[136] He was left to consider what to do with the new weapon and how to justify the billions that had been spent on it. Arthur Compton, sitting on the Scientific Advisory Committee that had recommended military use of the bomb, assented to its use against Japan. He signed the top-secret recommendation of June 16, 1945, on the immediate use of nuclear weapons. Neumann was chosen to help select the perfect cities to destroy. Truman approved their recommendation one day later. Szilard, who had developed misgivings about the weapon, tried unsuccessfully to prevent the dropping of the bomb by American bombers without prior warning to the Japanese. Hiroshima and Nagasaki were bombed on August 6 and 9, 1945.

Technical research into science's demons had paid off generously. Right after the destruction of the two Japanese cities, the US government released a single official report describing what up to then had been the secret technology that these scientists had helped create. The outcome—wholesale destruction of two large cities—appeared so demonic that the report considered whether some devil had inspired an evil scientist to create such a weapon. The answer was no: "This weapon has been created not by the devilish inspiration of some warped genius but by the arduous labor of thousands of normal men and women working for the safety of their country."[137] The report was authored by Karl Compton's Princeton colleague and the coauthor of their "Mechanical Maxwell Demon" paper, Henry De Wolf Smyth.

Smyth described the research effort in terms of the scientific topics that had interested him since those early years. The first page of the official report recounted the principle of conservation of energy, which since its formulation had "been the plague of inventors of perpetual-motion machines."[138] The discovery of nuclear power had cast doubt on the most cherished laws of the conservation of mass and energy, he continued, by showing how one could be transformed into the other. "If this is stated in actual numbers, its startling character is apparent."[139] One kilogram, in theory, could liberate 25 billion kilowatt hours of energy—enough to provide all of the electricity used by the entire United States for almost two months.

The bomb was frequently described in terms of perpetual motion and thermodynamics, but also with reference to ghosts, demons, and the devil.

These descriptions were not only metaphorically suggestive but technically correct. When Heisenberg heard the news of the bombings after his capture—he had been rounded up and taken to the Farm Hall M16 safe house in England, where the British secretly eavesdropped with hidden microphones—he "replied that had they produced and dropped such a bomb they would certainly have been executed as War Criminals having made the 'most devilish thing imaginable.'"[140]

Leó Szilard, in one of his most important public lectures on the bomb, explained his view of the history of its creation: "While the first successful alchemist was undoubtedly God, I sometimes wonder whether the second successful alchemist may not have been the Devil himself."[141] Richard Feynman, also employed at Los Alamos, would describe the Manhattan Project experiments with dangerous radioactive isotopes as "tickling the dragon's tail."[142] Back at Los Alamos, scientists kept experimenting with the nuclear material they had prepared in case a third bomb was needed. When an accident killed a second American (the first one had died from a previous accident), scientists started to informally refer to the fourteen-pound plutonium sphere of fissionable material left over at Los Alamos as the "demon core."

The year following the end of hostilities, Einstein referred to the scientific questions that kept him up at night as "problem demons" (*problemteufel*).[143] The physicist had been even less able to fend off the challenges that quantum mechanics posed to science than he was in trying to stall the challenges that atomic energy posed to the world.

In an article on "The Scattering of X-ray Photons" published in the *American Journal of Physics* (1946), Compton described a "typical example" of an event marked by quantum uncertainty: "the explosion of the atomic bomb." Quantum uncertainty, in his view, had affected not only microparticles but all of humanity. The bomb had not been created or used *ex nihilo*. Scientists and top-level government officials, including Compton and his brother, had been essential to the project. "Human actions are themselves examples of such events," Compton explained, and were no more predictable than the "direction [of] a particular x-ray photon" that scatters "when it falls upon a block of graphite."[144] Compton would continue to understand quantum mechanics in terms of the "vital human problem of freedom" and vice versa. Building the bomb had been hard. In his view, his perception of that "effort" was not an illusory side effect—it reflected the exercise of his free will. In his article, Compton chastised Laplace for launching an intellectual fashion that denied freedom of action. Yet, in light of the development of the bomb, he added: "It is perhaps pertinent to say, however, that no longer should a physicist try to argue, as Laplace might well have done, that effort to achieve a result can have no meaning."[145]

"Laplace," wrote Compton, "points out that a being with knowledge and understanding of the present state of the world, knowing the position and motion of each of its particles, would know both the past and the future as if they were present."[146] But his research had shown that the direction in which photons would scatter was essentially unpredictable. Where would they go? "No refinement of our experiments is in principle capable of making this direction predictable."

Laplace's demon, already nearly vanquished, suffered even more in the wake of the bomb's large mushroom clouds. Compton rehearsed the usual arguments that explained why quantum unpredictability was largely invisible to us, how it disappeared in the aggregate. Even if on the whole the disappearance of quantum effects left behind a pattern, its predictability was simply a cover for cloak-and-dagger operations at a fundamental level. For these reasons, quantum uncertainty should never be discounted as unimportant. "What is overlooked is that many large-scale events are based at some stage upon processes on a very minute scale." Something as dramatic as the atomic bomb could be set off by something tiny, such as "by the capture of a neutron that has in turn been released by some radioactive process." Moreover, these tiny effects could arise indeterministically throughout human activity. Freely created causes at the quantum scale could end up causing evil and good at the human level. What some considered inevitable might not be so.[147]

Back when he first wrote about demons, Compton had used the example of a single photon landing on a photocell and triggering sticks of dynamite to describe how microscopic laws could cause macroscopic effects. Little more than a decade later, his main example was no longer theoretical or speculative. Instead of photons landing on photocells, he described neutrons aimed at other neutrons in neutron-heavy elements that triggered a chain reaction setting off a nuclear bomb.

The theme of technology as a type of demon colored many scientific accounts during those years, particularly in Western Germany, in publications ranging from political commentary to apocalyptic fiction.[148] Max Born, in his recollections of those fateful years, explained that "since the days of Nagasaki and Hiroshima, the atom has become a ghost [*Gespenst*] threatening us with annihilation." He wrote: "We summoned the ghost ourselves. It served us faithfully for a while, but now it has become insubordinate."[149]

7
Cybernetic Metastable Demons

Most of the machines we take for granted today were developed in the immediate aftermath of World War II. They included "self-governing" electronic circuits that automatically took information from the environment and adjusted themselves to changes. They were central to the area of science and engineering called "cybernetics," a term derived from the Greek word for "governing" and popularized by the MIT mathematician Norbert Wiener in his best-selling book Cybernetics, or Control and Communication in the Animal and the Machine *(1948). In it, Wiener referred to these machines as "demons."*[1]

These demons could be organic or mechanical. Humans could act like them, or they could form part of larger systems of concatenated demons. The mathematician argued that their behavior was characterized by activities such as opening or closing switches, directing steering wheels, turning dials, and adjusting levers. Wiener described demons as "metastable," that is, they were sturdy enough to be practically useful but would one day cease to function. Some demons were as big as the economy, while others were as small as a single nucleotide inside a cell. They had the power to create productive imbalances by combining the rule-following powers of Laplace's demon with the ability of Maxwell's to adjust to new conditions. Innovators attempted to build more of them, furnishing computers with not only larger memories but also sensors, such as cameras and input devices (keyboards, mouses, and touchscreens), so that they could react, adapt, and learn.

These demons became prominent in the early days of electronic appliances, programmable computers, and robots, when scientists and engineers joined forces to enable humans and machines to interact more seamlessly with each other and with the world.

The decades after World War II proved to be particularly fertile for the study and development of demons. References to them became scientists' *soup du jour*, a veritable lingua franca used across fields. If the war had been the right time for studying quantum demons, the postwar period was the moment to build new demon-demon combinations.

"I am thinking of something much more important than bombs," announced John von Neumann soon after returning to Princeton from Los Alamos: "I am thinking about computers."[2] Computers had played essential roles in World War II. In the hands of cryptologists, among them Alan Turing, they had been used to crack the Germans' Enigma machine–generated cyphers. They were also useful for helping scientists with the complex calculations needed to model nuclear reactions and explosions. After the war, computers would take on even bigger roles.

To succeed, computers would have to change. One scientist in particular, the MIT mathematician Norbert Wiener, had a brilliant new plan for them. What if scientists started designing machines that would get the best out of Laplace's demon and combine it with the best qualities of Maxwell's creature? New robots and automata would be endowed with the abilities of both.

Wiener sketched out the main idea behind cybernetics in an ambitious lecture given in 1946. He considered the machines that had been built in the past with a certain disdain: "They are blind and deaf and perform their predestined dance."[3] The machines of the future, some of which Wiener helped invent, would be anything but deaf and blind. These machines would be furnished with "sense-organs" that could react. They would not simply follow instructions and produce predetermined outcomes. They had been developed and tested during World War II, with astounding success. "The proximity-fuse, which explodes a shell when the radar waves sent out by it are reflected by the target plane," explained Wiener, "is a sense-organ in the strictest meaning of the word."[4]

The war ended before most of Wiener's ideas could be successfully implemented, but the framework for new kinds of machines was in place. The Battle of Britain had tested the nerves of the bravest commanders. London burned, sirens wailed, and women and children took cover in subway stations and shelters. The superiority of German air power appeared to be undeniable from the sky, and to anyone looking up at German planes from the ground, it seemed like the war would be won only if someone took them down. As a young man during the war, Wiener had worked on ballistic calculations at the US Army Proving Grounds in Maryland. At that time, he noticed that a daredevil pilot's zigzagging motions resembled those of Brownian motion particles. Chasing a pilot with a gun was not easy, but it was slightly easier than trying to track Brownian motion particles. In tackling the movement of the world's most talented pilots as a mathematical problem with a theoretical limit, Wiener made important contributions to the mathematics of random motion and nonlinear dynamics.

Observing events in Europe unfolding from his MIT office across the Atlantic, Wiener nervously wondered how he could help. One winter day in 1940, he knocked on the door of his boss and offered him a brilliant idea. Soon thereafter, the National Defense Research Committee accepted his proposal and gave him a budget of $2,325 and the services of a talented graduate assistant, Julian Bigelow. A few months later, they submitted their first report, "Analysis of the Flight Path Prediction Problem," and started working on a new machine immodestly called "the predictor."

Wiener and Bigelow's work would become the foundation for cybernetics. Their new machine did more than just crunch out numbers given to it. It adapted according to the feedback it received. The machine would calculate, measure, and recalculate, before starting the process again until it got it right. It would learn by trial and error and improve every chance it got. To illustrate the talents of the "predictor," Wiener constructed a version of it that projected a light-spot moving randomly on a screen. He then created a "follower" to chase the light-spot automatically until it hit the moving target. His self-learning light-spot-chasing servomechanism, which in theory could go after any target, learned from its successes and failures. The following fall semester at MIT, Wiener demonstrated the new machine in a classroom, projecting on a screen a "follower" and a "target" engaging in a game of tag.

Wiener and his assistant gathered all the data they could get from the Antiaircraft Artillery Board at Camp Davis and from the Aberdeen Proving Grounds about the behavior of airplanes on bombing runs, then took it back to their lab. By February 1942, Wiener had written one of his most famous scientific papers, known colloquially as "the Yellow Peril." Owing to its military importance, its publication was restricted. Yet the contents of the paper circulated clandestinely in the long hallways of MIT, where it got many more researchers thinking about the future of smart weapons. Could an entire shooting battery become fully automated if a human gunner was replaced by a machine or drone? Could scientists construct an automatic machine to outgun the best gunner?

Wiener explained that "shooting at an airplane is like duck shooting. You do not aim at the plane itself, but at where the plane will be when the shell arrives." But shooting down a plane was obviously more difficult. Not only did that "duck" shoot back, but a gun mount on a naval AA battery swayed with the ship's motion. Wiener designed feedback systems that would react to these rapid changes by correcting the aim of the guns. Cybernetic "smart" guns would get better and better at predicting where to meet their target, and most eerily so, they appeared to act as if endowed with intelligence and *purposeful behavior*.

After the war, Wiener was exhausted, his nerves frayed by a diet of amphetamines. He took off for Mexico, where he worked at elaborating on the ideas for *Cybernetics, or Control and Communication in the Animal and the Machine*. Maxwell's demon occupied a central role in the book. When it was published in 1948, it caused such a splash that it soon ranked among the most influential books of the twentieth century.[5] The term "cybernetics" would come to refer to a new field of study that spread across traditional disciplines, from economics to philosophy. It encompassed a broad array of systems and technologies—both large and small technologies, and both old ones (like boat tillers) and new ones that reacted to new environmental conditions (like robots). "Economists, politicians, statesmen and businessmen cannot afford to overlook cybernetics and its tremendous, even terrifying implications," wrote a reviewer.[6]

Wiener started off his narrative by chastising those who brushed away Maxwell's demon for the sole reason that he was impossible. "Nothing is easier than to deny the possibility of such beings," he wrote. "It is simpler to repel" the creature than to take him seriously, and by taking such a route, "we shall miss the opportunity to learn something."[7] What most fascinated Wiener about the creature was how it could use information to produce energy efficiently. A working thermostat connected to a cooling device could act as a demon when it kept a room at an artificial temperature. By obtaining information from the environment and *feeding it back to itself*, it could adjust itself to keep some molecules in and others out. It could make a pocket of the world cool down, exactly as a Maxwell's demon would do.

The economic implications of tracing and even anticipating the future path of very fast and chaotic events were particularly intriguing. The skill required was as central for athletes and soldiers as it was for nerds and gamblers. From a pilot we might expect "the sort of intelligent use of his chances that we anticipate in a good poker player," wrote Wiener. "For example, he has so much opportunity to modify his expected position before the arrival of a shell."[8] Poker players beating the odds at Las Vegas, arbitrageurs and speculators gaming the stock market, and World War II pilots evading the barrage of bullets aimed at them were all successful when they used information about the new circumstance they were most likely to face next and quickly adjusted their next move accordingly.

Wiener did not believe that any cybernetic machines he described violated the second law of thermodynamics. Feedback was costly. Many other scientists like him who worked during the postwar period understood and accepted Szilard's argument that in order to work these demons needed to take a measurement, and that this measurement was information about a given situation, and that obtaining that information required energy. But while Szilard had been mostly concerned with engines that produced work, Wiener was hooked on improving communications systems and understanding them in terms of thermodynamics. He agreed wholeheartedly with Szilard's and Lewis's exorcisms of this being. "In the long run," he wrote, the demon "falls into a 'certain vertigo' and is incapable of clear perceptions." When that happens, "it ceases to act as a Maxwell demon." But the fatigue of the dizzy demon hardly mattered to him. One could employ him while he was still fresh. "There may be a quite appreciable interval of time before the demon is deconditioned," he wrote, "and this time may be so prolonged that we may speak of the active phase of the demon as metastable." The universe was so large that entire pockets of it could be (momentarily and locally) freed from the ravages of entropy. Where others saw entropy, fatalism, and defeat, Wiener saw an opportunity. His conclusion, and his belief in demons, was clearly stated: "There is no reason to suppose that metastable demons do not in fact exist." In fact, they were everywhere, even in our homes. They could be living or dead, human or not. These distinctions no longer mattered. Their unifying characteristic was their ability to reverse entropy, albeit momentarily and locally. Examples of metastable demons included "living organisms, such as Man himself," but were also nonliving elements, such as "enzymes" and other chemical "catalysts."[9]

A new interpretation of Maxwell's demon appeared at the intersection of two branches of science, thermodynamics and communications engineering. While engines pushed atoms from one place to another, communication technologies pushed electrons from one end of a channel to the other. In the latter, if Maxwell's demon let certain molecules pass and not others, a telephone filter let certain frequencies through and not others. Noise and entropy, Wiener insisted, were related. Both were a kind of disorder, and both should be eliminated for the purpose of prediction. The challenge of achieving perfect communication with each other and with our future, it turned out, was not so different from the challenge of building a perpetual motion machine. The demon, which had traditionally affected only engines and motors, was now seen impinging on information. Computers during this period were basically understood to be information engines. Computers got hot. When they did so, they started garbling their outputs. "In a large apparatus like Eniac or Edvac, the filaments of the tubes consume a quantity of energy," explained Wiener, "and unless adequate ventilation and cooling apparatus is provided, the system will suffer from what is the mechanical equivalent of pyrexia, until the constants of the machine are radically changed by the heat, and its performance breaks down."[10]

Research on "metastable demons" increased dramatically as researchers took notice of Maxwell's being's tricks on electronics and electromagnetic communications networks. MIT, IBM (International Business Machines), Bell Labs (the research branch of AT&T), and Western Electric were all ideal places to study information-sucking and message-altering demons. If a demon could act on the essential link of our social fabric, it could affect the political order as much as our private lives. Studying the relation of information to thermodynamics was key to the success of the emerging information economy.

When Wiener's *Cybernetics* appeared in print, the mathematician Claude Shannon reviewed it for the *Proceedings of the Institute of Radio Engineers*. "Wiener," explained Shannon, "makes the interesting conjecture that the paradoxes of the 'Maxwell demon' can be resolved by taking into account the information received by the 'demon' in entropy calculations." He concluded, "If this can be given experimental verification, it would be of considerable significance."[11] Shannon had moved to Bell Labs to join the team working on anti-aircraft technologies after studying at MIT, where he had taken a course with Wiener. At that time, he had already been looking forward to Wiener's forthcoming *Cybernetics*; he would acknowledge that "communication theory is heavily indebted to Wiener for much of its basic philosophy and theory."[12]

How efficient could an information engine get? How could scientists come as close as possible to the *non plus ultra* limits of the laws of thermodynamics? To answer those questions Shannon used the word "bit"; this portmanteau word, comprising "binary" and "digit," had been invented by his colleague John Tukey. "Bit" would become widely used in computer science, where it referred to Maxwell's demon's most elementary *materia operandi*. Theoretically, a bit was the smallest unit that could ever be used to convey something—the smallest quantity of information needed to sort and divide, a symbol for the tiniest bead in the tiniest abacus that could ever exist in the universe, the most basic tipping quantity, an entity so light that it could be insignificant. Yet the consequences of a bit could also be momentous—good if the bit was at the right time and place, and catastrophic if at the wrong time and place. The seemingly contrasting abilities of such a tiny unit made it the paragon of false modesty and understated elegance. The insights of Wiener's work and of Shannon's groundbreaking "Mathematical Theory of Communication" would not stay confined to the technical circles where they first appeared. Soon they were used to understand much larger systems, including the stock market, games of chance, artificial intelligence, government, and elections.

INFORMATION AND MEANING

Even in light of these technological advances, a few prominent holdouts continued to think of ways to return science to the path of determinism. Einstein's most faithful supporters, such as the physicist and philosopher Hans Reichenbach, had not lost their faith in the power of Laplace's demon. In his books *Elements of Symbolic Logic* (1947) and *The Rise of Scientific Philosophy* (1951), Reichenbach attributed to Laplace's creature the power of determining what was typically ascribed to chance. Where would a ball end up in a game of roulette? How would a die land? Reichenbach believed that the answer might be determined with exactitude by taking into consideration every little detail, from the dynamics of the forces of the person throwing the die to the air current affecting its fall, to the weight of each and every atom that might lead it to tilt and land on one of its six sides. "Laplace's superman" could counter otherwise intractable chance effects. "Such verification may be technically impossible," he wrote, "although in principle it should be possible to foretell the results of a throw of a die from the initial conditions, given the position of the die, the physiological status of the person considered, and other factors." Confidently, Reichenbach stated, "Laplace's superman could do it."[13] Could he, really? What gave Reichenbach such confidence?

Back in Germany, Grete Hermann was optimistic. She considered the possibility of using "Laplace's demon's assistants" to reinterpret quantum mechanics. Laplace's kindred spirits could work toward the goal of making things act in causal, determinate ways in time and space.

So much had changed in the intervening years since Hermann first wrote about Laplace's demon. Bremen lay in ruins. With Germany destroyed and Japan still reeling from defeat, she, along with other philosophers, mathematicians, and scientists, started to inquire about the limits of human action more generally. They plumbed the depths of a universe that had permitted the emergence of humans with the power to invent technologies to tick and explode for answers. Fixing the philosophical foundations on which quantum mechanics stood proved daunting. What did the atomic bomb reveal about how history developed? As an active member of the resistance during the war, Hermann knew firsthand how hard, perhaps even impossible, it had been to change the tragic course of events she had lived through.

Hermann, like Compton, used the example of the atomic bomb to analyze determinism, chance, and free will more generally. She accepted that microscopic events at the unpredictable quantum level could trigger macroscopic ones. Yet even "if the macrophysical effects triggered by this one electron were still far greater and more sustainable if, for example, it would be used to explode an atomic bomb," those effects, she argued, would not prove the fundamental indeterminacy of nature.[14] Scientists need not conclude from these examples that quantum uncertainty proved indeterminacy more generally, and it certainly did not prove the existence of free will.

After considering mainstream interpretations of quantum mechanics where "there is no room for the Laplacian demon," Hermann disagreed with them, especially when these interpretations were extended into the realm of ethics or when they were used to support a particular view of moral freedom. She was especially impatient with new quantum biological approaches developed by the German physicist Pascual Jordan and others. Their arguments, she claimed, were no better than the arguments of those who invoked "demon assistants" to prove determinism. "The quantum biologist," she concluded, "is no more competent than the assistants of Laplace's demons."[15]

Léon Brillouin disagreed. The war had not been won by letting entropy run its course. Brillouin was a renowned French scientist from a family of renowned scientists who was ready to put the final nail in Laplace's coffin. When the Nazis took over Paris, Brillouin was sent to Vichy with other French government officials. A few months into his new role, the Frenchman decided to switch sides. He emigrated to the United States to teach at the University of Wisconsin–Madison and at Brown University. Soon he was working for the Allies. After the war, Brillouin decided to stay in the United States, moving to the Cruft Laboratory at Harvard before finally leaving academia to work at IBM.

As soon as he heard about it, Brillouin became fascinated by Wiener's cybernetics, which he found to be "an entirely new field for investigation and a most revolutionary idea."[16] New research on Maxwell's demon and cybernetics showed that scientists could come very near to the dream of perfect efficiency. "Let us recall the paradox of Maxwell's demon, that submicroscopical being," he wrote in an article in *American Scientist*, "standing by a trapdoor and opening it only for fast molecules, thereby selecting molecules with highest energy and temperature."[17] How could its actions be harnessed for our purposes? "How does it become feasible on a large scale?" he asked.[18]

Brillouin was most impressed by how scientists had created nuclear power. He joined the ranks of many other thinkers who used the historical example of the atomic bomb and the science that led to it to think about science's demons and their demonic actions—that is, to consider how statistically improbable and minute local actions could connect and perhaps even bring about world historical transformations. He compared uranium to other standing reserves of power, such as a rock that, balancing on a cliff, could be easily tipped over, or an oil reserve that just "waits for prospectors to come."[19] The war had been won by exploiting preexisting natural resources in an unlikely and original manner. "Uranium remained stable and quiet for thousands of centuries; then came some scientists, who built a pile and a bomb and, like a naughty boy in the kitchen, blew up a whole city," he explained.[20] Continuing this line of thought, Brillouin asked how scientists could tap into the "negative entropy" of the universe. Minerals, water reservoirs, oil deposits, calories, and certain biological systems, such as ferments and yeasts, could all be used to produce it. Sometimes all that was needed to unleash the forces against entropy was a proper catalyst.

It was important, Brillouin argued, to pay attention to activities at the "fringe of the second principle." These demonic effects were felt in the macroscopic world and were everywhere. Humans engaged in these activities daily and without remark. "Man opens the window when the weather is hot and closes it on cold days!"[21] Because of the sheer size of the universe, "conditions forbidden on a small scale are permitted on a large one." With simple actions, the temperature of a room could momentarily be prevented from reaching equilibrium with the outside temperature.

Why would some demons eventually tire and fail? The main reason why Maxwell's "submicroscopical" demon could not go on with his mischief, argued Brillouin, was his inability to see when he was trapped in an enclosure. Brillouin noted that, in order to see molecules to manipulate them, Maxwell's creature needed some light. Without it, he was just hopeless, literally, in the dark. "The demon simply does not see the particles, unless we equip him with a torchlight," he explained. Exploring a clever suggestion of Wiener's that had "been generally systematically ignored," he speculated that, "once we equip the demon with a torchlight, we may also add some photoelectric cells and design an automatic system to do the work."[22] The setup was simple because "the demon need not be a living organism, and intelligence is not necessary either." Such a machine could reverse entropy, albeit locally and momentarily. What could be done to fortify this demon? His torchlight had a drawback: it injected energy into the system and eventually petered out. Any torch-carrying demon would fail in long-run missions, since "a torchlight is obviously a source of radiation. . . . It pours negative entropy into the system."[23] In years to come, physics textbooks would often include a picture of Maxwell's demon holding a flashlight.

Even if the fire of the demon's torch went out or his flashlight's batteries ran down, it was worth exploring all that he could do before darkness settled in. The venerated second law, Brillouin noted, was often not "strictly enforced." Exceptions might appear as "some sort of miracle," but they were hardly of mysterious or religious origin.[24] Momentary and local cases safe from the ravages of the second law were so numerous that Brillouin saw no benefit in thinking about the second law's universal validity. Such immense territories strained the intellect beyond reasonable measure. "It is better not to speak about 'the entropy of the universe,'" he stated. Why? Because "this, in my opinion, is very much beyond the limits of human knowledge." The subject led researchers into a hall of mirrors of unanswerable questions, such as: "Is the universe bounded or infinite?" "Do we know whether it is tight, or may it be leaking?" and "Do entropy and energy leak

Figure 2. Stanley W. Angrist and Loren G. Hepler, *Order and Chaos* (New York: Basic Books, 1967), 195, 197. Illustrations by Professor Ed Fischer Jr.

out or in?" He concluded firmly: "Needless to say, none of these questions can be answered."[25]

As research on demons in terms of information theory flourished, a new set of problems started bothering researchers. Did knowledge have to be placed on the same energy-accounting spreadsheet as any other bits of information? What made certain information more meaningful and relevant compared to other information? Could physics explain those differences? Brillouin illustrated the new problem: "Take an issue of the New York *Times*, the book on Cybernetics, and an equal weight of scrap paper. Do they have the same entropy? According to the usual physical definition, the answer is 'yes.'"[26] But answering yes, according to Brillouin, was an overly "hasty conclusion."[27] For him, "information value" could not be determined in exactly the same manner as other physical values.

The mischievous lawbreaker described by these scientists started to sow a rift between information theory in physics and communication theory in philosophy and in the humanities, with important consequences. Scientists such as Brillouin, Wiener, and Shannon did not believe that there was an exact one-to-one correspondence in the expenditure of energy and the gain in meaningful information. For them, and for others who continued to explore the possibility of interacting with nature in ways that would have unlikely results, the trick required reexamining the relation of physical information to meaning. By acting with particularly meaningful information, scientists might just about be able to momentarily stall entropic decline. Yet, for the most part, investigations about the meaning of information were separated from research that examined it as a physical quantity.[28]

By considering information as quantifiable energy, some scientists started to forget that information was only valuable if it was meaningful. At stake in these discussions was not only the nature of reality but also the possibilities and limitations of our capacity to understand the universe scientifically. In a universe considered in terms of bits of information, what was the meaning of meaning?

PHOTOCELLS AS MAGIC EYES

News of the possibility of constructing a new machine that would work at almost no expense gripped audiences during a lecture delivered at the University of Edinburgh on March 2, 1951. The speaker was Dennis Gabor, a Hungarian-born physicist and close friend of Szilard who would go on to win a Nobel Prize. He would have a demon, a quantum version of Maxwell's, named after him. The captivating lecturer described to his audience an imaginary machine, one that acted like a Maxwell's demon. Everyone wanted to know more about it. "The contents of the lecture," wrote Brillouin, "became known to a wider public through the distribution of a limited number of mimeographed notes, which have since become widely quoted in the literature."[29] MIT and Imperial College London invited Gabor to describe his machine in detail.

Gabor's "imaginary machine" appeared at first glance to be "a perpetuum mobile of the second kind." To help them understand it, Gabor summarized the previous research on demons from thermodynamics to Brownian motion, starting with Maxwell: "One of the most fruitful ideas in this direction came from Clerk Maxwell, who posed the question of demons opening a valve for fast molecules in a gas, and shutting it for slow ones." He continued by teaching them about Smoluchowski's research: "This led to the even simpler question: Why not spring-load the valve, so that only a fast molecule can open it?" This research, according to Gabor, opened up "the question of an 'intelligent demon,'" the question that fascinated Szilard. His conclusion, explained Gabor, was to simply assume that entropy increase had to accompany the act of selection. "Hence, in order to save the Second Principle, it must be assumed that such an observation could not be made by any 'demon,' intelligent or mechanical, without an entropy increase of at least this amount."[30]

In contrast to other such contraptions, Gabor's imaginary machine was equipped with state-of-the-art "photosensitive elements." So outfitted, it could determine when it should open or close a piston, like Maxwell's demon at his gate. Could it violate the laws of thermodynamics? Gabor concluded, predictably, that it could not. His *perpetuum mobile* would fail because of the particular behavior of the photons used to observe molecules. Although the science in his lectures was complex, his final lesson was prosaic: "We cannot get something for nothing, not even an observation, far less a law of nature!" Street-savvy, money-in-your-pocket wisdom and knowledge about demons was necessary to explain the behavior of light and information at the level of the quantum.[31]

Gabor explained clearly why Maxwell's demon could not work at human scale, but he also showed just how low the energy cost of obtaining information about a system really was. He was able to pinpoint the exact source of the cost. The photons necessary to "see" a particle interacted with it, adding entropy to the system. These photons were so tiny that it made a lot of sense to think about energy efficiency in those terms across fields. When information was particularly valuable, the price of obtaining it could fade into insignificance, as even a tiny amount of energy, such as $h\nu$, could be used to set off an avalanche, a chain reaction, or something even greater. Many other scientists in the 1950s soon reached the same conclusion as Gabor: even if a perfect Maxwell's demon could not be built at human scale, scientists could come pretty close to employing the ones they had.

In his MIT lectures on communication theory, Gabor reminded his students that certain information had much more value than other information. "To believe that this is what any 'yes' or 'no' is worth (for instance, the 'yes' or 'no" which decides a war) would be rather more than absurd."[32] They might not be able to eliminate all entropic decline from the universe or solve all energy and information problems, but if Gabor's demon was positioned at the right place at the right time, acting only when it really mattered, its actions could be extremely consequential. A bevy of new demonic devices placed where most necessary could accomplish transformational tasks. A demon in the age of photocell detectors was like "The Little Engine That Could." These tiny poltergeists, even with their limited powers, could be extremely useful when employed intelligently, though that was not easy work.

Arriving at answers required determining what the most meaningful points of intervention in the universe might be. Gabor's demon taught scientists valuable lessons about knowledge-gaining practices more generally, leading them to rethink the purpose, aim, and organization of science. Were scientists and intellectuals spending more energy than they were recovering, just like any old engine? Were they leaking more entropy into a system and getting out less than they put in? Were they impotent against the ravages of entropy in the long run? Or could they think themselves out of decay? Could information, knowledge, and intelligence be a source of salvation, especially in a world of scarce or limited resources? With his imaginary machine, Gabor showed how immensely powerful such technologies could be—if used in the right context and setting.

In covering the new adventures of demons who worked with photons, the *Journal of Applied Physics* featured an article by Brillouin titled "Maxwell's Demon Cannot Operate." The physicist reconsidered the topic by describing a machine equipped with a "photoelectric cell" equivalent to the demon's magical eyes: "We may replace the demon with an automatic device with a 'magic eye,' which opens the trap door at convenient instants of time."[33] Brillouin also concluded that the amount of energy wasted in the photon detection process was stunningly low, in the magnitude of Boltzmann's constant, and that this was true "for Maxwell's demon as well as for the scientist in his laboratory."[34]

"The physicist in his laboratory is no better off than the demon," warned Brillouin.[35] "The physicist making an observation," he insisted, "transforms negative entropy into information." The powers of the new engines furnished with photosensitive magic eyes were so stunning that Brillouin coined the word "negentropy," a neologism for negative entropy, or "entropy with the opposite sign" to explain them.[36] How could scientists create more of it? "We may then ask the question," continued Brillouin, "can the scientist change information into negentropy of some sort?" Brillouin suggested that scientists would create pockets of negentropy in the universe because they could create more and more machines with these powers. "With these laws he is able to design and build machines and equipment that nature never produced before; these machines represent unprobable [improbable] structures and low probability means negentropy?" By using their brains and "devising some ingenious gadget," they were able to curb the ravages of entropy. Brillouin entered here into complex philosophical territory. His analysis of contraptions that could almost break even led him to ask if there was something more than just breaking even in the innovation business. He was not sure. "Let us end here with this question mark, without attempting to analyze any further the process of scientific thinking."[37]

HIDDEN VARIABLES

Instead of focusing on all the "no can do" aspects of thermodynamics, many more scientists became concerned with the possibility of producing negative entropy offered by the small windows of opportunity available in the quantum world. For decades after he first published on the topic, Wiener continued to think about how to build "metastable demons" and where to find them. "One place to look for Maxwell demons may be in the phenomena of photosynthesis," he speculated.[38] Later some of his top candidates included viruses, genes, enzymes, amino acids, crystals, and other agents with self-organizing properties, such as snowflakes.

Wiener's most sustained discussion of the devil and entropy appeared in *The Human Use of Human Beings* (1950). He rehearsed the usual account of Maxwell's being as a "porter" handling a tiny gate, but what excited him the most was to see demonlike actions at human scale. Was a crowd of people on their morning or evening commute so unlike the particles handled by the demons of physics? "Perhaps I can illustrate this idea still further by considering a crowd milling around in a subway at two turnstiles," he explained. "One of which will only let people out if they are observed to be running at a certain speed, and the other of which will only let people out if they are moving slowly." What if the turnstiles acted like the demon charged with separating fast from slow molecules? "The fortuitous movement of the people in the subway," wrote Wiener, "will show itself as a stream of fast-moving people coming from the first turnstile, whereas the second turnstile will only let through slow-moving people." If these two turnstiles sent the fast-moving people in one direction and the slow people in the other, "we shall gather a source of useful energy in the fortuitous milling around of the crowd."[39] This simple scenario already represented a momentary violation of the second law.

By then, entropy, for Wiener, was the devil itself—not its sign, its manifestation, or its incarnation. It was *the* archenemy. The mathematician was downright apocalyptic in his understanding of entropy's ravages. "The scientist is always working to discover the order and organization of the universe, and is thus playing a game against the arch enemy, disorganization. Is this devil Manichaean or Augustinian?" Wiener sought to know these two devils with more precision in order to determine "the tactics to be used against them." The personality of the first one called for a specific strategy: "The Manichaean devil is an opponent, like any other opponent, who is determined on victory and will use any trick of craftiness or dissimulation to obtain this victory." This devil kept his cards close to his chest and changed his game plan midway: "In particular, he will keep his policy of confusion secret, and if we show any signs of beginning to discover his policy, he will change it in order to keep us in the dark." The other devil was simply "the measure of our own weakness." It "may require our full resources to uncover, but when we have uncovered it, we have in a certain sense exorcised it, and it will not alter its policy on a matter already decided with the mere intention of confounding us further." The first devil "will resort readily to bluffing," adding another level of complexity to the struggle "intended not merely to enable us to win on a bluff, but to prevent the other side from winning on the basis of a certainty that we will not bluff." The first devil possessed "refined malice," whereas the other "devil is stupid." "He plays a difficult game, but he may be defeated by our intelligence as thoroughly as by a sprinkle of holy water," Wiener concluded.[40]

Wiener, completely immersed in thinking about science and technology in terms of devils, genies, and ghosts, enthralled his audience at the New York Academy of Letters by reminding them that "the attitude of the fairy tale is very wise in many things relevant to modern life." The lines between sorcery and scientific work had become all too blurry for him. "Sorcery was condemned in the Middle Ages," he lectured, and "in those ages certain modern types of gadgeteer would have been hanged or burned as a sorcerer." Offering a reinterpretation of that time period, he argued that "sorcery was not the use of the supernatural, but the use of human power for other purposes than the greater glory of God." Lessons learned in medieval times were worth repeating, especially the tale of the fisherman and the genie. "The fisherman opens a bottle which he has found on the shore, and the genie appears." The adventure gets complicated as soon as "the genie threatens him with vengeance for his own imprisonment." In the fairy tale, all ends well. But in real life, Wiener warned, "when we get in trouble with the machine, we cannot talk the machine back into the bottle."[41]

Despite those risks, Wiener did not stop working or thinking about demons. One day he came across new research that excited him beyond belief. It had been done by the physicist and engineer Jerome Rothstein, who had worked on isotope separation for the atomic bombs. Like many others of his generation, he had become increasingly interested in quantum mechanics and was well aware of the enormous power that lay inside the atom. Rothstein was also interested in the new philosophical lessons that came with those discoveries. If theoretical physics was at a methodological impasse of sorts, it was a productive one.

In an article published in *Science* in 1951, Rothstein asked about the possibility of reinterpreting quantum mechanics in terms of a new demon: "A demon who can get physical information in other ways than by making measurements might then see a causal universe."[42] Limited beings like humans, who could only know by observing and measuring in ways that altered quantum systems, could not. But what about a demon who could see in a completely different way, in a manner alien to humans but not physically impossible? It would find a way to know whether "Schrödinger's cat" was going to die or live *before* the particle decayed, thereby either killing it or saving it. Rothstein believed that science had not ruled out this possibility. The fiction of a stealthier demon who could know more than anyone else because it would affect quantum systems was born.

After working on isotope separation, Rothstein joined the Army Research and Development Laboratories and the Solid State Electronics Branch of the Electron Division of the Signal Corps Engineering Laboratories. He then left the US Army and found employment at some of the most innovative technology companies in the United States.[43] He worked on masers, lasers, X-rays, medical electronics, and bionics, and he obtained joint appointments in the computing, information science, and biophysics departments at Ohio State University to research outer-space environments. Rothstein corresponded with Einstein, and the *New York Times* featured his work. The breadth of his training in science contrasted starkly with his humanistic training. In "areas often viewed with some suspicion by my brethren," such as "methodology, philosophy, and metaphysics," Rothstein was entirely self-taught.[44] He is virtually unknown today.[45]

When Wiener read Rothstein's new research on demons, he started to think about the future direction that quantum mechanics might take. The relation between light and entropy was most intriguing to him. "As a matter of fact, certain work now being done on the interaction between polarized light and matter," he stated, "is carrying us to a point at which we are constructing something very much like a Maxwell demon if not a Maxwell demon itself." All that scientists needed to circumvent the second law was to find a "mechanism of opening and closing the door" that "must depend on some signal going ahead of the particle and determining to some extent or other both its position and momentum." More research might show how this could be done. "As Mr. Röthstein has shown, this reaction goes much deeper," he wrote, "and penetrates even beyond the roots of quantum physics itself." It was there, he surmised, that one might "attempt to make the long-missing synthesis between quantum considerations and relativity considerations."[46]

If Rothstein's "demon who can get physical information in other ways than by making measurements" could be found, scientists would be able to work around the current limitations of thermodynamics in ever more productive ways.[47] "It is my intention to work on this field in the near future," Wiener stated confidently.[48]

Einstein needed allies. He had just turned seventy-one and was more famous than ever, appearing constantly on magazine covers, in newsreels, and even in television series. His autobiographical notes had recently appeared in print. That publication was something like "my own obituary," he wrote. The exercise had led him to reflect on how his contributions were holding up many decades after he first advanced them. He saw his work on relativity as analogous to that which others before him had done on perpetual motion machines. "Relativity," he argued, "is a restricting principle for natural laws, comparable to the restricting principle of the non-existence of the *perpetuum mobile* which underlies thermodynamics."[49] He had advanced the molecular understanding of nature, taking away most of the wind beneath the wings of Maxwell's demon. He had placed limits on the traveling speeds allowed in the universe, killing off the paradoxes attributed to Maxwell's demon colleague. Yet he now had to face the possible existence of new quantum demons that could "jump" at speeds faster than the speed of light and that threatened the possibility of ever knowing the world deterministically.

In the spring of 1950, Einstein, in response to a letter from Rothstein, explained what he thought was wrong with the direction that science was taking during those years. After a three-hour conversation followed up with more letter exchanges, Einstein and Rothstein came to the conclusion that *if* quantum mechanics was right, *then* the current understanding of physical reality had to be wrong.⁵⁰

The decade after Rothstein and Einstein's exchanges was marked by a flurry of new investigations not only by scientists but by philosophers as well. During the same years of his correspondence with Rothstein, Einstein, walking home with his colleague Abraham Pais, stopped suddenly. He asked his companion whether he "really believed that the moon exists only if I look at it." Einstein's question was an *argumentum ad absurdum*, a rhetorical strategy that he used to point out problems with certain interpretations of quantum mechanics. The two men "walked on and continued to talk about the moon and the meaning of the expression *to exist* as it refers to inanimate objects."⁵¹ "The quantum was his demon," concluded Pais.⁵²

Such a conversation would have been hardly memorable were it not for the fact that both men were discussing the foundations of physics. Using a hyperbolic example, Einstein was interested in a question much deeper and more detailed than that of regular object permanence. He sought to dispute the absurd consequences that might follow if certain claims made about quantum particles were true. Did nature change when it was looked at? Could these changes affect the world macroscopically, and if so, where would one draw the line separating quantum effects from those that affected our everyday lives? These were tricky problems, and experts were divided. While quantum mechanics was science's best bet for understanding the universe at its most elementary level, practitioners were conflicted about its implications for our general understanding of the universe, and remain so to this day. Relativity had given scientists wonderful insights into the cosmos at light-year scales, but the development of electronic and telecommunications technologies in the Cold War era had called for research into the atomic and even subatomic level of photons and electrons. No longer an arcane topic studied by the handful of scientists who had been involved in top-secret war work, quantum demons had moved to the forefront of physics.

BOHMIAN DEMONS

For a couple of decades already, Neumann's classic text on the foundations of quantum mechanics had been widely celebrated for its explanation of Maxwell's demon and used as evidence against the possible role of "hidden variables" in quantum mechanics. These variables, either as yet undiscovered or undiscoverable, might reveal aspects of nature that showed it to work in deterministically causal terms. If found, they might be used to prove that quantum indeterminacy was due to our incomplete knowledge of the universe, and that it was not a fundamental characteristic of nature. If found, most of the disturbing aspects of the quantum world (connected to nonlocality and uncertainty) would disappear, and Laplace's demon could get a new lease on life.

After starting his career as a graduate student at Berkeley under Robert Oppenheimer, the physicist David Bohm was barred from working at Los Alamos. His communist political affiliations had raised red flags among the establishment. Einstein, who shared some of his leftist sympathies, took Bohm under his wing at Princeton, giving him the support he needed to work on a book titled *Quantum Theory* (published in 1950). Bohm's book covered Szilard's exorcism in detail and restated the central conclusions of Neumann and other quantum physicists. "Irreversibility enters into the quantum theory in an integral way," he explained, and "this is in remarkable contrast to classical theory."[53] While his opinion on this matter would soon change, Bohm's politics remained very left of center. The year his book was published, Bohm was arrested for refusing to testify before the House Un-American Activities Committee for his alleged links to communism. Although he was acquitted a year later, he lost his job at Princeton and decided to move to Brazil, where he hoped to evade the long arm of McCarthyism.

In a controversial and much-discussed 1952 paper, Bohm sketched an alternative interpretation of quantum mechanics based on "hidden variables" or "pilot waves."[54] Although his research failed to gain much traction, it would lead some scientists to think that not knowing the results of certain measurements with utmost precision was no reason to abandon a belief in locality. Rather, it might still be assumed in good faith that something or someone hidden from view could be acting behind the scenes in mysterious ways to produce these outcomes. Other types of undercover commerce might be at work in the universe. In later discussions of Bohm's work, scientists started calling the suspected actor behind these possible relations "Bohm's demon," "Bohmian demons," "subquantum demons," and, in one particularly colorful phrase, "the surreal Bohm's witch." "What peculiar gifts belong to 'Bohm's witch'?" asked the physicist GianCarlo Ghirardi years later. "Well, she would simply be able to see the hidden variables of the system . . . the witch is able to know, system by system, exactly where the particle—let us say the one on the right—actually is."[55]

Einstein was pleased that a few remaining holdouts were not willing to give chance a chance. In 1954, a year before his death, he wrote to Louis de Broglie, one of the first scientists to have been seduced by the idea of "pilot waves," and told him, "I must look like an ostrich hiding always his head in the relativistic sand for not having to face those quanta villains."[56] For nearly three decades, de Broglie had abandoned the search for a way to reconcile the statistical indeterminism of quantum mechanics with traditional dynamics. Like most physicists of his generation and the generation that came after him, de Broglie had come to accept indeterminism as a fundamental aspect of nature.

By the middle of the decade, some scientists believed that discussions about demons were getting out of hand. Werner Heisenberg, who had discovered the uncertainty principle and had headed the German nuclear physics program during World War II, was not amused by those who tried to revive Laplace's demon. He would not tire of confronting others who, like Einstein, tried to save determinism. Scientists, he argued, should not be free to invoke willy-nilly imaginary creatures to explain away mysterious aspects of the universe. In *The Physicist's Conception of Nature* (originally *Das Naturbild der heutigen Physik*, 1955), he returned to the idea "expressed by Laplace in the fiction of a demon" to make a point against determinism.[57] In his Gifford Lectures at the University of St. Andrews in Scotland in the winter of 1955–1956, Heisenberg confronted those still hoping that a new strategy or creature "might possibly lead back" to a "completely objective description of nature." "There is no use in discussing what could be done if we were other beings than we are," he stated, trying to put a cap on the already long list of imaginary beings used by scientists.[58] Although he did not mention Hermann's use of "Laplace's demon assistants," he had her very much in mind.[59]

Brillouin agreed with Heisenberg that there was no point in trying to resuscitate Laplace's demon and the deterministic ideal it represented. The physicist had just secured funding from the National Science Foundation to publish the lectures he had given at Columbia, Berkeley, and IBM as *Science and Information Theory* (1956). Brillouin rallied readers around the possibility of successfully employing Maxwell's demons to reverse the natural flow of entropy by using simple electronic devices. In a chapter titled "The Demon Exorcised," he invited students to "investigate the possibilities of the demon" and show how it could be put to good use. Maxwell's demon functioned like an "ideal rectifier, acting on individual electrons" and "in contradiction with the second principle." Although the physicist warned that in practice "there is no ideal rectifier," the ones readily available to engineers came very close to it.[60]

Rothstein's *Communication, Organization, and Science* (1958) went further. He found it "amusing," he explained in the preface, that "it becomes possible to express all three laws of thermodynamics in demoniacal terms." The three main laws of physics curbed the powers of three demons, two of them named after physicists. "The first law excludes the existence of a demon who creates energy from nothing, the second does the same for Maxwell's demon, while the third disposes of Laplace's demon," he wrote.[61]

Rothstein's book, for the most part, received jarring reviews from his peers. One criticized it for being written "with a looseness and lack of rigour that must be outspokenly condemned as being inadmissible today."[62] A colleague who wrote the foreword for the book defended him by insisting that Rothstein should be congratulated for his unorthodox views. "Mr. Rothstein," he said, "is one of a growing number of refreshing and felicitous exceptions to a general reductivist tendency and tropism of twentieth century thinking today." Rothstein's brilliance, according to the author, resided in his refusal to equate nonmeasurability with nonexistence: "One of the greatest and most recurrent fallacies in twentieth century thinking has been to equate non-measurability with non-existence or insignificance," he claimed, then noted that this approach had backfired. "The attempt to relegate the allegedly nonmeasurable to the limbo of unreality"—whither "it stubbornly will not go"—had failed, he concluded.[63]

Despite the divisiveness caused by Rothstein's interpretative framework, there was no denying that he was an experienced scientist who would not back down after acerbic criticisms were leveled at him. The American Physical Society's meeting in Pasadena in January 1959 afforded Rothstein a golden opportunity to present his demonological approach. Several hundred scientists were in attendance. The *New York Times* covered his lecture with stunning headlines: "A SCIENTIST GIVES DEMONS THEIR DUE; 'Aladdin's Demon' Added to List of Impossible Imps That Help Physicists." The article reported that "demonology was elevated to a place beside far more concrete topics yesterday in a report to the American Physical Society." The *Times* reporter quoted the presenter, who had said that "even in his paper the demons have the last word" and that he believed that the announced

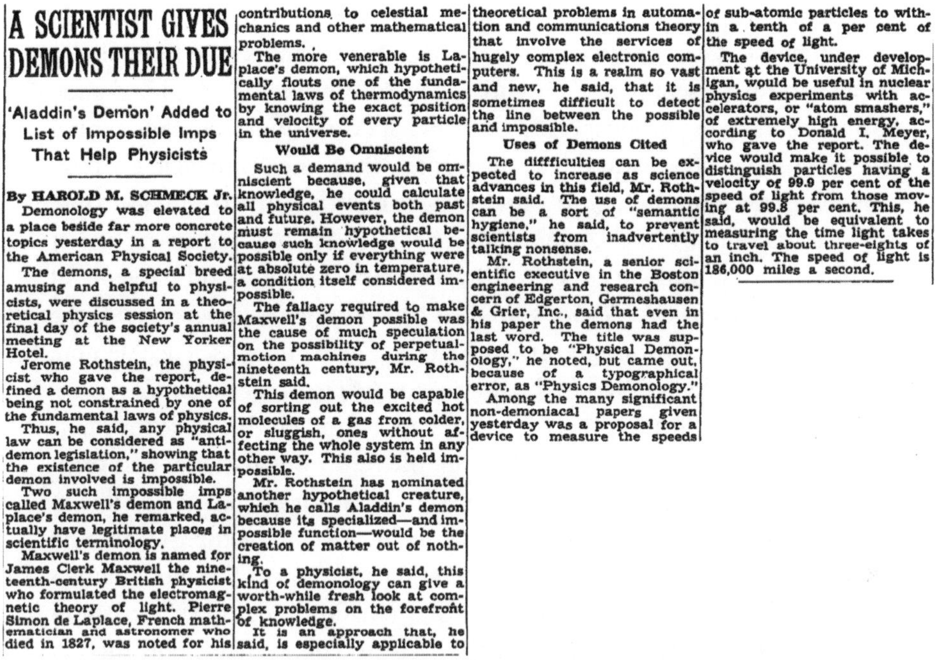

Figure 3. Harold M. Schmeck Jr., "A Scientist Gives Demons Their Due: 'Aladdin's Demon' Added to List," New York Times (February 1, 1959), 2. ©1959 The New York Times Company. All rights reserved. Used under license.

title was changed by one of them: "The title was supposed to be 'Physical Demonology,' he noted, but came out, because of a typographical error, as 'Physics Demonology.'"[64] Rothstein reportedly found demon-talk useful because it forced physicists away from jargon and back to substance: "The use of demons can be a sort of 'semantic hygiene,' he said, to prevent scientists from inadvertently talking nonsense." The article conveyed Rothstein's reasons for believing that the time was ripe for reintroducing demons into scientific thought: "It is an approach that, he said, is especially applicable to theoretical problems in automation and communications theory that involve the services of hugely complex electronic computers." Because "this is a realm so new, he said, that it is sometimes difficult to detect the line between the possible and impossible."[65]

Despite having no formal training in philosophy, Rothstein would manage to get an article published in the prestigious *Philosophy of Science* journal. His article "Thermodynamics and Some Undecidable Physical Questions" once again explained the progress in science in terms of demons: "The subject of 'physical demonology' is not without interest in its own right, and is related to the present paper." Inventing demons was useful. "For if a demon be defined as a hypothetical entity capable of doing things humans cannot do because of some natural law," he wrote, "then one can invent a demon capable of deciding a physically undecidable question, and the undecidability is tantamount to outlawing such a demon." He referred to his earlier work, where he had added a third hypothetical being: Aladdin's genie of *One Thousand and One Nights*, who does the bidding of whoever holds the lamp he inhabits. Rothstein turned Aladdin's genie into a science demon, who could also be referred to, he suggested, as "First Law demon."[66]

Rothstein told readers that he had been working on this topic for many years, citing his own prior work. "The inherent informational and organizational nature of physical laws," he wrote, "always makes it possible to express a law as anti-demon legislation." So far, "two of the three (Maxwell's and Laplace's, corresponding to second and third laws of thermodynamics respectively) are informational, the other (Aladdin's demon, first law of thermodynamics) having the ability to create energy from nothing." A published version of his American Physical Society talk appeared under the title "Physical Demonology" in an international journal on cybernetics and linguistics titled *Methodos*. "Any law can be formulated as nonexistence of some demon," he explained. The most interesting of them were the ones that had done great service in particular scientific fields, those that "were invented, not ad hoc, but before their corresponding laws were well understood."[67]

Modern science "killed" and "destroyed" demons, sometimes mercifully, other times cruelly:

> Aladdin's genie, who could create matter and energy from nothing, is killed by the first law. Maxwell's demon, who acquired physical information (results of measurement) without paying an equivalent entropy increase, is destroyed by the second law. Laplace's demon (either the original or a quantum version), who, knowing the state of the universe at any one time, knows all of the past and future, receives his coup de grace from the third law (unattainability of absolute zero temperature).

The distinction between different demons was not hard and fast, since they all shared a certain lineage: "Most of the ad hoc demons are either equivalent to Aladdin's genie, hybrids of Aladdin's and Maxwell's or Laplace's demons, or more or less sophisticated variations or specializations of the last two." Aladdin's genie and Maxwell's demon had an especially complicated relationship to each other. "The outlawing of Aladdin's genie does not necessarily administer the coup de grace to Maxwell's demon, even though, as we have seen, the former can masquerade as the latter." The relation of Maxwell's to Laplace's demon was basically one of size: "Laplace's demon is Maxwell's demon writ large," he explained. And Aladdin's genie, or the "First Law demon," lorded over them and could take over the work of the other two: "The first law demon is the most potent of all thermodynamic demons. He can simulate the activities of Maxwell's and Laplace's demons."[68] In addition to "the hardiest and most interesting of the species," Rothstein found "junior Laplace demons" (which he also referred to as "L.D. Jr.," using the demon's initials) that were "quantum" and "relativity" demons, but at this stage in his investigations he was still unsure as to their real nature. "We must therefore withhold judgment for the time being as to whether the quantum and junior Laplace demons are distinct, related, or identical."[69] He had a hunch about quantum and relativity demons: "One suspects—and we emphasize that it is only a suspicion—that when a relativistic quantum mechanics is formulated, the quantum and relativity demons will turn out to be one and the same."[70]

For Rothstein, ever attentive to the wiles of science demons, the most important lesson derived from his study was that they were powerful, but not all-powerful: "Omnipotent and omniscient demons are incompatible with thermodynamics," he explained.[71] The study of demons could show scientists which kinds of machines were impossible to construct, thus paving the way for the construction of those that were not impossible. By studying demons, researchers could pinpoint which aspects of machines were worth tweaking in order to come as close as possible to the demon-ideal, while knowing that certain characteristics would never be imitated in full.

"The demonological formulation of thermodynamics and other fields," explained Rothstein, "may thus have more than historical, philosophical, or humorous import—it might prevent trying to design a machine as impossible to make as a perpetual motion device."[72] With the assistance of these creatures, the mental and physical energy of innovators would not be wasted chasing away empty fantasies, but in the pursuit of attainable goals. The coupling "of measuring apparatus, computers, and machines controlled by the computers (which make their decisions in accordance with the information supplied by the measuring equipment) has resulted in apparatus so versatile and powerful that it can be compared to a demon." These innovations were so magical that "the impact of these machines on people with a medieval outlook," surmised Rothstein, "would surely be no less than that of a real demon."[73]

Demon-work attracted journalistic attention, raised the public's eyebrows, and rattled the scruples of some of Rothstein's peers. Yet in technical circles, research continued. The Austrian-American scientist Heinz von Foerster, who was then working for the Information Systems Branch of the Office of Naval Research at the University of Illinois at Urbana, was among many others who reconsidered Laplace's and Maxwell's demons by reference to quantum mechanics during those years. "My demonology," he explained at a May 1959 symposium, was based on interlocking systems of "internal" and "external" demons. "If the two demons are permitted to work together, they will have a disproportionately easier life," he explained, compared to when they were "forced to work alone." Concatenated in such a way, they could come close to matching their former glory. "To-day these fellows don't come as good as they used to come," he said, "because before 1927 they could watch an arbitrary[ily] small hole through which the newcomer had to pass and test with arbitrary[ily] high accuracy his momentum." Now they "are, alas, restricted by Heisenberg's uncertainty principle."[74]

Contributing to *Nature* magazine's coverage of the new research on demons and quantum mechanics, Brillouin wrote in an early 1959 article, "Laplace invented, more than a century ago a demon [that] the present discussion constitutes an exorcism of Laplace's demon." Brillouin concluded that "the negentropy principle of information had already provided a definite answer to the paradox of Maxwell's demon; it also eliminates Laplace's demon without further discussion."[75] While quantum uncertainty had shown the limitations of Maxwell's demon, the death of Laplace's creation had been announced thrice.

After reading the latest work of "a French friend of mine, Léon Brillouin," Max Born, widely celebrated for his statistical interpretation of quantum mechanics, referred to Laplace's demon once again in a lecture. He first explained why the need for Laplace's demon had arisen.[76] "Since the arts of observation and arithmetic required" by classical physics "seem to exceed human capacities, the astronomer Laplace (at the end of the eighteenth century) spoke of a spirit of the demon who can do everything and emulate the ideal of the physicist."[77] Born recounted how the idea then spread quickly, and in his opinion without justification, to other areas, including history, sociology, and economics, influencing policy and even politics. "There were and are schools such as Marxian materialism," he wrote, "which claim to be able to correctly and infallibly predict the social and political development of humanity." This demon's enthusiasts forgot that "the Laplacian demon [*der Laplacesche Dämon*] can only do his job, when he can measure with absolute precision," which was impossible in many cases.[78]

"We are men and not demons," Born concluded, and "the demon is already only a distant ideal."[79] Because the possibility of absolute measurement worked only for "a demonic, but not human power," he argued that the concept of determinism, along with the demon associated with it, was no longer valid, even if researchers such as Planck and Einstein did not want to let go of it.[80] He shared none of their compunctions. "If one wants to characterize determinism literally, I would call it a fantastic novel; what is called in English 'fiction.'" Born was ready to rescind his belief in such a fiction. "I, too, took delight in this novel for a long time, until I realized that it is not a picture of reality."[81]

Nevertheless, the "novel" that Born had been able to put down for the simple reason that it had been exposed as fictional continued to hold others in thrall. The nuclear physicist John Bell tried harder than most to solve the challenges that quantum mechanics posed to determinism. He developed the famous "Bell inequality" in response to Einstein's criticisms of quantum mechanics. Initially, Bell had been prominent among those who turned their attention to the idea that there might be other forces at work in quantum mechanics. In light of his research, Bohm's work and the smart objections of Hermann—which had remained marginal in the larger community of physicists—received renewed attention. With support from the US Atomic Energy Commission, Bell believed that Neumann's conclusions were "found wanting."[82] In response, Bell considered the possibility that hidden variables might give scientists a fuller—even superdeterministic—understanding of quantum effects. While previous attempts to explain the entanglement of quantum particles had required that "the signal involved must propagate instantaneously," Bell instead considered the possibility that the entanglement might be due to instructions set at much earlier times that were as yet undisclosed. Causation need not be found in real time. It could have been established "sufficiently in advance to allow them [particles] to reach some mutual rapport by exchange of signals with velocity less than or equal to that of light."[83] Perhaps what most physicists ascribed to acausal statistics was really caused by something else, such as a preestablished code that could have been set eons before. Bell would never be able to find these causes, or what he described years later as particles that "carry with them programs, which have been correlated in advance, telling them how to behave." "It is a pity that Einstein's idea doesn't work," he lamented. Despite his best efforts, Laplace's demon's heartbeat had almost gone silent. "The reasonable thing just doesn't work."[84]

Energized by these investigations, Bohm organized a colloquium at Cambridge University in 1967, believing that "a paradigm change in quantum theory may be imminent."[85] Bohm and his allies continued to look for a demon who would bring science back to the path of determinism. Physical demonology did not scare Bohm one bit; on the contrary, he had become convinced that Eastern mysticism might provide valuable lessons to physicists. Bohm invited Rothstein as one of twenty-five participants in his colloquium, seeking to destabilize what by then had solidified as the mainstream interpretation of quantum mechanics. In the edited volume, Rothstein focused on the paradoxes of statistical mechanics, which he understood as "Loschmidt, Zermelo, irreversibility, and assorted demons."[86] Even though quantum physics was based on the assumption that no such beings could ever be found, Rothstein insisted that scientists' lack of success in detecting them *so far* did not mean that they would *never* detect them.

A handful of other scientists during those years considered the attempts to restore Laplace's demon to his supreme position a worthy pursuit. At the same time, however, most agreed that the manner in which Bohm proposed to go about the task and the way Rothstein actually did so (by explicitly invoking demons) would require more changes in metaphysics, epistemology, and logic than was allowed for in science, even in the countercultural 1960s. But what was no longer in doubt for many prominent scientists was that traditional materialism, which had emerged from the old Cartesian mind-body division, no longer seemed like a viable framework for understanding our place in the universe. Eugene Wigner explained the new paradigm in "Remarks on the Mind-Body Question," written a few years before he was awarded the Nobel Prize in Physics. "The epitome of this belief was the conviction," he explained, "that, if we knew the positions and velocities of all atoms at one instant of time, we could compute the fate of the universe for all future."[87] That conviction had been thrown into doubt and with it the belief that a firm line could be drawn between mind and matter. What or who would fill this power vacuum?

8
Computer Daemons

Starting in the 1950s, artificial intelligence (AI) pioneers developed a new creative programming strategy to breach the limits of what hardware could accomplish. The cryptologist and mathematician Alan Turing articulated the new approach. He argued that Laplace's view should be abandoned as a model for computer programming. Instead of designing computers to follow algorithms and reach previously determined outcomes, they should be programmed to learn by themselves. These machines, designed to be smarter than us, could arrive at results "that we cannot make sense of at all."[1]

The MIT computer scientist Oliver Selfridge successfully implemented the new strategy by using subroutines called "demons." Computer demons of this kind started to become more common in the 1960s, when "a 'daemon' process" was defined as "a system (not a user) process, the operation of which is automatic."[2] *By the 1970s, Marvin Minsky and other artificial intelligence pioneers tasked graduate students at MIT with programming more and more "demons" or "daemons" to help computers gain general understanding. In years to come, email "mailer demons," "internet daemons," and other demons running background chores enabled communication between humans and machines. Dictionaries would soon add a new definition to the word "demon" and include its variant "daemon." Both usages referred to a kind of computer program that "runs in the background without intervention by the user, either continuously or only when automatically activated by a particular event or condition." Recently, a historian of the internet calculated that "daemons are now the products of hackers, free software developers, telecommunication companies, and the $41 billion networking infrastructure industry."*[3]

These kinds of programs characterize the new open-ended programming approach that now powers most AI systems and that elicit widespread fears, ranging from concerns about the introduction of subtle bias into automated systems to fears of a total AI takeover of our world and universe.

The question "Can man build a superman?" was printed across the cover of the January 23, 1950, issue of *Time* magazine. The illustration by Boris Artzybasheff, an artist known for his *diablerie*, depicted Harvard's new Mark III computer as a bony-handed part-machine, part-human bespectacled cyclops. These supercomputers, some weighing as much as seven tons, sparked fears about their demonic potential.

That same year the mathematician Alan Turing published an essay that would forever change the history of computing. Once the war ended, Turing found the time to think carefully about the future of computers. In "Computing Machinery and Intelligence," published in the scholarly journal *Mind*, he considered Laplace's idea carefully, noting that Laplace's conception of intelligence had already been practically realized in the computers he worked with himself, such as the Manchester Mark I.

Turing admired the scientific ambitions of "Laplace's view," since "from the complete state of the universe at one moment of time, as described by the positions and velocities of all particles, it should be possible to predict future states." Computers could now perform those calculations. "Even when we consider the actual physical machines instead of the idealized machines," he wrote, "reasonably accurate knowledge of the state at one moment yields reasonably accurate knowledge any number of steps later." With these new machines, "the prediction which we are considering is, however, rather nearer to practicability than that considered by Laplace," he concluded. But these powers had brought about unforeseen problems. In the "universe as a whole," it was impossible to calculate all consequences from initial conditions, as Laplace had imagined. A minuscule change in any single condition could "have an overwhelming effect at a later time." The slightest missing detail could throw off the future calculations of any such pettifogger, with life-and-death consequences. "The displacement of a single electron by a billionth of a centimeter at one moment might make the difference between a man being killed by an avalanche a year later, or escaping."[4] Laplace's brilliant idea was imperfect because of just how perfect it was. Could computer scientists do better?

Turing proposed a different strategy. "Many people" would consider that a test—"like the playing of chess"—could be used to determine intelligence, whether in humans or in computers. But perhaps that criterion was also wrong, Turing suggested. He proposed an original new test for intelligence based on an entirely different criterion: a well-known parlor game of those years known as "the imitation game."[5] This game, it could be argued, resonated more closely with our general assessment of intelligence than even chess. Surprisingly, too, it might be easier to construct a machine that excelled at it.

The imitation game usually had three players. One player was the interrogator; this individual, without seeing the other two players, would ask questions to try to guess the gender of each one, male or female. One of the two would try to get the interrogator to come to the wrong conclusion. Could a computer be programmed to take on the role of one of the two players and "provide answers that would be naturally given by a man" to deceive the interrogator and win the game?[6] If so, this machine would have effectively fooled the interrogator into thinking it was human, regardless of gender. The "Turing test" would soon emerge as a criterion for artificial intelligence.

If a computer was practically indistinguishable from an intelligent being such as a human, why should it be denied that status? Turing calculated that it would take sixty workers coding as fast as he himself could work, steadily, for fifty years to be able to program a computer to play the imitation game competently. Turing's article went on to systematically take down renowned detractors of the idea that computers could possess intelligence, among them "Ada Lovelace's objection." With these new programming techniques, computers would not follow predetermined orders, but instead would "[do] something that we cannot make sense of at all."[7] If scientists then provide "the machine with the best sense organs that money can buy, and then teach it to understand and speak English," the capabilities of computers could be limitless.[8]

Across the Atlantic, computer scientists were pursuing a different strategy to reach a similar goal. A breakthrough arrived in the fall of 1958, when for four intense days leading scientists from across the world met at the National Physical Laboratory (NPL) in Britain to discuss the mechanization of thought processes. The NPL director's welcoming remarks the first morning referred to Descartes: "The thinking process brings to mind Descartes' famous remark. 'Cogito ergo sum'—I think, therefore I exist." But recent events, he noted, required a different maxim. "Let me invert Descartes' proposition and say that, since we exist, we can think and let us now think to some purpose."[9] A brilliant presentation delivered toward the end of the conference would completely revolutionize the field and, as part of the symposium's published proceedings, become a foundational text for computer science and artificial intelligence.

Oliver Selfridge, a scientist who had traveled to the conference from MIT, argued that further advances in computing would not be forthcoming unless scientists changed how they thought about thinking. His proposal was similar to Turing's. The reason why the "thinking machines" had been unable to advance beyond an elementary level was simple: computers were only doing what they were told to do. They were only spitting out what was given to them beforehand. They were pumping electrons down a fixed set of wires just as the ancients had pumped water down their fixed set of ducts and canals. In both instances, the path of an algorithm was being slavishly followed.

Selfridge argued that computers' limits lay not only in the algorithm itself but in the very conceptualization of their capabilities. They would never become smarter if they were given only a sequence of rules to follow. The scientist knew a thing or two about the topic. As the illegitimate grandson of the millionaire owner of the Selfridges & Co. department store in Britain, where his mother had been a clerk, Selfridge came from a family that had broken with tradition and eschewed convention.

Selfridge traveled back to the old continent where he had been born in secrecy to deliver a paper titled "Pandemonium: A Paradigm for Learning." From the podium, and in front of some of the most prominent researchers in the field of artificial intelligence, he argued that progress would come if computers were programed to act like "demons." Instead of having computers follow rules set in advance, programs involving demons would try out different strategies and options and adjust themselves on the fly, depending on the success or failure in achieving the task at hand. A host of "demons" and "subdemons" would work in parallel and be organized hierarchically. Those at the very bottom would labor behind the scenes like store clerks, doing repetitive simple tasks before summoning others higher up in the computer program's hierarchy to take on the next steps.

Selfridge's presentation created quite a stir at the conference. His idea for computer programming would become so effective at making machines appear intelligent that scientists would begin to wonder whether the human mind itself worked in the same way. Cognitive psychologists soon started to refer to the processes in intelligent minds and computers interchangeably as "daemons" and "demons." While Selfridge had initially

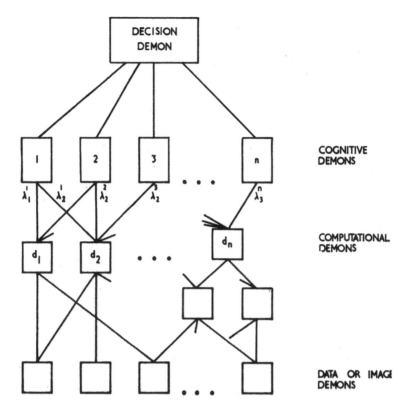

Figure 4. Oliver G. Selfridge, "Pandemonium: A Paradigm for Learning." In *Mechanisation of Thought Processes* (London: Her Majesty's Stationery Office, 1961), 517.

referred to his work as a "paradigm for machine learning," it would soon become a model for intelligence more generally, in humans as much as in machines.

Selfridge's proposal shared many similarities with one made by Frank Rosenblatt from the Cornell Aeronautical Laboratory. Rosenblatt stressed that the world needed machines that could learn by themselves for reasons to be found in the laws of thermodynamics." No single algorithm could once and for all beat this law, but in specific cases intelligent machines or beings could adapt themselves to their environment to circumvent it momentarily and produce handsome profits. Thermodynamics showed Rosenblatt the need for building a machine that could "spontaneously improve in its ability to organize, and to draw valid conclusions from information."

Such machines were the reason why scientists like him had to develop "a system which is capable of reorganizing its own logic."[10] Instead of demons, Rosenblatt proposed "neural" elements that could work in parallel and whose output could be used as an input for another neural element charged with a different task so that the computer could land on "original ideas."[11] These two approaches, at times competing and at times complementary, inaugurated the "neural networks" approach to machine and human intelligence that would dominate the rest of the century. Cognitive psychologists and neuroscientists, who had long thought of the brain as a type of machine, would soon start to think about it as a computer that could run *these kinds of programs*.

In the third session, Selfridge introduced demons directly and unapologetically. "We are not going to apologize for a frequent use of anthropomorphic or biomorphic terminology," he stated, as "they seem to be useful to describe our notions."[12] Selfridge proceeded to propose a model for artificial intelligence based on a hierarchy of shrieking demons working behind the scenes and getting noticed by those higher up in the spirit hierarchy depending on how loudly they screamed: "Each cognitive demon computes a shriek, and from all the shrieks the highest level demon of all, the decision demon, merely selects the loudest." These "cognitive demons" were tasked with making simple comparisons by guessing. If a demon's guess matched positively against the target, it was instructed to shriek. The better the match, the louder it hollered and bellowed.[13]

Selfridge worked with a vacuum-tube programmable IBM 704, one of the first mass-produced computers to hit the market. As the leader of Group 34 at the MIT Lincoln Laboratory, he was engaged in the practical task of teaching computers how to read Morse code that had been manually transmitted by a human typist and to translate it automatically into typewritten messages. If successful, these computers could replace thousands of telegraphists and typists by streamlining communications around the globe. Selfridge dreamed of replacing many other secretarial jobs by, for instance, constructing a voice-to-text machine and improving on computer pattern recognition more generally so that machines could take in messages automatically, with no need for a person's manual input.

Demons in Selfridge's pandemonium program would have to adapt and evolve to survive. These "adaptive changes," he hoped, would "promote a kind of evolution in our Pandemonium." A "natural selection" process could be designed inside the computer to be cut-throat. "If they serve a useful function they survive, and perhaps are even the source for other subdemons who are themselves judged on their merits." No mediocre demons would be spared or allowed to reproduce. "Eliminate the relatively poor," he wrote, "and encourage the rest to generate new machines in their own images."[14]

The physicist and neuroscientist Donald MacKay praised Selfridge's presentation, which led him to rethink his own work, he admitted, since he "had never suspected its demonic implications!" He could now see demons in other examples far beyond those covered in the paper: "Dr. Selfridge's 'pandemonium' has a certain family resemblance to a class of mechanism [sic] considered in some earlier papers."[15]

John McCarthy, another of the luminaries present that day, was in awe. What intrigued him about the new model was more than just its usefulness for machine learning. Selfridge's model seemed to describe aptly how the human brain worked. McCarthy decided "to speak briefly about some of the advantages of the pandemonium model as an actual model of conscious behavior."[16] He had only recently coined the term "artificial intelligence" at the Dartmouth conferences of 1956. McCarthy's goal at the time was to teach computers common sense, a preliminary step to computers eventually evolving intelligence of a human order.[17] He was an evangelist of "strong AI," a term representing the view that computer programs were capable of thinking. In later years, McCarthy would argue that machines could not only think but also hold beliefs: "Machines as simple as thermostats can be said to have beliefs, and having beliefs seems to be a characteristic of most machines capable of problem-solving performance."[18]

What struck McCarthy most about Selfridge's presentation was the similarity between the behind-the-scenes work undertaken by the demons and our unconscious thinking processes: "If one conceives of the brain as a pandemonium—a collection of demons—perhaps what is going on with the demons can be regarded as the unconscious part of thought, and what the demons are publicly shouting for each other to hear, as the conscious part of thought."[19] McCarthy's comments that evening proved prescient.

By the dawn of the 1960s, Selfridge's pandemonium was successfully implemented. For the first time, computers at MIT's Lincoln Labs were able to recognize handwritten letters. Scientists were well aware that the basic operation of computer demons programmed to follow "if-then" rules were similar to those performed by Maxwell's: *if* hot molecule, *then* stop;

if cold, *then* let pass. This activity—having microprocessors handle bits to go in one direction instead of others—worked at the level of software as much as in hardware. Just how much smarter could they get? Could they be employed to reduce entropy or even to circumvent the second law of thermodynamics?

"A Pandemonium system that learns from experience has been tested," explained Selfridge excitedly in "Pattern Recognition by Machine," published in *Scientific American* in 1960.[20] Until then, machines had been extremely smart in some respects, but exceedingly dumb in others. Yes, some computers of those years could play chess and checkers better than their programmers, but these machines, Selfridge lamented, "are not well equipped to select from their environment the things, or the relations, they are going to think about."[21] They could not even recognize the numbers that "a child recognizes before he learns to add them." And even worse, "they cannot understand the simplest spoken instructions." If machines were to *think*, these inabilities had to be corrected. "Understanding speech and reading print are examples of a basic intellectual skill that can variously be called cognition, abstraction or perception; perhaps the best general term for it is pattern recognition."[22]

The trick was to have many demons look at the same input at once: "One might think of the various features as being inspected by little demons, all of whom then shout the answers in concert to a decision-making demon," explained Selfridge. The demons were organized to forestall the possibility of internecine conflict. "A Pandemonium program," he wrote, "can handle the situation by having the demons shout more or less loudly," thereby permitting repetitive tasks to be carried out in parallel, maximizing efficiency. "From this conceit," he explained, "comes the name 'Pandemonium' for parallel processing."[23]

The artificial intelligence pioneer Marvin Minsky also delivered a paper at the "Mechanisation of Thought Processes" symposium. He loved the new approach. Selfridge hired Minsky to work at MIT's Lincoln Labs, and the two would become some of the strongest contributors to the field. Minsky's groundbreaking article "Steps toward Artificial Intelligence," published in the *Proceedings of the Institute of Radio Engineers* in 1961, described the benefits of the "Pandemonium" computer-programming model in detail. "It is proposed there that some intellectual processes might be carried out by a hierarchy of simultaneously functioning submachines called 'demons,'" he explained.[24]

AI demons kept improving. Math PhD dropout Allen Newell was prominent among the many other scientists who realized how promising Selfridge's pandemonium model really was. In years to come, he would become one of the most influential voices in the field of AI and cognitive psychology, practically merging the two disciplines into one. Drafted into the US Navy as a nineteen-year-old, Newell had watched the atomic bomb nearly destroy the South Pacific Bikini atoll. He was charged with making maps of the ensuing radiation around the island. After leaving the military, Newell decided to study physics at Stanford and mathematics at Princeton, but then left graduate school to work for RAND in Santa Monica.

One day Newell had "a *conversion experience* while he attended a talk by Oliver Selfridge" and became convinced that programming demons was the best route to achieving artificial intelligence.[25] A few months after hearing Selfridge's talk, Newell published his groundbreaking article "The Chess Machine," which outlined a new way to teach computers how to play the game. Years later, he would succeed in programming a computer that found proofs for many of the logical theorems of Alfred North Whitehead and Bertrand Russell's *Principia Mathematica*.

Just as two brains can be smarter than one, two demons can be smarter than one. What they could do appeared to be nearly infinite, but only if they were properly schooled; the problem of too many chefs in the kitchen applied to too many demons as well. In "Some Problems of Basic Organization in Problem-Solving Programs" for the book *Self-Organizing Systems*, Newell explained that computer demons could be efficiently employed only if they were well organized. "Metaphorically we can think of a set of workers," he explained, "all looking at the same blackboard: each is able to read everything that is on it, and to judge when he has something worthwhile to add to it." This common classroom and office setting was modeled in code for the benefit of computer demons.[26] The way demons were organized mirrored the hierarchical organization of cut-throat and even hellish modern workplaces designed to streamline a business's efficiency to maximize profits. Only the fittest could survive in the corporate jungle. Stronger demons used weaker ones and weaker ones used even weaker ones until the last ones down the ladder were culled and spat out. The ones at the top shrieking success gained at the expense of the others. Those demons near the bottom of the ladder were thought to act like our unconscious "working memory," busybodies so absorbed with simple and repetitive tasks that they did not even know what they were doing. The world inside the universe of computer demons designed to compete with each other mimicked the world outside, where exploitation was the rule rather than the exception, and where engineers worked hard to find ever more successful ways of replacing others with machines.

TEXTBOOK DEMONS

By the beginning of the decade, science demons and their exorcisms were standard topics in physics textbooks. *Symbols, Signals, and Noise* (1961) by the communications engineer John Pierce included a drawing and detailed description of the "hypothetical and impossible creature" known as Maxwell's demon. A new edition of Norbert Wiener's *Cybernetics* appeared that same year. A second edition of Léon Brillouin's *Science and Information Theory* arrived in 1962.[27] In his new preface, Wiener noted that "the automata which the first edition of this book barely forecast have come into their own," and "the related social dangers" he had warned about had also "risen well above the horizon."[28] Brillouin added a new section to the new edition, aptly titled "A Simple Example for Discussion: Laplace's Demon Exorcised," as he continued to show that recent advances in electronics and computing could be used to momentarily plug the leaky boat of our entropic universe.[29]

Most successful among these were *The Feynman Lectures on Physics*, published to great public acclaim from 1963 to 1965. "Half genius and half buffoon" was how a colleague described the unorthodox physicist.[30] That epi-

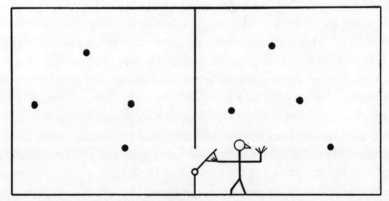

Figure 5. John R. Pierce, *Symbols, Signals, and Noise: The Nature and Process of Communication* (New York: Harper & Row, 1961), 199.

thet might also fit the demon that would later bear Richard Feynman's name. If there was someone who could write the ultimate exorcism, it was probably Feynman, widely respected as one of the most brilliant physicists of the century. He tried, twice—first while lecturing to impressionable undergraduates about the impossibility of producing work from nothing, and later when he dedicated himself to understanding the physics of computing. Brilliant and unconventional, Feynman held even more brilliant and unconventional views about Maxwell's demon. His "little demon" had eyes that made him extra special.

One of Feynman's most celebrated lectures started by describing "an extra special demon" who "can have eyes at the back of his head."[31] Feynman explained to eager undergraduates why no one, no matter how smart they were, could ever succeed in building a perfect Maxwell demon that could violate the second law. "We have two boxes of gas at the same temperature, with a little hole between them," he explained. "At the hole sits a little demon (who may be a machine of course!). There is a door on the hole, which can be opened or closed by the demon."[32] Feynman's model for a demon-operated machine was similar to models for traditional watermills or windmills built with cogwheels, ratchets, and pawls. The cogwheel or ratchet was used to move the mill. The pawl was used so that it could only advance in one direction. If currents changed and the vane was pushed in the opposite direction than desired, the pawl would halt it. Feynman used this concrete example to pose a more general theoretical question: when and how can the movement of particles going back and forth in relatively random ways be used to harness energy and produce useful work?

Ratchets and pawls are widely used for winding all sorts of mechanical clocks, toys, and other instruments. Similar contraptions are employed to pull up shades, tighten belts and ropes, and trap air in the pneumatic pumps used to lift heavy weights. They work wonders, but always break eventually, because of stress and friction. Wear-and-tear causes ropes to burst, strings to snap, and hinges to bend. Screws have to be tightened or they pop out. Vanes need to be readjusted, and gears need to be replaced.

What about a mini-molecular windmill? Would it eventually break down as well? Quantum mechanics offered new opportunities for speculation. Feynman's demon was tiny, as tiny as anything that had been realistically imagined thus far. Could a tiny ratchet and pawl take advantage of Brownian molecular motion to lift a flea? The physicist described how such a wind-powered flea-lifting contraption might work:

Let us say we have a box of gas at a certain temperature, and inside there is an axle with vanes in it. Because of the bombardments of gas molecules on the vane, the vane oscillates and jiggles. All we have to do is to hook onto the other end of the axle a wheel which can turn only one way—the ratchet and pawl. Then when the shaft tries to jiggle one way, it will not turn, and when it jiggles the other, it will turn. Then the wheel will slowly turn, and perhaps we might even tie a flea onto a string hanging from a drum on the shaft, and lift the flea![33]

Feynman easily proved that such a machine would also eventually stop working, breaking down from the uneven pressure of the molecules. The physicist explained that the root of the problem lay at the level of the "little demon," who would begin to tire and shake until it no longer knew what it was doing.

"It turns out, if we build a finite-sized demon, that the demon himself gets so warm that he cannot see very well after a while." After operating for a limited amount of time, "the demon . . . must heat up." "Soon," Feynman argued, "it cannot tell whether it is coming or going, much less whether the molecules are coming or going, so it does not work."[34]

Feynman's explanation was accepted wisdom by the late 1960s. Electrical circuits were not so different from filters or mechanical ratchets and pawls. "We find the same thing in an *electrical rectifier*," he explained.[35] Except for momentary exceptions, the universe flowed and evolved in a certain direction because of the demon's congenital weaknesses. His limitations produced the arrow of time. "Its one-way behavior is tied to the one-way behavior of the entire universe," explained Feynman. This unidirectionality was an undeniable characteristic of the universe: "Are all the laws of physics reversible? Evidently not! Just try to unscramble an egg! Run a moving picture backwards, and it takes only a few minutes for everybody to start laughing." Could we fight this directionality? We do it all the time: "We pull up the shades and let the light out."[36] With effort and ingenuity, we can make *our* universe seem different from *the* universe, but only for a while.

Feynman's demon raised more questions than it answered. Science did not rule out the possibility that we could learn how to get something from very little effort most of the time, though not all of the time. The physicist admitted that he could not completely answer all questions pertaining to these demons and the universe in which they operated. Why were universal laws suspended in delimited enclosures? Why did the universe permit the existence of pockets of momentary exceptionality? How many of these exceptions could there be? To answer such questions required knowing more about the universe and the circumstances around its emergence. "One-way behavior," concluded the physicist, "cannot be completely understood until the mystery of the beginnings of the history of the universe [is] reduced still further from speculation to scientific understanding."[37]

During these years, the benefits of computing technologies were such that every advanced nation and institute in the world aspired to equip itself with the best machine money could buy. Israel would not be left out. Conflict with Palestinians did not deter scientists in Rehovot from unveiling the "most modern computer based on ultra high speed electronics," with a "tenfold improvement in speed over the latest model, already among the world's fastest."[38] Gershom Scholem, the world's foremost expert on Jewish mysticism, offered an original way of thinking about computer demons by reference to the Golem, a clay creature from late-medieval Jewish lore who followed orders until one day he went berserk. In his dedicatory remarks at the Weizmann Institute, he lamented, "I have been complaining [that the institute] has not mobilized the funds to build up the Institute for Experimental Demonology and Magic which I have for so long proposed." He explained that "they preferred what they call Applied Mathematics and its sinister possibilities to my more direct magical approach."[39] Scholem's comparison was not just metaphorical.

THE MICROCHIP

The year 1967 was a good one for science's demons. Maxwell's character was the subject of a feature-length article in *Scientific American* that opened with an intriguing subtitle: "This hypothetical being, invoked by James Clerk Maxwell nearly a century ago as a violator of the second law of thermodynamics, has occupied the minds of many prominent physicists ever since."[40] Its author, Werner Ehrenberg, was chair of the Experimental Physics Department at Birkbeck College. His interest in these questions dated back to the 1940s, when he showed a new way to derive the second law in "A Note on Entropy and Irreversible Processes," published in the *London, Edinburgh, and Dublin Philosophical Magazine and Journal of Science*.[41] At Birkbeck College, he worked next to David Bohm, who had been hired back from Brazil as chair in theoretical physics. Decades earlier, Ehrenberg had made important contributions to quantum mechanics with his colleague Reymond

Siday. The two were among the first to discover that electromagnetic potentials could affect electrically charged particles even when the electric and magnetic fields in their proximity were null (now known as the Aharanov-Bohm or Ehrenberg-Siday-Aharanov-Bohm effect). Years earlier, in his attempts to peer further into the structure of the atom, Ehrenberg had developed the fine-focus X-ray tube that would later be central to the discovery of DNA. Rosalind Franklin and Maurice Wilkins used an instrument based on Ehrenberg's design to obtain and compile the data essential for James Watson and Francis Crick's discovery of the helical structure of DNA.

"For nearly a century the 'demon' invoked by Maxwell in the foregoing passage has haunted the world of physics," explained Ehrenberg. Interest in demons only increased with the use and development of microchip-based electronics. To understand how solid-state electronic devices such as diodes, semiconductors, and transistors worked and how they could be improved, scientists dug deep into the literature of Maxwell's demon, Szilard's demon, Gabor's demon, and other quantum creatures. "These devices," argued Ehrenberg, "are analogous to the hypothetical gadgets that translate the up-and-down movement of a Brownian particle into a purely upward motion, or that perform the old demon's trick of permitting only fast molecules to go from left to right."[42]

In the 1960s, more and more microchips were programmed to direct a stream of electrons one way or the other by opening or closing semiconductor gates, working away as tireless Maxwell's demons. What most excited researchers about these microelectronic components was the possibility of coupling them with older and larger engineering systems, such as cars and airplanes, to lower the ratio of energy obtained versus energy spent. "A demon, or another intelligent being, may first meditate about improving the ratio of entropy lost to entropy gained by using large fluctuations," wrote Ehrenberg.[43]

Ehrenberg asked readers to consider a simple fishing net. How similar to or different from rectifiers, filters, diodes, and semiconductors was it? "Similar devices on a macroscopic scale are well known and sometimes useful for example in catching fish," he explained.[44] Like any net, valve, membrane, or filter, these would eventually bend or break under stress. For this reason, scientists had "informed Maxwell's little demon that in reality he and his wall are only a semipermeable membrane of a type that does not exist" except when idealized. Yet little did it matter that the finite-sized demon tended to heat up if its life could be extended by keeping it at a cool temperature or if it could be easily and cheaply replaced by a young and fresh one.

As long as limits could be pushed, limits *were* pushed. Ehrenberg concluded his *Scientific American* article by suggesting that the creative powers of the human mind might conquer these practical challenges: "So the demon has become *Homo sapiens*, and the ball is thrown to the biological sciences, in particular to the study of man."[45]

Even as new electronic microchip-based computer technologies seemed to offer the most promising possibilities, the scientific literature on demons led some scientists such as Ehrenberg to warn against overly optimistic expectations of science-based high-tech society and culture. Not everyone seemed to be benefiting equally from advances in science and technology. Extreme poverty and wealth, deadly accidents, and unexpected windfalls seemed to be spreading around the globe despite being caused by extremely improbable events. The physicist discussed a deadly avalanche of coal waste that descended on the Welsh village of Aberfan in 1966, killing 116 children and 28 adults. One dislodged pebble caused tons of coal mining debris to suddenly slip. A liquefied mass of rock and shale buried first a school and then the village. Could scientists tame computer demons to understand and possibly prevent such tragedies?

Increased use of electronic components that imitated the abilities of Maxwell's demon risked skewing the ratio between low-probability and high-probability effects in nature even further. With such demons, it made sense to bet on exceptions to the second law. The demon in the enclosure "must expect to expend more photons than the jackpot is worth, but what if he takes courage, trusts his luck and succeeds the first time? After all, some people do win the Irish Sweepstakes."[46]

Ehrenberg considered boons and disasters to be examples of Maxwell's demon–type actions. A minuscule event could lead to a major tragedy. "Think of ball lightning or catastrophes such as the Aberfan disaster, in which a heap of mine refuse suddenly began to move and engulfed many houses," he wrote. But the benefits were clear as well. "Even if he does not assist in providing power for a submarine," this demon's small actions could add up significantly: "Let us stop here and be grateful to the good old Maxwellian demon."[47]

Besides, scientists might be able to find yet another demon with enhanced powers. After discussing the ongoing debates about "hidden variables" in quantum mechanics, Ehrenberg asked: "Shall we then find a new, smaller and better demon, who interferes with the hidden variables?"[48]

Ehrenberg would die less than a decade after writing his Maxwell's demon article, leaving behind a draft manuscript for a book on causality, necessity, and chance. Published posthumously, the book introduced a creature named "Born's demon" in its discussion of the paradoxes of quantum mechanics. This creature could see determinism hiding behind quantum mechanics' apparent indeterminism. His double vision allowed the possibility of developing an "ensemble interpretation" of quantum effects that could reconcile dynamical *and* probabilistic explanations at the level of quanta. "Born's demon," explained Ehrenberg, "is then confronted by the ensemble, and not first with an undisturbed, uncharted sea and then with one in which the island has just been found." Born's creature was able and fast enough to persistently chase subatomic particles. Thanks to its speed and capabilities, it could connect statistical values to determinable outcomes. The task of chasing electrons in their orbits around nuclei was no doubt difficult, since "an electron, taken as a real particle, would spin around the nucleus at about 10^{16} times a second." Under such circumstance, "even a demon might give up." Yet if found, it could reconcile quantum mechanics with the "belief that probability spreads in space (multi-dimensional) and time according to deterministic laws in the form of differential equations."[49] It would explain why although "in the hostelry of life we are offered a limited menu; our deliberate choice is pure chance for the cook."[50] By the time Ehrenberg wrote these lines, he believed that the chances that this demon existed were remote. No experiments had come even close to revealing him. "So it comes to this: physics is not demonology," he concluded.[51]

MIT AND PROJECT MAC

While physicists grew increasingly divided about how to answer some of their discipline's most fundamental questions, computer scientists kept building demons. The student protests of 1968 kept some radical students out of their classrooms. Others burrowed more deeply than ever in their studies, becoming practically glued to their laboratories and computer monitors. A new generation of scientists and engineers were seduced by the utopian hope that new computer-human partnerships and unimpeded flows of information could lead to a better and more peaceful world.

At MIT, the ambitious Project MAC (acronym for Men and Computers) was in full swing in its attempt to "bring about a partnership between men and computers" and to create the right technologies "for men and computers to work together directly and effectively."[52] It was progressing rapidly, attracting some of the most talented researchers of those years.

The Project MAC progress report of 1968–1969 defined "a 'daemon' process" as "a system (not a user) process, the operation of which is automatic."[53] Fernando Corbató, who led the initiative to build a daemon process, later explained that the developers were inspired by Maxwell's demon.[54] Since then, the term has been used to designate a software program that runs silently in the background and stands ready to answer requests from other daemons.

Project MAC's daemons were essential for new "computer networks," which at that time were "a new field of very great potential." "Several daemons have been developed during the year," according to the Project MAC report, and were almost ready to be released.[55] With them, researchers could develop the "general-purpose multi-access computer systems" that became a model for ARPANET and then for the internet. "We are looking forward to participation in the experimental ARPA network, which will link multi-access computers in several universities," stated the 1969 report. Abhay Bhushan, a student who had moved from India to earn a master's degree in electrical engineering, was tapped from the team to develop more daemons, since he "has already published two papers in the new field and is planning thesis research in it."[56]

A few years after MIT's Project MAC first announced that its computer communications infrastructure would rely on daemons to run background processes, Bhushan launched one of the first effective computer file exchange systems, the file transfer protocol (FTP). It allowed users to exchange files by using daemons (or FTPds, short for FTP daemons) without having to "explicitly log into a remote system or even know how to 'use' the remote system." To receive a file, the computer user no longer needed to follow a set of instructions in real time. The FTPd did that automatically with the help of "'daemon' processes which 'listen' to agreed-upon sockets, and follow the initial connection protocol much in the same way as a 'logger' does."[57] Bhushan described these processes as powering a "user-level protocol that will permit users and using programs to make indirect use of remote host computers." Daemons could run "an intermediate process [that] makes most of the differences in commands and conventions invisible to you."[58] They were so helpful that, with them, scientists succeeded for the first time in connecting two computers at MIT (the GE645/Multics with the PDP-10/DM/CG-ITS), and they hoped to join these two computers with Harvard's PDP-10 soon.

PROGRAMMING DEMONS

As computer scientists created daemons to facilitate networking between computers and between computers and humans, they concocted other demons to make computers smarter. Computers could excel at lightning-fast calculations, but engineering them to gain "general understanding" was much harder. Students at MIT's Artificial Intelligence Laboratory were tasked with the challenge. Eugene Charniak, who studied with Minsky, Seymour Papert, and Terry Winograd, was particularly successful in developing demons that taught computers how to understand stories. "Charniak's demons" soon became an insider term used by practitioners to designate computer demons of this type.[59] The US Department of Defense eagerly funded his project. Charniak started a successful career in the fields of artificial intelligence, deep learning, and neural networks by first focusing on teaching computers to "understand" children's books. His PhD dissertation for MIT's Department of Electrical Engineering was submitted with the ambitious title "Toward a Model of Children's Story Comprehension" (1972).[60] Its technique for developing "deep semantic processing" in machines required working with "demons" to "model the effect of 'context.'" The author explained the use of the term "demon" as "first coming into computer literature with" Selfridge and Minsky.[61] The approach was so novel that it was hard to explain. "In trying to summarize this thesis we might look at some of the threads which wander through many of the chapters," Charniak noted. "Probably," he concluded, "the most persistent thread is the idea of 'context' as implemented through demons."[62]

In teaching computers how to answer questions about simple stories, scientists learned about how children gained understanding and knowledge more generally. AI research advanced by thinking of computers as simple and ignorant little children, and children, in turn, were increasingly considered to be like not-so-smart computers. Eventually, the science of education would be reoriented toward a view of learning as a sort of programming of the human mind.

Programming demons was hard work. To begin with, "there were several problems which had to be ironed out," explained Charniak. "A demon should embody a somewhat general fact about a situation independent of the particular people or objects involved, so we need machinery to specify demons to specific situations."[63] Programmers needed not only to write down everything they knew about a particular word but to define "demon demon interactions" (cautioning that "we should add that demon demon interaction is probably more complex than we have indicated so far").[64] They also had to construct instructions to "destroy demons," for "if demons are to represent the current context of the story, we need machinery for removing demons when no longer needed." The underlying risk, of course, was the possibility that "we would destroy the demon when we shouldn't."[65]

Charniak's first programmed demons were used to create software that could answer questions about *Up and Away*, the simple first-grade reader published by Houghton Mifflin. That story proved most challenging for neophyte machines. The project was going well, until Charniak had to deal with a "piggy bank" that came up in the narrative. The phrase threw the program for a loop, setting it off on a futile and circular attempt to parse out relations between banks and pigs. To prevent this from happening and to teach a computer how to answer a question about a piggy bank, Minsky's star student programmed an appropriate demon (called the DEMON PB) to provide the context for the appearance of the term. The programmer first wrote down everything they knew about piggy banks: all of this information fit on a single page. For the program to work with other stories, the same programming effort would be required for every other object that ever appeared in another story. This particular graduate student was neither intimidated nor deterred. Many others would soon join him in similar efforts to program demons, then sharing them as open-source code. Computers were thus able to slowly answer more and more questions about more and more things.

Another graduate student of Minsky's who worked at the Artificial Intelligence Laboratory took on the same challenge: to "construct a program that will understand stories that children will understand."[66] Not knowing much about children himself, he enlisted his wife's help and that of a few children, constructing "DEMONs that correspond to contextual information" in the story. The stories he chose to tackle involved characters who were babies, so he proceeded to "write DEMONs for understanding infants." He started by drafting a list of everything he thought children might know

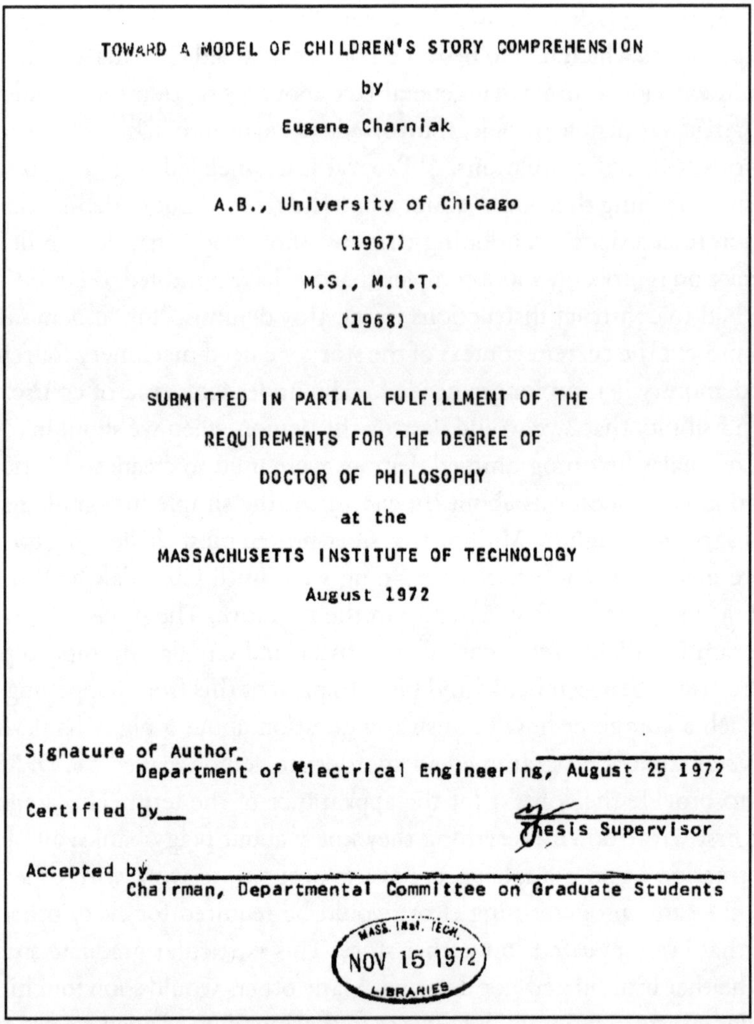

> TOWARD A MODEL OF CHILDREN'S STORY COMPREHENSION
>
> by
>
> Eugene Charniak
>
> A.B., University of Chicago
>
> (1967)
>
> M.S., M.I.T.
>
> (1968)
>
> SUBMITTED IN PARTIAL FULFILLMENT OF THE
>
> REQUIREMENTS FOR THE DEGREE OF
>
> DOCTOR OF PHILOSOPHY
>
> at the
>
> MASSACHUSETTS INSTITUTE OF TECHNOLOGY
>
> August 1972
>
> Signature of Author_____
> Department of Electrical Engineering, August 25 1972
>
> Certified by_____
> Thesis Supervisor
>
> Accepted by_____
> Chairman, Departmental Committee on Graduate Students
>
> NOV 15 1972

Figure 6. Eugene Charniak, "Toward a Model of Children's Story Comprehension" (PhD dissertation, MIT, 1972), 1, 79. Reproduced with permission of Massachusetts Institute of Technology.

about babies, before continuing to write about other topics that might appear in the story. In a little bit more than two pages, he laid down most of the basic information that would be used by the program's BABY DEMONs. After writing his first demon, he wrote yet another page about everything he knew about baby bottles for the next one, BABY BOTTLE DEMON. The student continued to code—for BABY HUNGRY DEMON, BABY-GIVE-

```
                                                            79

4.2  What Demons Look Like

    At the very end of section 3.5 we used a simplified
diagram to described a demon which connected money coming out
of a piggy bank with the person who shook the PB now having
the money.  In this section we give a more detailed notation.
    The basic form of a demon is:
        (DEMON  <demon's name>
                <list of variables>
                <pattern the demon is looking for>
                <e1>
                <e2>
                  .         Program to be run if the
                  .         proper assertion is found
                  .
                <en>)
So our PB-OUT-OF demons would have the outline
        (DEMON  PB-OUT-OF
                (<variables>)
                (?N OUT-OF ?M ?PB)
                  .
                  .
                  .
                )
To fill in the rest of the demon we will need two primitives,
GOAL and ASSERT.

The Primitive GOAL

    GOAL corresponds exactly to THGOAL in Micro Planner, so
those who know the language can skip this description.  GOAL
is primarily an information obtaining primitive.  So for
example:
        (GOAL (IN JACK HOUSE))
```

Figure 6 (continued).

FOOD DEMON, BABY-CRY DEMON, BABY-MAKES-NOISE DEMON, and so on—hoping to "make it easier to build a children story comprehender, which may in fact be necessary if we want to understand general discourse."[67]

A slight mistake in the interpretation of a single word could cause general havoc throughout the program. Minsky's student developed worrisome

```
                MASSACHUSETTS INSTITUTE OF TECHNOLOGY
                   ARTIFICIAL INTELLIGENCE LABORATORY


Artificial Intelligence                              August 1972
Memo #265                                      Reprinted January 1974




                       INFANTS IN CHILDREN STORIES -
                TOWARD A MODEL OF NATURAL LANGUAGE COMPREHENSION

                              Garry S. Meyer


                                 ABSTRACT

         How can we construct a program that will understand stories that
    children would understand? By understand we mean the ability to answer
    questions about that story.  We are interested here with understanding
    natural language in a very broad area.  In particular how does one under-
    stand stories about infants? We propose a system which answers such
    questions by relating the story to background real world knowledge.  We
    make use of the general model proposed by Eugene Charniak in his Ph.D.
    thesis (Charniak 72).  The model sets up expectations which can be used
    to help answer questions about the story.  There is a set of routines
    called BASE-routines that correspond to our "real world knowledge" and
    routines that are "put-in" which are called DEMONs that correspond to
    contextual information.  Context can help to assign a particular meaning
    to an ambiguous word, or pronoun.


    Work reported herein was conducted at the Artificial Intelligence
    Laboratory, a Massachusetts Institute of Technology research program
    supported in part by the Advanced Research Projects Agency of the
    Department of Defense and monitored by the Office of Naval Research
    under Contract Number N00014-70-A-0362-0003.


    Reproduction of this document, in whole or in part, is permitted for
    any purpose of the United States Government.
```

Figure 7. Garry S. Meyer, "Infants in Children's Stories: Toward a Model of Natural Language Comprehension" (master's thesis, Artificial Intelligence Laboratory, MIT, August 1972, memo 265), 1, 30, 31, 32.

doubts about the claims of his professors, namely, their assertion that they were creating machines with the capacity to *learn* and act *intelligently*. Rarely did hardworking students like him get any credit, and in his thesis he confessed that, in all honesty, he did not believe that he was creating what they claimed. "We are forced to do all the work when we want to add knowledge to the system," he concluded.[68] Nevertheless, although this graduate student was exhausted by the end of the project, others labored away.

> ### A First Look at Infants
>
> The initial task I attempted was to write DEMONs for understanding infants in the context of the Charniak model. The first thing I did was to look at several stories that were either about infants or had them in them. From this experience, plus the knowledge gained by talking with three children, ages 2, 6, and 8, and with my wife's help, I produced the following summary of the knowledge that I think children have about infants:
>
>> Infants are happy to sleep most of the time, which they do either in a crib, bassinet, or any comfortable and "safe" place (mother's arms). When they are not sleeping they are either playing, eating, or relieving their bodily functions (described by children with euphemisms like: poo-poo, bunny, wee-wee, etc.). Infants usually wear diapers because they don't control their body movements very well. Dad or Mother must change the diaper after this happens. Excrement causes irritation which leads to pain and will thus cause crying. Babies also have all of the common characteristics we attribute to most humans (i.e. 2 arms, 2 legs, 2 eyes, 10 fingers, etc.), except they are proportionally smaller. Their size being bigger than a Tiny Tears doll (or any small doll for that matter) and smaller than a large stuffed animal. Infants are "new" (although not necessarily improved) so they can not do many of the things that older humans do. That is, they don't know how to walk, talk, play with most games, or dress, wash, or feed themselves. Infants cry when hungry. Crying has been shown to be inhibited both by feeding and by nonnutritive

Figure 7 (continued).

The challenge to develop AI systems based on computer demons spread from MIT to Stanford on the West Coast. A student of the AI pioneer Roger Schank, having read everything there was to learn about demons, explained that, "basically, a demon is a process which can be activated by certain combinations of situations." He laid down his plan for his project:

> The idea is to spawn demons at each point in, say, a story. The demons will "lie in wait" until they detect that they are applicable to some later event or situation, at which time they become active, releasing their

> sucking. You can reduce crying by holding a baby or by supplying a continuous auditory stimulation (singing a lullaby). Young infants are burped after feeding, this is to bring up trapped air that they may take in while feeding. To burp, hold the baby over your shoulder and pat it on the back gently. They are fed liquid food (usually milk or a formula) in a bottle or from a mother's breast. When they get a little older they move up to soft foods called "baby food." When little, they eat frequently and at regular intervals. Milk or formula must be warmed to take the chill off. When the baby starts cutting teeth, called teething, the baby gets cranky and cries easily due to its gums being tender and sore. Baby is washed by mother in a layette or small tub possibly with the help of older children or father. Infants are too little to splash and play with toys when they bathe as older children often do. Infants' skin are very tender so they are oiled and powdered. This is done after a bath and is accomplished by sprinkling a little on and rubbing it in "very" gently. They feel small and soft, and they are therefore "nice" to hold, but they wiggle a lot so you must be very careful or they may fall. Dropping a baby is considered "bad form" and can be very serious, as are other actions that cause harm. Pinching, hitting, pushing, and kicking are some of the ways in which you can hurt a baby. If you intentionally hurt a baby then that is reason for your being reprimanded. There are other ways of causing a baby to cry, like making a loud noise, taking away something it thinks it owns, or frightening the baby by holding it the wrong way. Very young infants grab onto things like your finger or eyeglasses and smile at you. They make sounds when happy, like goo-goo. They are also highly susceptible to diseases so you must stay away from them when you are sick. Babies need a lot of attention from parents, causing other older children to feel "out of it" and thus to become jealous. This jealous reaction is also caused by the fact that baby is not responsible for its actions and therefore is

Figure 7 (continued, above and opposite).

potential to influence the interpretation of the pattern which has activated them. . . . Demons come out of suspended animation when patterns with which they are equipped to deal are detected.

Lamenting that "demons are often guilty of 'playing their cards too close to their faces,'" this student, like many others during those years, sought ways to widen their applicability.[69]

> 32
>
> not punished for the same things that an older child might be punished for. Babies don't always know what is bad for them so you must watch them. This appears to the older child to be "unfair" treatment. Another cause for hostility is the fact that people bring presents to the new baby and not to the older children. Having a baby is a "Joyous Event" and is usually celebrated. Mother has to go to the hospital to have the baby. Pregnancy precedes giving birth.
>
> The amount of information here is considerable larger and more complex than the knowledge about objects like piggy bank or baby bottle. Much of the information is of the form of "babies can not ____ ". This implies that children have a good idea of what they themselves can do. Many of the stories are based on "can not do" sort of facts and many are centered around the ways in which a baby acts differently from "us" (children). We can not use the information in the form that it is above. It is not clear at all how we would use a fact like "Babies feel small and soft." I have only been able to formalize a small subset of the facts above. The task of formalizing all of it in the context of Charniak's model may in fact be very difficult. I will present what little I have done here and then move on to baby bottle.

THE PATRON SAINT OF CYBERNETICS

Demons were everywhere in the age of computers, proving to be necessary for understanding software as much as hardware. Physics students of the 1970s were instructed to add a demon to the equations of statistical mechanics. If they left it out, they might be left with the wrong impression about what was possible in this universe. In his textbook on the *Foundations of Statistical Mechanics* (1970), the physicist Oliver Penrose described why it was necessary to include the demon. "If we combine the demon (denoted by D) and the system he acts on (denoted by S') into a composite system $D+S'$," the entropy of the entire system always increases, he explained.[70]

Penrose started by modeling "Sherlock Holmes's brain" as a thermodynamical system. "A Maxwell demon with this amount of memory capacity could bring about entropy decreases not exceeding 10^{10} k ln 2m which is roughly 10^{-11} of the entropy decrease when 1 g of water is cooled from 300°K to 299°K." Penrose's Maxwell's-demon-with-Sherlock's-memory chimera showed that memory limitations ultimately affected the thinking capacity or calculating power of anything in the world. By writing a symbol for this creature into the equations of thermodynamics, he discovered one of the most important laws about the limitations of all kinds of computing systems, including biological ones: forgetting—clearing out memories—was both necessary and thermodynamically costly. Once a brain filled up with memories, it could not work anymore. To convince his readers, the physicist cited an insight from Sir Arthur Conan Doyle's detective stories: keeping useless facts in memory was harmful. "I consider that a man's brain originally is like a little empty attic, and you have to stock it with furniture as you choose," said Sherlock. "It is a mistake to think that that little room has elastic walls and can distend to any extent. Depend upon it, there comes a time when for every addition to knowledge you forget something that you knew before. It is of the highest importance, therefore, not to have useless fact elbowing out the useful ones," he concluded.[71] For solving problems, crimes, and riddles, forgetting was as important as remembering.

Maxwell's demon was crowned as the patron saint of cybernetics during the annual conference of the American Society for Cybernetics in the fall of 1972. Heinz von Foerster, cybernetician and professor of electrical engineering at the University of Illinois in Urbana-Champaign, delivered the keynote lecture. He had studied Maxwell's demon for decades as part of his work running the Biological Computer Laboratory, where he had come to practical, mathematical, and philosophical understandings of self-organizing systems.[72] Computers, in his view, were "functional isomorphs" of this being. Addressing his colleagues, he asked them to ponder what had brought them together that evening. More importantly, which saintly figure looked after them?

Foerster longed for a benign overseer who could unite the disparate fields comprising cybernetics and help its practitioners develop it further. "Why then, unlike most of our sister sciences, do we not have a patron saint or a deity to bestow favors on us in our search for new insights, and who protects our society from evils from without as well as from within?" he asked.

Figure 8. A. Y. Lerner, *Fundamentals of Cybernetics* (New York: Plenum, 1975), 257. Reprinted by permission of Springer Nature.

"Astronomers and physicists are looked after by Urania; Demeter patronizes agriculture; and various Muses help the various arts and sciences. But who helps Cybernetics?" Luckily, he found an ally in Maxwell's demon. "Nobody else but this respectable demon could be our patron, for Maxwell's Demon is *the paradigm for regulation*," he argued.[73]

"Regulation," in Foerster's view, was a "general and all-pervasive notion." The demon knew how it functioned better than anyone. "Maxwell's Demon is not only an entropy retarder and a paradigm for regulation, but he is also a functional isomorph of a Universal Turing Machine," he explained. This able patron brought together "the three concepts of regulation, entropy retardation, and computation." They "constitute an interlaced conceptual network which, for me, is indeed the essence of Cybernetics," he explained.[74]

Foerster recalled Maxwell's demon appearing to him one evening and suggesting himself as an ally. The scientist described the encounter humorously during his keynote. "One night when I was pondering this cosmic question I suddenly had an apparition." His reaction to the visitor was not initially positive. "Alas, it was not one of the charming goddesses who bless the other arts and sciences," he lamented. This one was comical. "Clearly, that funny little creature sitting on my desk must be a demon." The scientist and the demon soon struck up a conversation, but then the creature vanished: "After a while he started to talk. I was right. 'I am Maxwell's Demon,' he said. And then he disappeared."[75]

The idea that most of the actions of computers were basically identical to Maxwell's demons proved to be popular and long-lasting. For the rest of his life, Foerster would consider "the machine's computational competence and the demon's ordering talents" to be "equivalent."[76] The computer scientist Richard Laing from the University of Michigan also promoted this idea, arguing that Maxwell's demons were really just "computing automata" that could be seen to operate across a wide variety of contexts. "Ordinary Maxwellian Demon systems," he wrote, "may be rigorously interpreted as computing automata."[77]

By then, demons were central to a number of disciplines beyond physics that sought to explain general thinking processes. Computer scientists were invited to publish in journals of linguistics and psychology and were welcomed at conferences in these fields. A popular introductory psychology textbook written by professors at the University of California–San Diego was illustrated with "delightful demons" that showed the processes through which computers and people were able to recognize the shapes of the letters of the alphabet.

One take-home lesson about Maxwell's demons was that in the macroscopic world these creatures acted partly handicapped if tired, drunk, or suffering from illness. "To use a medical analogy, the demon who wants to operate the molecular trap is like a patient with a severe case of Parkinson's disease trying to thread a fast-vibrating needle!" wrote a well-known scientist.[78] Yet once the string was threaded, everything could change—for the better—in a heartbeat.

Just as they spread to other disciplines, demons also moved geographically. By the 1970s, the West Coast was ready to embrace science demons, even as it pushed away their relation to older forms of religion and superstition. Asked in 1974 to give a commencement address to the graduating students of the California Institute of Technology, Feynman explained that science eliminated superstition. "During the Middle Ages there were all kinds of crazy ideas, such as that a piece of rhinoceros horn would increase potency," he told students. His words that day would be frequently repeated. Feynman explained that after those dark times "a method was discovered for separating the ideas—which was to try one to see if it worked, and if it didn't work, to eliminate it." "This method," he lectured, "became

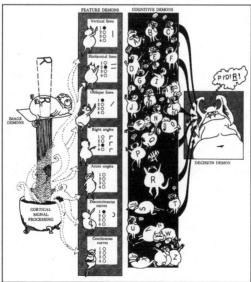

Figure 9. Peter H. Lindsay and Donald A. Norman, *Human Information Processing: An Introduction to Psychology* (New York: Academic Press, 1972), front cover and p. 121. Reprinted with permission from Elsevier.

organized, of course, into science."[79] Even by the time of Feynman's 1974 address, demons still had not gone the way of crazy ideas.

WHEELER'S DEMON AND BLACK HOLES

In light of recent research on demons and information theory, cosmologists reworked their theories of the universe. To answer some of the most important riddles of the universe pertaining to its origins and history, the place of human intelligence in it, and the possibility of harnessing its energy content, they started to wonder what would happen to demons in the vicinity of a black hole.

"Wheeler's demon" was first invoked during a conversation between Jacob Bekenstein, a young PhD physics student at Princeton University, and his adviser, the physicist John Archibald Wheeler. Wheeler was a towering figure in the field. He had worked on the Manhattan Project during World War II before moving on to work on the H-bomb. At Princeton, he had become interested in "black holes," a term he coined. Black holes and H-bombs were similar in that central aspects of each could be understood in terms of thermonuclear reactions.

However remote at the time, the possibility of extracting energy from black holes fascinated Wheeler and a small circle of researchers around him. So did the idea that these strange singularities could be home to a wholly new set of demons, some of which might swallow information and energy, decrease entropy, and stop time like nothing yet known to humankind.

Bekenstein climbed the academic science ladder rapidly, arriving at Princeton after a stint at the Polytechnic Institute of Brooklyn. During the fateful conversation, Wheeler told his student that existing explanations of black holes allowed "a wicked creature—call it Wheeler's demon—to commit the perfect crime against the second law of thermodynamics."[80] The possible existence of such "wickedness" in the universe proved to be "distressing" for the young man, who proceeded to combat this creature in his dissertation. Bekenstein vowed "to defeat the schemes of Wheeler's demon and make black hole physics consistent with thermodynamics."[81]

The enormous gravitational pressure that gave rise to black holes crunched up everything inside them. Would the neat-fingered and delicate Maxwell's demon be able to survive under those extreme conditions? A black hole was challenging for even the lithest of imaginary beings, to say the least. Would the demon be able to operate under this tremendous pressure? He might not make it. The weighty molecules he was used to handling would not be avail-

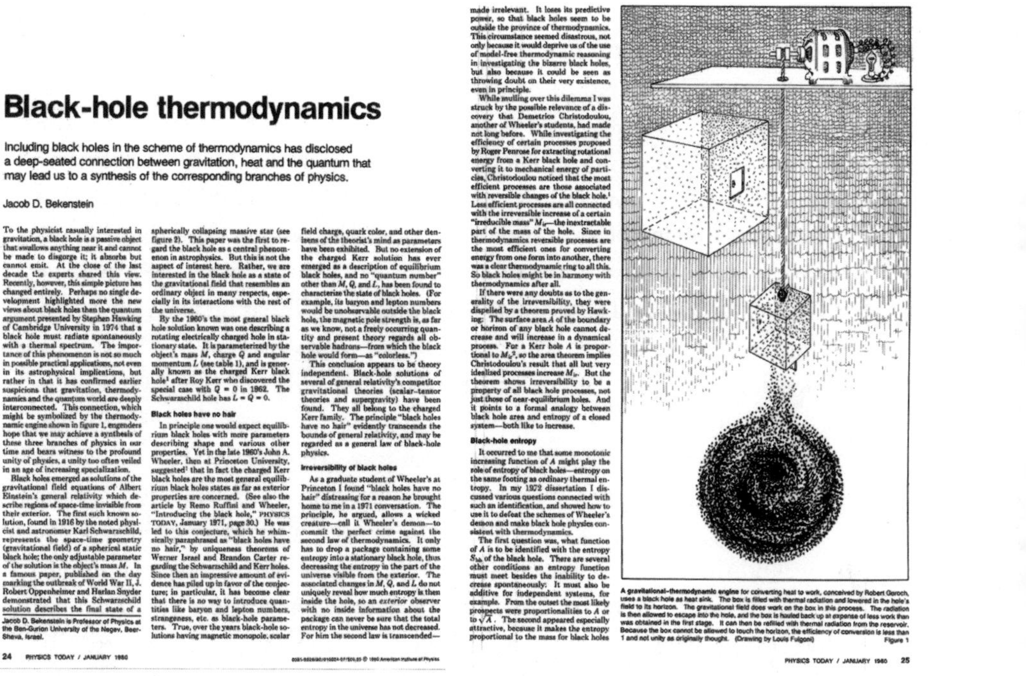

Black-hole thermodynamics

Including black holes in the scheme of thermodynamics has disclosed a deep-seated connection between gravitation, heat and the quantum that may lead us to a synthesis of the corresponding branches of physics.

Jacob D. Bekenstein

To the physicist casually interested in gravitation, a black hole is a passive object that swallows anything near it and cannot be made to disgorge it; it absorbs but cannot emit. At the close of the last decade the experts shared this view. Recently, however, this simple picture has changed entirely. Perhaps no single development highlighted more the new views about black holes than the quantum argument presented by Stephen Hawking of Cambridge University in 1974 that a black hole must radiate spontaneously with a thermal spectrum. The importance of this phenomenon is not so much in possible practical applications, not even in its astrophysical implications, but rather in that it has confirmed earlier suspicions that gravitation, thermodynamics and the quantum world are deeply interconnected. This connection, which might be symbolized by the thermodynamic engine shown in figure 1, engenders hope that we may achieve a synthesis of these three branches of physics in our time and bears witness to the profound unity of physics, a unity too often veiled in an age of increasing specialization.

Black holes emerged as solutions of the gravitational field equations of Albert Einstein's general relativity which describe regions of space-time invisible from their exterior. The first such known solution, found in 1916 by the noted physicist and astronomer Karl Schwarzschild, represents the space-time geometry (gravitational field) of a spherical static black hole; the only adjustable parameter of the solution is the object's mass M. In a famous paper, published on the day marking the outbreak of World War II, J. Robert Oppenheimer and Harlan Snyder demonstrated that this Schwarzschild solution describes the final state of a spherically collapsing massive star (see figure 2). This paper was the first to regard the black hole as a central phenomenon in astrophysics. But this is not the aspect of interest here. Rather, we are interested in the black hole as a state of the gravitational field that resembles an ordinary object in many respects, especially in its interactions with the rest of the universe.

By the 1960's the most general black hole solution known was one describing a rotating electrically charged hole in stationary state. It is parameterized by the object's mass M, charge Q and angular momentum L (see table 1), and is generally known as the charged Kerr black hole[1] after Roy Kerr who discovered the special case with $Q = 0$ in 1962. The Schwarzschild hole has $L = Q = 0$.

Jacob D. Bekenstein is Professor of Physics at the Ben-Gurion University of the Negev, Beer-Sheva, Israel.

Black holes have no hair

In principle one would expect equilibrium black holes with more parameters describing shape and various other properties. Yet in the late 1960's John A. Wheeler, then at Princeton University, suggested[1] that in fact the charged Kerr black holes are the most general equilibrium black holes states as far as exterior properties are concerned. (See also the article by Remo Ruffini and Wheeler, "Introducing the black hole," PHYSICS TODAY, January 1971, page 30.) He was led to this conjecture, which he whimsically paraphrased as "black holes have no hair," by uniqueness theorems of Werner Israel and Brandon Carter regarding the Schwarzschild and Kerr holes. Since then an impressive amount of evidence has piled up in favor of the conjecture; in particular, it has become clear that there is no way to introduce quantities like baryon and lepton numbers, strangeness, etc. as black-hole parameters. True, over the years black-hole solutions having magnetic monopole, scalar field charge, quark color, and other denizens of the theorist's mind as parameters have been exhibited. But no extension of the charged Kerr solution has ever emerged as a description of equilibrium black holes, and no "quantum number" other than M, Q, and L, has been found to characterize the state of black holes. (For example, its baryon and lepton numbers would be unobservable outside the black hole, the magnetic pole strength is, as far as we know, not a freely occurring quantity and present theory regards all observable hadrons—from which the black hole would form—as "colorless.")

This conclusion appears to be theory independent. Black-hole solutions of several of general relativity's competitor gravitational theories (scalar-tensor theories and supergravity) have been found. They all belong to the charged Kerr family. The principle "black holes have no hair" evidently transcends the bounds of general relativity, and may be regarded as a general law of black-hole physics.

Irreversibility of black holes

As a graduate student of Wheeler's at Princeton I found "black holes have no hair" distressing for a reason he brought home to me in a 1971 conversation. The principle, he argued, allows a wicked creature—call it Wheeler's demon—to commit the perfect crime against the second law of thermodynamics. It only has to drop a package containing some entropy into a stationary black hole, thus decreasing the entropy in the part of the universe visible from the exterior. The associated changes in M, Q, and L do not uniquely reveal how much entropy is then inside the hole, so an *exterior* observer with no inside information about the package can never be sure that the total entropy in the universe has not decreased. For him the second law is transcended—made irrelevant. It loses its predictive power, so that black holes seem to be outside the province of thermodynamics. This circumstance seemed disastrous, not only because it would deprive us of the use of model-free thermodynamic reasoning in investigating the bizarre black holes, but also because it could be seen as throwing doubt on their very existence, even in principle.

While mulling over this dilemma I was struck by the possible relevance of a discovery that Demetrios Christodoulou, another of Wheeler's students, had made not long before. While investigating the efficiency of certain processes proposed by Roger Penrose for extracting rotational energy from a Kerr black hole and converting it to mechanical energy of particles, Christodoulou noticed that the most efficient processes are those associated with reversible changes of the black hole.[1] Less efficient processes are all connected with the irreversible increase of a certain "irreducible mass" M_{ir}—the inextractable part of the mass of the hole. Since in thermodynamics reversible processes are the most efficient ones for converting energy from one form into another, there was a clear thermodynamic ring to all this. So black holes might be in harmony with thermodynamics after all.

If there were any doubts as to the generality of the irreversibility, they were dispelled by a theorem proved by Hawking: The surface area A of the boundary or horizon of any black hole cannot decrease and will increase in a dynamical process. For a Kerr hole A is proportional to M_{ir}^2, so the area theorem implies Christodoulou's result that all but very idealized processes increase M_{ir}. But the theorem shows irreversibility to be a property of all black hole processes, not just those of near-equilibrium holes. And it points to a formal analogy between black hole area and entropy of a closed system—both like to increase.

Black-hole entropy

It occurred to me that some monotonic increasing function of A might play the role of entropy of black holes—entropy on the same footing as ordinary thermal entropy. In my 1972 dissertation I discussed various questions connected with such an identification, and showed how to use it to defeat the schemes of Wheeler's demon and make black hole physics consistent with thermodynamics.

The first question was, what function of A is to be identified with the entropy S_{bh} of the black hole. There are several other conditions an entropy function must meet besides the inability to decrease spontaneously: It must also be additive for independent systems, for example. From the outset the most likely prospects were proportionalities to A or to $\sqrt{A}$. The second appeared especially attractive, because it makes the entropy proportional to the mass for black holes

A gravitational–thermodynamic engine for converting heat to work, conceived by Robert Geroch, uses a black hole as heat sink. The box is filled with thermal radiation and lowered in the hole's field to its horizon. The gravitational field does work on the box in this process. The radiation is then allowed to escape into the hole, and the box is hauled back up at expense of less work than was obtained in the first stage. It can then be refilled with thermal radiation from the reservoir. Because the box cannot be allowed to touch the horizon, the efficiency of conversion is less than 1 and not unity as originally thought. (Drawing by Louis Fulgoni) Figure 1

24 PHYSICS TODAY / JANUARY 1980

0031-9228/80/010024-07/$00.50 © 1980 American Institute of Physics

PHYSICS TODAY / JANUARY 1980 25

Figure 10. Jacob D. Bekenstein, "Black-Hole Thermodynamics," *Physics Today* (January 1980): 24–25. Reproduced by permission of the American Institute of Physics.

able to him. The standard model used to understand the subatomic world broke down under those conditions. He could have no torch, no flashlight, and no light. Scientists had pictured Maxwell's being as a cricket player—or, as George Gamow saw him, a tennis master—but the challenges that this sportsman now had to face were greater than catching or hitting balls. Inside such a hole, or even in its vicinity, the being was not only out of his league but in an entirely different ball game, one played with perhaps no balls at all.

If Maxwell's demon might not make it, however, a creature quite similar to him might. When it came to black hole research, the devil was in the details—the very, very tiny details. Bekenstein combed through the existing literature on the topic. In a section of his doctoral dissertation titled "Of Demons, Entropy, and Black Holes," he explained that, "to the author's knowledge, the only satisfactory proposal [about Maxwell's demon] made to date is that of Brillouin."[82] But Brillouin's analysis needed to be expanded to be useful under such extreme conditions.

After referencing the "classic example" of Maxwell's being, Bekenstein conjured a new creature named "Wheeler's demon." This creature could make the entropy created in a thermodynamic process disappear by dropping it into a black hole. If so, it could stop time and its ravages. Could it exist? "We shall give Wheeler's demon the benefit of the doubt," Bekenstein wrote in his dissertation, following up with a section titled "Demon Drops Entropy into a Black Hole." Until these questions were solved, black holes offered the universe a window of opportunity: they could be the most perfect trash cans that could ever exist—so good at annihilating what they swallowed that they could even disappear entropy and decay.

Bekenstein pictured "Wheeler's demon" dropping entropy "freely" into a black hole and then, "by means of a string," examining what happens when "the demon drops the apparatus and withdraws the string."[83] He also wondered what would happen when the demon changed its strategy by using "a spring wound about a rigid bar which originally does not touch the spring" and then "slowly withdraw[ing] the bar."[84] With these imaginings, Bekenstein tried to answer a difficult question: "Can Wheeler's demon send entropy into a black hole without causing the black hole to change irreversibly?" In conclusion, he answered with a firm no. "The second law is valid in black hole physics in a generalized form: that there exists a well-defined black hole entropy (defined from the area) which increases so as to compensate for the loss of thermal entropy from the black hole exterior."[85]

The cosmologist and theoretical physicist Stephen Hawking worked on black holes at the same time as Bekenstein. He agreed with the student's conclusions but would soon provide a different explanation for Wheeler's demon's powers and for Bekenstein's contributions. "He was basically correct," wrote Hawking years later about his colleague, "though in a manner he had not expected."[86]

After reading Bekenstein's work on Wheeler's demon, Hawking embarked on one of his most celebrated scientific projects. A proper investigation into what might occur inside a black hole required that extreme gravitational forces be included in thermodynamic and quantum dynamic explanations of Maxwell's demon. When this task was done, Hawking predicted that black holes would have to radiate particles, regurgitating what they had swallowed.[87]

Hawking considered the possible actions of "gnomes" in his 1978 "Black Holes and Unpredictability" article for *Physics Bulletin*, picturing "a race of little gnomes stationed just outside the event horizon, the boundary of the black hole," watching closely what fell into the hole and sending light signals to each other to communicate. According to Hawking, when the black hole swallowed one particle in a particle-antiparticle pair, the little gnomes witnessing the act would not be able to signal "accurately the time at which each particle fell into the black hole" because "they would have to use photons of the same wavelength and therefore the same energy as that of the infalling particle." As a result, "a large amount of information about the system is irretrievably lost in the formation of a black hole and the number of bits of information lost can be identified with the entropy of the black hole."[88]

For cosmologists like Hawking, a black hole was a kind of outer space information destroyer, and depending on what it was eating up, it could do bizarre things to us. "Our region of the universe," he wrote, "is receiving unpredictable influences from a region about which we have no knowledge."[89]

Hawking's best-selling *A Brief History of Time*, which appeared to great success in 1988, described the boundary around a black hole as acting like a Maxwell's demon. "The event horizon, the boundary of the region of spacetime from which it is not possible to escape," he wrote, "acts rather like a one-way membrane around the black hole: objects, such as unwary astronauts, can fall through the event horizon into the black hole, but nothing can ever get out of the black hole through the event horizon." To describe these effects, Hawking quoted a line from Dante's *Divine Comedy*: "One could well say of the event horizon what the poet Dante said of the entrance to Hell: 'All hope abandon, ye who enter here.' Anything or anyone who falls through the event horizon will soon reach the region of infinite density and the end of time."[90]

Could something simply vanish? Without a trace? What if the stuff sent down a black hole contained particularly valuable information about our world and ourselves? Could it ever be found and recovered? Could a black hole demon stash information away in the darkest corners of the universe, or could it be made to disappear completely? Nothing is as lost as something misfiled. Out of sight, out of mind. To understand the universe, its history and origins, it was important for scientists to determine what happened inside those holes and to clarify the difference between gone and *gone* gone. For that, they needed to know more about how demons might react under such conditions. Bekenstein's and Hawking's research expanded physicists' understanding of black hole cosmology, but it also opened up the as-yet-unresolved information paradox that would be debated for decades to come.

In the years that followed, Hawking continued to work hard to investigate certain properties of black holes which would "protect chronology" and make "the universe safe for historians."[91] As his illness worsened—Hawking suffered from amyotrophic lateral sclerosis (ALS)—he grew ever more nostalgic about the concept of "Laplace's determinism," in which "one needed to know the positions and speeds of all particles at one time in order to predict the future." It could be salvaged if only black holes were discovered to emit something, perhaps some kind of radiation, back to space. "If information were lost in black holes, we wouldn't be able to predict the future," he lectured to a packed audience at Harvard University's Sanders Theater. "It might seem that it wouldn't matter very much if we couldn't predict what comes out of black holes. There aren't any black holes near us. But it is a matter of principle," he insisted. "If determinism breaks down with black holes, it could break down in other situations," and "even worse, if determinism breaks down, we can't be sure of our past history either. The history books and our memories could just be illusions. It is the past that tells us who we are. Without it, we lose our identity."[92] Black holes worried Hawking—who had dedicated his life to searching for evidence of determinism—because otherwise these holes would introduce "a level of unpredictability in physics over and above the usual uncertainty associated with quantum mechanics."[93]

SEARLE'S DEMON

As personal computers started to show up in more and more American households in the 1980s, the idea of a demon who could fiddle with neural synapses was born. "Searle's demon" was a philosophical creation designed to take the wind out of the sails of scientific and philosophical theories that considered the brain in terms of computer software. This hypothetical demon would allegedly sit in your brain, intercepting neuronal synapses momentarily before passing them on again. Nothing would change outwardly. You would act *exactly as you would have without him*. Your behavior and thoughts would be identical to your behavior and thoughts without him. The thought experiment fascinated philosophers and scientists alike. The mere possibility that a central aspect of thinking processes could be taken over by this demon might show that a simple mechanism could be used to perform basic thinking operations. If true, simple AI mechanisms could be considered to be identical, in essence, to living intelligent organisms.

John Searle, a Berkeley philosophy professor who forcefully contested the idea that computer programs possessed a kind of intelligence, did not know when he first published "Minds, Brains, and Programs" (1980) that the paper would bring him the honor of having a demon named after him. As soon as the paper appeared, it was recognized as one that "surely is destined to become, almost immediately, a classic."[94] In spite of its academic style, it clearly conveyed a sense of exasperation.[95] It named names. It was nasty. It was brilliant. The philosopher went after "strong AI," a label for the headline-grabbing school of thought that was making huge headway in academia by claiming that computers could think and perhaps do so better than us. Wounds from the ensuing conflicts festered, with no signs of healing.

Searle had no problem with the idea that certain very special kinds of machines could think. Our own brain, he admitted, was just such a machine. His point of disagreement was subtler. "According to strong AI," Searle explained, "the computer is not merely a tool in the study of the mind; rather the appropriately programmed computer really is a mind, the sense that computers given the right programs can be literally said to *understand* and have other cognitive states." He combated the claim that programs inside computers were responsible for that thinking. "No program by itself is sufficient for thinking," he categorically concluded.[96]

Searle took issue with most interpretations of the famous Turing test, lambasting it for "being unashamedly behavioristic and operationalistic."[97] In preemptive attacks, Searle systematically went after colleagues at Berkeley, Yale, MIT, and Stanford by taking down a long series of anticipated objections to his claims one by one. The neurophysiologist John Eccles, who counted as his achievements winning the Nobel Prize for his study of synapses in the nervous system, applauded Searle: "It is high time that strong AI was discredited."[98] Among many others, Searle targeted John McCarthy for attributing beliefs to such things as thermostats. Searle was exasperated with the rhetoric surrounding new computer programs that could handle natural language. One of them, computer scientist Terry Winograd's SHRDLU, could take requests from a user to manipulate virtual objects on a computer screen, such as blocks and cones. Another one, Joseph Weizenbaum's early chatbot ELIZA, could engage in casual conversation and even masquerade as a psychoanalyst. Searle disputed the "strong AI" interpretation given to some of these innovations, as advanced primarily by Schank at Yale University. "The claims made by strong AI are that the programmed computer understands the stories and that the program in some sense explains human understanding," Schank explained. Searle would have none of it. After taking down Schank, he went after Allen Newell and Herbert Simon, both winners of the Turing Award given by the Association for Computing Machinery. They "write that the kind of cognition they claim for computers is exactly the same as for human beings," Searle protested.[99]

Searle's targets struck back in the open peer commentary following his *Behavioral and Brain Sciences* paper. "Sophistry," replied the philosopher and cognitive scientist Daniel Dennett, arguing that the author was using verbal "tricks with mirrors that give his case a certain spurious plausibility." His evidence was "misleadingly presented" to lead readers "on a tour of red herrings."[100] Dennett capped his response by comparing Searle to the *enfant terrible* philosopher Henri Bergson. Searle's critique was nothing other than "an updated version of *élan vital*," stated Dennett, who, like Bergson, thought that there was a vital force that was irreducible to matter.[101]

Searle had struck sensitive chords. "We may worry about whether any of our fellows is an automaton," wrote the philosopher Arthur Danto, who was one of the few commentators who appeared more amused than irritated.[102] The philosopher Richard Rorty claimed that Searle "sets up the issues as would a fundamentalist Catholic defending transubstantiation" and wondered whether "the money spent playing Turing games with expensive computers should have been used to subsidize relatively cheap philosophers like Searle and me."[103] Funding streams and entire careers were suddenly in jeopardy.

Did this mean that computers could never be considered intelligent? Marvin Minsky was exasperated. "I just can't see why Searle is so opposed to the idea that a *really* big pile of junk might have feelings like ours," he lamented.[104] The cognitive scientist Douglas Hofstadter was irate and had some of the harshest things to say. Searle's provocation was "a religious diatribe masquerading as a serious scientific argument," and "one of the wrongest, most infuriating articles I have ever read in my life." "What Searle and I have is, at the deepest level a religious disagreement," he wrote. Dialogue between Searle and his critics was as copious as agreement was scarce. "I doubt that anything I say could ever change his mind," concluded Hofstadter.[105]

The *Gedankenexperiment* that Searle conceived became known as the "Chinese room argument." Searle asked his readers to imagine a person "locked in a room and given a large batch of Chinese writing." Could an English-language speaker with no knowledge whatsoever of Chinese simulate a native Chinese speaker by secretly reading notes of what he should say? To someone outside the locked room, this person might seem to understand the language, but Searle showed that such an appearance was deceptive. All this person needed to do was to have access to a set of instructions in English for correlating a set of questions to the Chinese symbols he was handed. With no one witnessing the paper shuffling required by the translation process, this person could well impersonate a native Chinese speaker who ostensibly "understood" a story in Chinese without actually understanding a single word of it. One set of strange symbols would come in through one slot of the box and another set of symbols with the correct response would come out on the other side. Searle used this example to prove that the person in between the input and output processes could appear to understand without really understanding anything.

Searle dismissed almost in their entirety the acerbic criticisms of the multitude of renowned scientists who responded, except for one. It came from the philosopher John Haugeland, whose response Searle considered "genuinely original."[106] Haugeland explained that Searle's example was effective because it essentially made readers consider the parallel case of a person doing their job but having no understanding of what was meant by the instructions they followed. Software running inside a computer might have equally little or no understanding of anything it did. "The crucial move," he explained, "is replacing the central processor (c.p.u.) with a superfast person—whom we might as well call 'Searle's demon.'" In that way Searle tried to show that "an English-speaking demon could perfectly well follow a program for simulating a Chinese speaker, without itself understanding a word of Chinese."[107] But another lesson could be drawn by taking the thought experiment a bit further. What if a person's thought processes were replaced by the travails of Searle's synapse-intercepting demon? In Haugeland's new example, the demon's victim was Chinese and female, perhaps even a "Chinese criminal." While it was hard for Searle's critics to argue that the locked-in respondent could understand Chinese (even if it appeared so from the outside), once reframed in terms of Searle's demon, the *Gedankenexperiment* raised a much scarier possibility. What if the locked-in being was locked inside the brain of a living person?

"Now imagine covering each of the neurons of a Chinese criminal with a thin coating," explained Haugeland, "which has no effect, except that it is impervious to neurotransmitters." The brain of such a criminal would suddenly stop working. "And imagine further that Searle's demon can see the problem, and comes to the rescue." The demon could serve as a link between synapses, taking over "her neurons" and thus reconstituting the "victim's behavior" in its entirety. The victim would then act in exactly the same way as if the neurons had not been intercepted. Nobody would notice anything odd. In fact, Haugeland proposed, the woman would act exactly as she would have before the demon's intervention. Would we then have to admit that the criminal, technically, was no longer thinking because she could not do so without help from the demon? Who was doing the thinking, the Chinese criminal or the demon? Who would be the intellectual author of a crime if she went so far as to commit one? Was there an accomplice or a conspirator? Who, in short, was to blame?

Searle restated Haugeland's argument before replying. What if a woman had "her neurons coated with a thin coating that prevents neuron firing," he asked? And what if we "suppose 'Searle's demon' fills the gap by stimulating the neurons as if they had been fired"?[108] Did that then mean that the demon should be thought of as possessing thinking powers? Searle responded to Haugeland's reply by saying that the demon would have to act exactly as the criminal would in order for her behavior to remain "unchanged." If that was the case, then her demon would have to have thinking powers identical to hers, and if so, she should still be considered as doing the thinking. In his view, the demon was simply helping her trivially. Searle closed his reply confidently: "I conclude that the Chinese room has survived the assaults of its critics."[109]

Disagreements about "strong AI" revealed deeper disagreements between Searle and his interlocutors. The possible attribution of intentionality to machines or programs opened up a new universe of murky ethics. Could an AI be blamed if its actions led to a crime or a disaster? How would cases involving AIs compare to those involving intoxication, insanity, or unconscious behavior? Yet another complex quandary surfaced in discussions of Searle's demon: how much could we replace or enhance our thinking processes and still remain ourselves? Something similar to a Cartesian evil demon could be operating *within* our brains, where not even ratiocination could save us.

Searle was right in his criticisms of strong AI, argued Haugeland, but not for the reasons he thought. The claims of AI advocates failed, "not because there's no such demon, but because there's no such program," he argued.[110]

Finding one would be equivalent to finding the other. What would be needed to debunk Searle was to build such a thing. If a program for simulating speaking Chinese could indeed be perfectly used by a monolingual English speaker, then Haugeland might be ready to concede to Searle. The terms of the debate, he argued, should be changed to "*whether* there is such program and *if not, why not*." These "are in my view, the important questions."[111] In Haugeland's view, practical developments had to be connected to theoretical answers.

So much had changed from Turing's Manchester Mark I programs of the 1950s to the PDP-10 computer programs of the 1980s. Could researchers expect to see an actual implementation of Searle's case example with new automatic translation software or better chatbots? Research marched on. The debate grew. Haugeland would later have a demon named after him, "Haugeland's demon," aka "the H-Demon."

Searle's demon was not alone. In the responses to his article, Searle's commentators started referring to a host of colorful beings to help them better understand what *thinking* really meant. For Schank, the question of who understood thinking had a clear answer: "Why, the person who wrote the rules of course. And who is he? He is what is called an AI researcher."[112] But understanding, just as much as intelligence, was a slippery concept. Another commentator went after the presumed intelligence of AI programmers, noting that their notion of intelligence would inevitably shift with time: "The ineluctable core of intelligence is always the next thing which hasn't yet been programmed."[113] These commentators discussed the essence of "understanding" by reference to common delegation and exploitation practices. Various replies to Searle were based on examples in which a person was replaced with something or someone else to determine whether the essential neurological process counted as a legitimately intelligent action. They discussed replacing functions usually associated with human thought with work performed by "green aliens," "homunculi," "elves," a "robot," a "poor slave of a human," "museum monkeys" who produced great literature by chance, a lonely computer "adventitiously manipulated by uncaring programmers," a "parasite," "a piece of stale cheese," "an intelligent human from a chess-free culture, say in Central Africa," or "a Chadian in the chess room" who "hasn't the foggiest notion what he is doing."[114] Hofstadter's reply to Searle involved fiddling with the gender of the characters introduced in the Chinese room thought experiment. "Let us add a little color to this drab experiment and say that the simulated Chinese speaker involved is a woman and that the demons (if animate) are always male," he wrote. Hofstadter also proposed making distinctions between beings that were traditionally deemed to possess understanding and those that did not. "Parrots who parrot English do not understand English," he insisted, adding that "the recorded voice of a woman announcing the exact time of day on the telephone time service is not the mouthpiece of a system that understands English."[115]

It is a poor carpenter, the saying goes, who blames his tools. But these studies showed just how difficult it was to draw a firm distinction between ourselves and our tools. If there was one main lesson to be drawn from the discussion of Searle's demon, it was about the difficulty of determining who was manipulating whom. Credit-taking competed with victim-blaming, and determining who was a perpetrator required knowing how agency operated across systems made up of humans, things, and machines.

One commentator added Descartes's demon to the discussion. While this demon's original powers of deception were limited to the realm of the senses, Searle's could go deeper into our brains. But what if Descartes's demon could manipulate computers' input devices just as it could manipulate our senses? A "smart" program that could seem to understand instructions to manipulate a world of blocks displayed graphically on a computer monitor, such as SHRDLU, could be interpreted as "the victim of a Cartesian evil demon," since the blocks that could be moved by the instructions given to the SHRDLU program "do not exist in reality."[116] Searle's demon revived fears about Descartes's aging being; the view of him as having the ability to hack computer vision and install an alternative reality transfused him with new blood for the next millennium.

Searle's strongest critics, Dennett and Hofstadter, continued the polemic around Searle's demon in a later book compiling a collection of essays and commentaries to explore the issue in more detail. The authors hoped that once our minds—which they referred to as the *"minimal victim* of the [Descartes's] demon's deceit"—were analyzed scientifically, complex thinking could be broken down into much simpler mechanical processes.[117] To clarify the issues pertaining to Searle's demon, Hofstadter introduced yet another demon, called "Haugeland's demon" after the philosopher who had first started to think about the experiment in those terms. Hofstadter took Searle's arguments in a new direction. "If it is a normal-sized human being, we shall call it a 'Searle's demon,'" and "if it is a tiny elflike creature that can sit inside neurons or on particles, then we shall call it a 'Haugeland's demon.'" In Hofstadter's retelling, the "S-demon," envisioned as a life-sized person, competed for credit with the little "H-demon" who was making the connections behind the scenes. By dividing the two in this way, he aimed to show that most readers would attribute the capacity to think to beings with which they would *more easily identify*. They would tend to consider the human-sized demon as a biological thinking being, and they would tend to consider the small one a mostly mechanical helper. Hofstadter argued that Searle had constructed his argument in such a way that most readers identified with the more humanlike demon and thus became convinced, with Searle, of the inanity of "strong AI" claims. But this interpretation resulted only from their particular *identification at the expense of the little guy*. Hofstadter accused Searle of attributing understanding to the life-sized person only. He portrayed Searle as answering, "Fools! Don't listen to her!" and, "She's merely a puppet whose every action is caused by my tickling, and by the program embedded in these many neurons that I zip around among." If readers would identify instead with the smaller demon, then they might reconsider the claims of "strong AI." Hofstadter aimed to show that Searle's ability to persuade readers of the particular thinking powers of some entities but not others relied on nothing more than anthropomorphic identification.[118]

Identity politics, racism, and sexism entered into discussions of who was *perceived* to possess thinking abilities, displaying just how difficult it was to distinguish between having smarts and seeming smart. Searle responded with a scathing review in the *New York Review of Books*, titled "The Myth of the Computer," about the "spectacular failure" and "exaggerated hopes" of the ambitious AI program. It was as ridiculous to attribute understanding to a machine as it was to attribute thirst to it, no matter how loudly and frequently it told us it was thirsty. Strong AI, he continued, was "nonsense," and Dennett and Hofstadter were just taking advantage of people who did not know better: "A lot of the nonsense talked about computers nowadays stems from their relative rarity and hence mystery." "As computers and robots become more common," he insisted, "as common as telephones, washing machines and forklift trucks, it seems likely that this aura will disappear and people will take computers for what they are, namely useful machines."[119] Searle then imagined a program for simulating thirst that "run[s] on a computer made entirely of old beer cans, millions (or billions) of old beer cans that are rigged up to levers and powered by windmills." After being programmed to be thirsty, "at the end of the sequence a beer can pops up on which is written 'I am thirsty.'" He concluded his philosophical argument sarcastically: "Does anyone suppose this Rube Goldberg apparatus is literally thirsty in the sense in which you and I are?"[120]

Searle referred readers to the second chapter of Weizenbaum's classic book *Computer Power and Human Reason*, where he "shows in detail how to construct a computer using a roll of toilet paper and a pile of small stones," thereby illustrating how ridiculous it was to attribute intentionality and the capacity to think to a machine program.[121] Early on, Weizenbaum had claimed that computers and certain computer programs were considered intelligent, not because they were so, but because they had the ability to deceive their users. He interpreted ELIZA, the chatbot that he created, in ways that were very different from typical "strong AI" interpretations. His conversational program showed more than the blatant superficiality of most of our human interactions. It revealed the hopeless naïveté of so many who engaged in comparable interactions with machines and who lost sight of who their interlocutors really were. Those who believed in strong AI similarly tended to forget that they were interacting with machines programmed to seem intelligent. Weizenbaum noted that many users "soon forgot that fact," "just as theatergoers, in the grip of suspended belief, soon forget that the action they are witnessing is not 'real.'"[122]

Searle's demon and Haugeland's demon neither proved nor disproved the dream of AI. For some scientists and philosophers, Searle's *Gedankenexperiment* showed that only machines as special as us could traffic in meaningful discourse. For Searle's critics, it proved that machines could do so too, and could potentially beat us at the game. These demons exacerbated the disagreement between AI's critics and defenders, widening the already badly strained relationship between scientists and philosophers. As more and more thinking machines were built, more and more researchers started to debate what thinking really meant. What was artificial in it? What was natural? Who excelled in it and who did not?

FEYNMAN'S DEMON

Theoretical physicists soon delved into the debate about the limits and possibilities of intelligence and computation more generally. New research published in the *International Journal of Theoretical Physics* under the unassuming title "The Thermodynamics of Computation: A Review" would change existing views about "the old Maxwell's demon problem" for decades to

Figure 11. *Abacus: Magazine for the Computer Professional* 3 no. 4 (Summer 1986): front cover image by Ed Soyka. Republished by permission of Springer; permission conveyed through Copyright Clearance Center, Inc.

come. "Many people don't like their scientific ideas overthrown," recalled the author, Charles Bennett, referring to the conclusions he was about to question. Those ideas had been advanced decades earlier by his senior colleague Rolf Landauer, one of the most important theoretical physicists of his generation.[123] Landauer held an undergraduate degree and a PhD from Harvard, where he had worked with Léon Brillouin. After graduating, he started working at IBM and became an expert on microchips. "Rolf was an old IBM type who did things on the straight and narrow," recalled Bennett. "I was a scruffy hippie. He endured that." Landauer magnanimously arranged to get his contradictor hired at IBM Research.[124]

Richard Feynman was fascinated by the bold conclusions of the rookie. "Surprisingly," he wrote, "Bennett has shown that Maxwell's demon can actually make its measurements with zero energy expenditure, providing it follows certain rules for recording and erasing whatever information it obtains."[125] The insight was nothing short of revolutionary, and Feynman was forced to revise his previous conclusions about Maxwell's creature and change his views about physics and cosmology. "When Bennett discovered all this, no one knew it could be done," he explained, adding that "there was a lot of prejudice around that had to be argued against."[126]

Feynman was as fascinated by Bennett's research as he was by the potential of new microchips and semiconductor technologies. From 1983 to 1986, he taught a series of courses on the physics of computers at Caltech, interrupted by returning bouts of cancer. His lectures from these courses would be published posthumously as the *Feynman Lectures on Computation*. In them, Feynman returned to considering the possible actions of "our little friend" who could "shine a demonic torch" on molecules in order to see them and manipulate them. Feynman believed that his willingness and ability to momentarily violate the second law suggested that "there is probably something fishy about him." He proceeded to investigate Maxwell's demon in detail.[127]

"The diode," wrote Feynman, "is a cunning device which allows current to flow in one direction only" and as such could take on the role of a Maxwell demon.[128] What could be achieved with these little devices? What did they contribute to what could and could not be done with computers, and why?[129] Feynman was already thinking ahead to the time "when we come to design the Ultimate Computer of the far future." To prepare for machines "which might have 'transistors' that are atom-sized, we will want to know how the fundamental physical laws will limit us."[130] Feynman recommended that readers brush up on the existing literature on demons. "Those of you who wish to take your study of the physics of information further," he said, "could do no better than check out many references to a nineteenth century paradox discovered by the great Scottish physicist James Clerk Maxwell," adding that "*Maxwell's Demon*, as it is known, resulted in a controversy that raged among physicists for a century, and the matter has only recently been resolved." Once students had done some homework on that creature, they could catch up with the most recent research to properly understand the current resolution: "In fact, it was contemplation of Maxwell's demon that partly led workers such as Charles Bennett and Rolf Landauer to their conclusions about reversible computing, the energy of computation, and clarified the link between information and entropy." Feynman acknowledged that it would be impossible to cover everything that had already been dis-

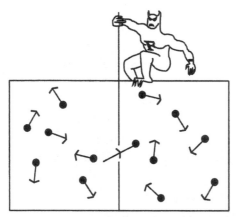

Figure 12. Richard Feynman, *Feynman Lectures on Computation*, ed. Anthony J. G. Hey and Robin W. Allen (Cambridge, MA: Perseus, 1999), 149.

covered about the demon, but he promised to "supply" his students "with enough tidbits to at least arouse your interest."[131]

"It is easy to imagine that if we built a big enough computer, then it could compute anything we wanted it to," explained Feynman.[132] When a computer was conceived as an information engine, the fuel that could keep it going was the size of its memory. Feynman explained how a supercomputer could be realized practically by summarizing Bennett's "subtle" idea of "using a message tape as fuel" for a computer and relating "the information in the tape to its fuel."[133] While fuel would inevitably be spent in steam and coal engines, all that information engines needed to calculate anything was a longer and longer memory tape. The moment when it seemed that a computer would start to run out of power, it could be given more memory to keep it going. How far could the analogy be taken? Scientists of the Industrial Revolution had been concerned with reversible heat engines, which could reuse spent fuel, and they were now concerned with reversible computation—reharnessing used-up energy as a source of power. If not a source of perpetual *motion* in the classical sense, this energy could become a source of perpetual *computation*.

As a young student, Bennett had heard Landauer lecture on the topic. It took him decades to challenge the senior scientist's conclusions. The living memory of organisms seemed much more powerful than any computer memory. The genetic code, all those bits and bits of information that shaped biological organisms, were stored in a tiny part of every cell, which was somehow one of nature's most impressive information storage technologies. "The existence of a storage medium as compact as the genetic one indicates that one can go very far in the direction of compactness," argued Landauer.[134] What is more, this genetic memory replicated itself in stunning ways. Why not use that as a model to build better memories for computers? Bennett, following a cue from Landauer, compared the storage and copying mechanisms of computers to those inside biological cells, "in the biochemical apparatus responsible for the replication, transcription and translation of the genetic code."[135] The process of DNA transmission and replication could serve as a model for breaking through current limitations on the memory storage capacity of computers and lowering their energy consumption. In work funded by the US Atomic Energy Commission at the Argonne National Laboratory, Bennett considered "the biosynthesis of messenger RNA" "a physical example of reversible computation."[136] Comparing cells, computers, and engines would prove extremely productive.

"The digital computer may be thought of as an engine that dissipates energy in order to perform mathematical work," he explained.[137] But in contrast to old engines, Bennett believed that energy dissipation in computers could be minimized to the point of negligibility. The energy dissipated by computers might not be brought all the way down to zero, but it could be brought close to nothing.

Bennett agreed that there was only one limit on computers' theoretically infinite power: memory. What if they could be furnished with more and more of it? "The essential irreversible step, which prevents the demon from breaking the second law, is not the making of a measurement (which in principle can be done reversibly) but rather the logically irreversible act of erasing the record of one measurement to make room for the next."[138] The physicist landed on these conclusions by going further than anyone in inquiring into "the physical state of the demon's 'mind.'"[139]

Landauer disagreed with Bennett's conclusions. What good was it to think of the marvelous qualities of infinite memory capacity and infinite velocities if they did not—and could not—exist? Landauer grew more and more exasperated at scientists' attempts to build the perfect computer and their claims that such a project was even possible. Herein lay the crux of their disagreement, and the reasons for the birth of Landauer's principle and, later, Landauer's demon.[140] Even if the laws of physics dictated that we might not be able to *do* everything we ever dreamed of, some scientists still believed that one day we might be able to *know* everything. Landauer, in contrast, argued that the same limitations that affected the energy content in the universe dampened the possibilities of our knowledge of it.

Was there a physical or even cosmological limit to human ingenuity? Landauer and Bennett agreed that the bottleneck preventing computers from getting smarter was their limited memory. Whoever could figure out a way for computers to become more efficient and expend less energy would be handsomely rewarded. Engineers far beyond IBM raced to make computer memories and microchips more efficient by making them smaller, thinner, faster, and cooler. The challenge had to be approached both practically and theoretically. These questions went far beyond even physics and touched on biology as well. Yet even as Landauer stressed that ever more powerful memory storage devices could be found in the future, he stressed that none would ever be able to store *infinite* amounts of information. At some point, they would have to clear their data, and in the erasure of "the unnecessary details of the computer's past history," there would be additional "energy dissipation."[141]

Computer technologies had changed radically since Feynman first started considering Maxwell's demon, and miniaturization was advancing quickly. The idea that computers could become as small as atoms fascinated the physicist: "So, in 2050, or before, we may have computers that we can't even see!"[142] His excitement about their possibly diminutive dimensions was tied to an even more momentous possibility: that they could be designed to expend almost no energy. Feynman described computers as expert shufflers, shifting binary "pebbles" from one end to the other. They were nothing but superfast sorters. After all, computers only needed to distinguish between a 1 and a 0 and send the digit left or right. The simplicity of that task took away neither their magic nor their intelligence. In fact, it was what made them so smart. Feynman highlighted this aspect of intelligent machines in his *Lectures on Computation*. "The inside of a computer is as dumb as hell but it goes like mad!" Its tricks were due to the speed at which it could accumulate the computation of very simple tasks, "just like a very dumb file clerk." The final result looked like intelligence: "It is only because it is able to do things so fast that we do not notice that it is doing things very stupidly," Feynman concluded.[143] Computers, in his estimation, were very fast idiots.

"The marvelous thing is, with sufficiently detailed rules this 'idiot' is able to add two numbers of any size!" wrote Feynman with amazement. All that this idiot required was some programming to enable him to know how to shuffle binary digits around, which could be compared to moving simple pebbles around, or the beads of "the abacus of old." But be careful what you wished for, he warned: "What you tell an apparent idiot, who can do no more than shuffle pebbles around, is enough for him to tackle the evaluation of hypergeometric functions and the like." Add speed, and he would surely beat you: "If he shifts the pebbles quickly enough," Feynman explained, "he is justified in thinking himself smarter than you!"[144]

The physicist was among many others who were fascinated by how DNA copying processes in cells seemed like a "living computer" that expended almost no energy and were able to reproduce. In a footnote to his published lectures, Feynman described computing machines modeled after these processes. "Bennett has nicely christened machines like this 'Brownian computers,'" he wrote, "to capture the manner in which their behavior is essentially random but in which they nevertheless progress due to some weak direction of drift imposed on their operation."[145] Decades later, a host of living and nonliving machines that could produce work from nearly random fluctuations would come to be understood by reference to "Feynman's demon."

As physicists continued to push the boundaries of what demons could do within hardware, computer scientists charged on with developing them using software. Software demons were designed to act like a group of expert spectators: "A demon is a routine of some kind that watches for an event to happen, then 'spontaneously' starts executing when that watched-for event occurs." Anyone who had ever watched an event carefully in order to spring into action based on the outcome, from buyers at an auction to stock traders to bookies, could relate. "An analogy is an arena of experts sitting watching the center of the arena for information they can use. When a piece of information appears, experts jump up and say 'Yes, I can do something with that!,' and would immediately start performing functions with or upon that information."[146] "Demons fire (or become invoked)," he explained, just like a group of experts waiting attentively to act on information. Although the term "watcher" could be considered a "more appropriate, less sinister name," the term "demon" was deemed most appropriate.[147]

By then, successive improvements to Selfridge's pandemonium model had proved the value of the parallel processing demons that became central to parallel distributed processing (PDP) cognitive models and computer architectures. Demons arranged in parallel featured prominently in new artificial intelligence programs that created rules as they went along and did not need to be programmed beforehand. Computers running the new AI programs learned by themselves, ostensibly like children. "In contrast" to other architectures, explained Allen Newell, the instructions in them "fire in parallel whenever their conditions are satisfied, acting as demons."[148]

Advances in programming techniques and the storage memory capacities of computers made the goal of AI seem ever more attainable. "In the Middle Ages scholars wondered how many angels would fit in the head of a pin," started a *Los Angeles Times* article in 1986. "As the result of recent technological advance at Stanford University we now know that the entire Encyclopedia Britannica would comfortably fit there," it concluded.[149] Feynman, who had funded a prize to reward such improvements, was ecstatic. "We can now easily get a whole book, such as an encyclopedia or the Bible, onto a pinhead—rather than angels!" he wrote.[150] Scientists started imagining the possibility of fitting entire libraries into a few square feet.

Marvin Minsky was so optimistic about these new developments that he started to imagine how to program computers to be able to understand jokes. For machines, differentiating comedy from tragedy was notoriously difficult, and identifying irony and sarcasm was even harder. Simple stories posed formidable challenges for computer programmers who wanted to design computers to answer simple questions about them. Minsky recalled that his student Charniak had proposed using demons to help computers understand "context." Charniak had "suggested that whenever we hear about a particular event, specific recognition-agents are thereby aroused." These demon agents "then proceed actively to watch and wait for other related types of events," which then set off other demons, and so on. "Because these recognition-agents lurk silently, to intervene only in certain circumstances, they are sometimes called 'demons,'" Minsky explained. Their employment "raised many questions," such as: "How easy should it be to activate demons? How long should they then remain active?" The risks were evident: "If too few demons are aroused, we'll be slow to understand what's happening. But if too many become active, we'll get confused by false alarms." While "you might understand certain parts of a story by using separate, isolated demons," Minsky argued, "there are no simple solutions to these problems, and what we call 'understanding' is a huge accumulation of skills." If computers could start to gain "understanding" by activating demons in succession, our own understanding might be subjected to their whims too. Minsky wondered if we too might be understanding simply by virtue of their manipulations. "How much of the fascination in telling a story, or in listening to one," he asked, "comes from the manipulations of our demons' expectations?"[151]

Maxwell's demon stole scientific headlines in 1987. In October, the financial markets crashed alongside the economic framework that had sustained them since the Great Depression. In November, *Scientific American* ran an article titled "Demons, Engines, and the Second Law," authored by Bennett. "The real reason Maxwell's demon cannot violate the second law," he argued, "has been uncovered only recently." The "unexpected result" came from "a very different line of research: research on the energy requirements of computers."[152] The culprit was not the high price of information, as was commonly thought; it was the cost of forgetting.

To anticipate the surprise of readers learning that physicists thought about the universe in terms of the possible action of a demon, Bennett explained that a demon's utility was due to "its far-reaching subversive effects on the natural order of things." This demon, like the others, affected the most important aspects of our economy. "Chief among these effects," he wrote, "would be to abolish the need for energy sources such as oil, uranium and sunlight." The demon had the potential to liberate us from fuel dependency. "Machines of all kinds," he marveled, "could be operated without batteries, fuel tanks or power cords. For example, the demon would enable one to run a steam engine continuously without fuel, by keeping the engine's boiler perpetually hot and its condenser perpetually cold." Could this liberator ever be found? "One way to uncover the reasons Maxwell's demon cannot work is to analyze and refute various simple inanimate devices that might function as demons," wrote Bennett.[153] After considering the usual contraptions that had all shown how Maxwell's demon would eventually tire, cease to work, and start expending energy, he considered yet another one: computers based on small molecular fluctuations.[154]

Bennett determined that Maxwell's demon's problems were exacerbated when he filled his memory capacity, then had to erase what he knew in order to make room for more information. Erasing took energy: "Forgetting results, or discarding information, is thermodynamically costly."[155] That insight pointed out a possible solution. "If the demon had a very large memory," it might save itself the thermodynamic expense of erasing it.

Broad transformations in the media landscape of those years helped Bennett's conclusions appear less counterintuitive. "We do not generally think of information as a liability," he noted. "We pay to have newspapers delivered, not taken away." But the exponential increase of information during those years was creating a dystopian problem as well, one that affected many, including demons. "Yesterday's newspaper" was a problem for demons because it "takes up valuable space, and the cost of clearing the space neutralizes the benefit the demon derived from the newspaper when it was fresh." Bennett speculated that "perhaps the increasing awareness of environmental pollution and the information explosion brought on by computers have made the idea that information can have a negative value seem more natural now than it would have seemed earlier in this century."[156] With Bennett's research, science gained a new insight into the problem of having too much information.

Decades after Bennett and Landauer first started investigating these topics, a few key questions remained unanswered. Both men agreed that one limit to the powers of computation lay in the minimum amount of energy needed to erase bits of information. For Bennett the problem of memory was largely technical, but for Landauer it was essential. While Landauer focused on the limitations that computers faced due to basic physical laws, Bennett continued to think about ways to bypass those limitations. Landauer argued against drawing conclusions about physics based on a platonic view of the power of mathematics, while Bennett had no problem extrapolating beyond practical limits and calculations. In his first articles on Maxwell's demon, Bennett had placed no limits on "the size of the parts" that would enable him to function. Landauer, in contrast, liked to point out that those scenarios were unrealizable "in the real universe, with a limited selection of parts."

Even the conclusions that physicists drew from mathematics and logic, Landauer argued, should be subject to the limits dictated by the laws of physics. "About a century ago," he complained, "mathematicians declared their independence of the real physical universe." Since then, they could not care less if something "does not even exist" and "scorn questions about physical executability."[157] To those who dreamed that computers embodied a world of limitless possibilities and that with the right program anything could be accomplished, he delivered a pessimistic lesson: "There is really no software, in the strict sense of disembodied information, but only inactive and relatively static hardware."[158] Practical limitations should not be brushed away, even by theoreticians, he argued. To back up his argument, Landauer cited the mathematician Herman Weyl's complaints about his colleagues' proclivity to believe uncritically in the transcendental world of infinite lengths and perfect continua, a belief that "taxes the strength of our faith hardly less than the doctrines of the early Fathers of the Church or of the scholastic philosophers of the Middle Ages."[159] Physics had to contend with issues pertaining not only to energy but to meaning and knowledge. Was the cosmos bounded or unbounded, finite or infinite? What was the relation of abstract mathematics to concrete reality? As these questions remained unanswered, scientists continued to debate what demons could or could not do. "Then where are the real limits?" asked Landauer, ever ready to call attention to the *non plus ultra*. "They are, presumably, related to more cosmological questions, and unlikely to be answered by those who started the field."[160]

By the time Bennett published "The Thermodynamics of Computation," Maxwell's demon had been through a lot. One of his biographers listed at least "four obituaries" in succession: "'1914: the Mechanical Demon (Age Forty-seven) Succumbs to Heat'; '1929: Measurements Fell the Intelligent Demon (Age Sixty-two)'; '1950: The Demon (Age Eighty-three) Is Crushed by the Cost of Information Acquisition'; '1982: The Demon (Age 115) Drowns in Garbage.'"[161] Bennett was most responsible for the last one.

"A SOCIETY OF LITTLE DEMONS"

While cosmologists were preoccupied with questions pertaining to the information and energy content of the universe, computer scientists continued to push these limits practically by trying to make computers more efficient and smarter. As personal computers became more popular and demon-based computer programming techniques spread, more and more scientists started to understand the workings of our minds by reference to them. By the mid-1980s, computer scientists, physicists, biologists, and psychologists had joined forces to understand intelligence, genetics, and computers by reference to Maxwell's demon.[162] The director of the Institute for Cognitive Science (ICS) at the University of California–San Diego became a leader of this multidisciplinary approach, gathering under the umbrella of the PDP (Parallel Distributed Processing) Research Group top researchers from different disciplines, including biophysicists such as John Hopfield, computer scientists such as Eugene Charniak, and geneticists such as Francis Crick.

In 1988, *Science* published an article describing the new "style of learning" of the best AIs.[163] These machines employed "a society of little demons" with excellent "manners" for the purposes of imitating intelligence.[164] These helpful pucks escaped from the constraints of rule-following, algorithm-based computer programs. The reviewer explained that they were not "a conventional programming algorithm," and certainly "not if 'algorithm' means an unambiguous, precisely defined procedure that tells the computer what to do at every step along the way." New AIs achieved "something a little more subtle."[165] They could learn from experience.

Subtle thinking came with not-so-subtle problems. The risks of such open-ended AI architectures were clear. Demons could fight with each other inside programs. "But what happens if things do not go well?" "What happens if none of the demons has anything to say—that is, if none of the rules apply to the situation at hand?" Or "worse, what happens if several demons are activated simultaneously and start struggling for power?"[166] This particular architecture had distinct benefits, as well as significant drawbacks, since the program shared with humans the perilous capability "of drawing the wrong lessons from its experience." In imitating human intelligence so well, open-ended AI architectures also imitated human weaknesses; they had "no way to delete an incorrect rule from memory, any more than a human can forget his or her name just by deciding to." Just as we may tend to draw incorrect lessons from the environment, so a computer programmed on that model could do the same. The computer might even be drawn to madness. "The potential is catastrophic," warned the article. "One can all too easily imagine the program piling up errors until its reasoning resembles a kind of hopelessly muddled schizophrenia."[167]

Newell and his collaborators' first step to avoiding these pitfalls "was to teach the demons some manners. Instead of having each one shriek out a command when its conditions were met, he had them express opinions such as 'operator Q1 (take your opponent's queen) is better than operator Q2 (take your opponent's pawn),' or 'operator Q7 (sacrifice your bishop) is best.'" But good manners were not enough. Recognizing the need to install a demon as a conference moderator, programmers "modified the production system as a whole to operate like an exceedingly polite business conference." The demon-moderator played a critical role in this arrangement: "instead of letting the activated demons fight over who gets to give the orders," it would "let all of them have their say. Only when everyone's opinions were on the table would the program decide what to do."[168]

The main benefit of employing these demons came from their wait-and-see, contemplative attitude: "These demons spend most of their time in quiet contemplation of something called 'working memory,' which is a kind of internal blackboard that records data about the current situation." Remaining in such a state was an ideal way of implementing if-then orders and optimizing memory use, because "when one of them sees something it likes—that is, when the conditions on the IF side of the rule match the current situation in working memory—it jumps up and shrieks out the command listed on its THEN side: 'DO this.'" Moments later the demons could return to their inactive-yet-vigilant sentinel mode, "and everyone settles down again to wait for another demon to jump up."[169]

FROM MICROCHIPS TO UNIX

A few months before the Berlin Wall came tumbling down in 1989, former president Ronald Reagan heaped praise on the microchip in his first speech after leaving office. The occasion was a reunion in honor of Winston Churchill. "The Goliath of totalitarianism," stated Reagan, "will be brought down by the David of the microchip." The *New York Times* reported that, in the ex-president's view, "the international revolution in communications technology had unleashed an irrepressible march toward democracy in the Communist world, bringing nearer the collapse of totalitarianism." "Electronic beams blow through the Iron Curtain as if it were lace," Reagan stated.[170]

The usefulness of these tiny pieces of silicon-based hardware, so often understood by reference to Maxwell's and quantum demons, grew alongside innovations in software programming techniques. Demons and subdemons programmed to fulfill essential background tasks were regularly coded into operating systems. The UNIX manual of that year, known colloquially as the "Devil Book," included the term in its glossary, where it explained why the term should not be interpreted negatively: "The old English term, daemon, means a 'deified being,' as distinguished from the term, demon, which means an 'evil spirit.'"[171]

That same year Evi Nemeth, a programmer and operating system expert, explained the use of the term "daemon" in her field. "'Daemon' is actually a much older form of 'demon'; daemons have no particular bias towards good or evil, but rather serve to help define a person's character or personality," she wrote in one of the bibles of computing, the *Unix System Administration Handbook* (1989). She elaborated on the use of the term: "The ancient Greeks' concept of a 'personal daemon' was similar to the modern concept of a 'guardian angel'—*eudaemonia* is the state of being helped or protected by a kindly spirit." Nemeth concluded: "As a rule, UNIX systems seem to be infested with both daemons and demons."[172]

While physicists continued to gain understanding of the limitations of Maxwell's demon, computer demons took the 1990s by a storm. *The New Hacker's Dictionary* included the term in one of its early editions:

daemon /day'mn/ or /dee'mn/

[from the mythological meaning, later rationalized as the acronym "Disk And Execution MONitor"] n. A program that is not invoked explicitly, but lies dormant waiting for some condition(s) to occur. The idea is that the perpetrator of the condition need not be aware that a daemon is lurking (though often a program will commit an action only because it knows that it will implicitly invoke a daemon) . . . writing a file on the . . . spooler's directory would invoke the spooling daemon, which would then print the file. The advantage is that programs wanting (in this example) files printed need neither compete for access to nor understand any idiosyncrasies. . . . They simply enter their implicit requests and let the daemon decide what to do with them. Daemons are usually spawned automatically by the system, and may either live forever or be regenerated at intervals. Daemon and demon are often used interchangeably, but seem to have distinct connotations. . . . Although the meaning and pronunciation have drifted, we think this glossary reflects current (1991) usage.

The entry further explained that "the term 'daemon' was introduced to computing" by people working on a compatible time-sharing system.[173] The difference between them was slight but important:

demon

n. 1. [MIT] A portion of a program that is not invoked explicitly, but that lies dormant waiting for some condition(s) to occur. See [daemon]. The distinction is that demons are usually processes within a program, while daemons are usually programs running on an operating system. Demons are particularly common in AI programs. For example, a knowledge-manipulation program might implement inference rules as demons. Whenever a new piece of knowledge was added, various demons would activate (which demons depends on the particular piece of data) and would create additional pieces of knowledge by applying their respective inference rules to the original piece. These new pieces could in turn activate more demons as the inferences filtered down through chains of logic. Meanwhile, the main program could continue with whatever its primary task was. 2. [outside MIT] Often used equivalently to daemon—especially in the UNIX world, where the latter spelling and pronunciation is considered mildly archaic.[174]

What are computer demons, actually? "Even if their physical manifestation is no more than the flicker of electric current through a silicon computer chip," argued the technology correspondent Andrew Leonard in his book *Bots: The Origin of New Species*, bots "do exist in the real world."[175] By the late 1990s, daemons were not only a subset category of many other kinds of bots but software versions of mechanical robots that "explore the hyperlinked reality of cyberspace, mapping out and indexing the vast quantities of information available through the World Wide Web."[176] They served as models for many daemons to come. "Corbato's daemon belongs at the top of the great Tree of Bots," noted Leonard. "Call it the ur-bot, the primeval form to which all present and future bots owe ancestry."[177] With the rise of programmable personal computers and the internet, the reach of these bots increased exponentially. "Today daemons are a programmer's best friend, one of the lubricants that keeps the electronic machinery of bits and bytes moving smoothly."[178] In later decades, they would make up most of the traffic across the internet, masquerading as human users and surpassing them in numbers. Even regular users of email "meet up constantly with the vigilant mailer daemon, a vital cog in the electronic post office that deals with a host of mailing problems (messages sent to nonexistent addresses, overflowing mailboxes, forwarding and readdressing needs.)"[179]

As UNIX-based computer systems conquered the world, the term "daemons" became abbreviated as a simple "d" at the end of other technical terms, the most popular ones being "inetd," "nfsd," "sshd," "named," and "lpd" (for line printer daemon). They generally acted as they should. "Printer-spool daemons monitor computer file servers to see if any files are waiting to be sent to the printer," and "clock daemons" execute "instructions at specific times." These daemons played key roles in technologies that allowed users to communicate via computers connected to a network. They performed scheduled tasks on networks, answered and directed emails auto-

Figure 13. *Unix System Manager's Manual* (1984), front cover image by John Lasseter, Lucasfilm, Ltd.; UNIX T-shirt designed by comic artist Phil Foglio based on Polaroid photograph of the Chicago Circle's DEC PDP-11/40 computer. Reproduced by permission of Phil Foglio.

matically, and helped configure software. They were invisible (out of sight, out of mind), often doing menial tasks in the background, and they were attendant, waiting patiently for the right moment to act. By then, the figure of the daemon was firmly ensconced as a mascot for UNIX.

HTTPd, the first daemon-based web server protocol allowing computers to communicate with each other, was soon made available as free software. "An HTTP (hypertext transfer protocol) daemon," explained Leonard, years after its first release, "resides in a computer that stores World Wide Web homepages, always on the alert to respond to requests from other computers asking for access to local Web documents."[180] The World Wide Web and the development of the internet depended on HTTPd.

Computer science's reliance on demons proved long-lasting. "The internet is alive with daemons of all kinds," explained a historian of the internet. "Their inhuman intelligence spread through the global infrastructure."[181] "The internet is now an infrastructure possessed. Embedded in every router, switch, and hub—almost every part of the internet—is a daemon listening, observing, and communicating."[182] By the end of the century, a full "catalogue" of smart demons was available to cognitive and computer scientists.[183]

SMART DEMONS

During the summer of 1990, the editor of *Nature* published an article describing a groundbreaking paper that got scientists one step closer to reaching the dream of creating work without the requisite expenditure. Most science journals, including *Scientific American* and *New Scientist*, covered the new research with dramatic headlines. According to *New Scientist*, "Maxwell's demon can, in certain circumstances, do the impossible: transfer energy from a cool body to a warmer one." Some scientists were enthralled, but others were unconvinced. New exorcisms were thought up. A scientist from the Department of Electrical Engineering at the University of Southampton explained why the article was so timely: "The question is important because our search for a pollution-free energy source would soon be solved for ever if an army of Maxwell's demons could be put to such use on a commercial scale."[184] The article in *Nature* described a "small gang" of Maxwell's demons guarding an "orifice." Its author explained that, after fantasizing about well-oiled and nearly frictionless shutters and doors for centuries, scientists had discovered that "the demon's Achilles' heel" was "cognitive, not mechanical."[185]

Previous researchers had noted that the information-erasure actions of Maxwell's demon led to entropy increase, but Carlton Caves of the University of Southern California found a way to make these energy losses almost negligible. "Maxwell's demon," explained Caves, "has been both a celebrity and a conundrum for succeeding generations of physicists." He read up on its history and learned that "early work in 'demonology' showed that a one-way valve does not work, because it heats up to ambient temperature and ceases to function."[186] The trick to bypassing this limitation was simple: instead of using one engine that would create work at one minimum expense per engine, use many engines. Instead of letting each engine spend the least bit of information possible by considering them separately, scientists could operate an engine system only when they saw a group of molecules coincide on one side in every engine. They could then use this least possible bit of information to extract work from several engines at once. "In the rare, 'simple' occasion when all 10 molecules are in the same half of their boxes, one bit of information only is needed to describe them," explained the editor of *Nature*.[187] In consequence, "this means that only one-tenth as much energy is needed to describe the position of the molecules." N number of engines could potentially lower costs n number of times. This new proof showed that "a demon can achieve a net gain in energy by choosing to operate only in such cases."[188] Caves named this new "sophisticated" demon the "Unruh demon," after his colleague William Unruh.[189] In comparison to these new connected systems of engines, the single-molecule contraption thought up by Szilard back in 1929 looked like "a crude, and quite theoretical, type of demon, called a Szilard engine."[190]

Caves had been inspired by an article published in *Nature* a year earlier by the theoretical physicist Wojciech Zurek of the Los Alamos National Laboratory, who had tried to figure out if there was a "smart" way of erasing the demon's memory that would take less energy than the "dumb" proposals made thus far. Zurek had been mulling over that challenge for decades. In the 1980s, he had sat in on Feynman's courses on computation and worked closely with John Wheeler. The "recurring themes" that arose for Zurek while working with Wheeler and Feynman became "somewhat of an obsession" for him. Zurek would go on to study the demons named after his two mentors. When one of the biographers of Maxwell's demon visited Zurek at his Los Alamos office to "learn about the Demon's latest adventures," he found, to his great delight, that Zurek was "someone who truly understood" Maxwell's creature. "Instead of threatening to throw the poor little goblin out the window," Zurek "shooed him out of his gloomy box, and gave him a fresh purpose in life."[191]

The excitement around a new "sophisticated" demon that acted only when most profitable did not last long. A few days after the new research appeared in print, the authors sent a note to *Physical Review Letters* explaining why their demon could not create energy without expenditure.[192] In a quick follow-up article in the *New Scientist* with the title "Sacred Law of Physics Is Safe After All," a journalist who looked into the matter reported that, "at the seminar at the Santa Fe Institute, Caves and other 'demonologists' found a hidden energy cost in the technique."[193] It was indeed safe, but scientists had shown yet another way to make dissipation more negligible than ever.

For years to come, Zurek continued to think about demons. In an article written for *Feynman and Computation* (1999), the physicist said that the old Maxwell's demon who would tire from taking measurements was no longer challenging or that interesting. What fascinated researchers now was its smart version, "a complete Maxwell's demon" who "should be able to measure, and it (. . . ?; he? she?!) should be of course intelligent." Zurek returned to previous attempts to make a Maxwell's demon smarter by pairing him with a program that instructed him to act intelligently. This "demon of choice" was an "intelligent and selective version of Maxwell's demon."[194] He/she/it could work with groups of molecules by waiting patiently until rare thermal fluctuations took place and a certain number of molecules gathered at the right place. Only then would the demon act. This new demon combined the traditional powers of Maxwell's demons with those coming from a procedure that instructed him to only act in specific favorable cases and wait out the rest by doing nothing and sitting still.

"The aim of this paper," Zurek wrote, is "to exorcise the demon of choice." Zurek figured out exactly where its "cost to doing business" lay. Its energy gains were not "illicit," because the demon had to pay an additional price for keeping more complex results, which had to involve more molecules, in its memory: "When the demon forgets the measurement outcome, it will repeat the measurement and remain stuck forever in the unprofitable cycle." To completely forget the past, explained Zurek, was as unproductive as remembering it in its entirety. The "famous saying that *'those who forget their history are doomed to relive it'* applies to demons with a vengeance!" he concluded.[195] It also applied to scientists.

QUANTUM MAXWELL'S DEMONS

Another idea soon took hold of scientists' imaginations: quantum mechanical Maxwell's demons (QMMD), or more simply, quantum Maxwell's demons (QMD). These new demons, conceived at MIT and elsewhere, were an innovative class of nanoscale demons, or nanobots (also referred to as microbots or microrobots), that performed work at almost no expense. Some of them were based on manipulating entangled quantum properties, like electron spin. Scientists surmised that quantum entanglement might provide a way to take a measurement inexpensively by acting on one system while leaving another one untouched. Would additional costs emerge when these quantum motors produced macroscopic amounts of work? Researchers raced to find out.

"Up to now, Maxwell's demon has functioned primarily as a thought experiment that allows the exploration of theoretical issues," explained Seth Lloyd, the principal investigator, who proposed "a model of a Maxwell's demon that could be realized experimentally using magnetic or optical resonance techniques." The "NMR demon" (short for nuclear magnetic resonance demons) was one of a larger "collection of many demons of the sort described [that] could be used to pump macroscopic amounts of heat and do macroscopic amounts of work."[196] They took advantage of quantum entanglement to obtain information from a distant system, so "essentially any quantum system that can obtain information about another quantum system can form the basis for a quantum demon." These demons were not hypothetical. "The systems discussed here," wrote Lloyd, "can be realized experimentally."[197]

By the next millennium, QMDs were no longer a fiction. "Feynman's Demon has recently been built as a nanoscale Brownian motor," stated an article around the turn of the century.[198] Articles covering these advances in the international scientific press (such as *Chemistry World* or *Anales de Química*) and in mainstream news media outlets (including the BBC, CNN, MSNBC, the *Washington Post*, the *Daily Telegraph*, *La Repubblica*, and others) bore headlines such as: "Maxwell's Demon Goes Quantum," "A Demon of a Device," "Molecular Machines: Facing Their Demons," "Maxwell's Demon Tamed," "Demon Ratchets Up Nanotech Revolution," "Demonic Chemistry," "Laws of Nature Survive Attack by Nano Demon," "Opening the Door to Maxwell's Demons," "Maxwell's Devilishly Clever Idea Moves Closer to Reality," "Scientists Build Maxwell's Demon," "Nanomachines: Maxwell's Demon Becomes a Reality," "Maxwell's Demon Created by Scottish Scientists," and "140 Years Later, James Clerk Maxwell's 'Demon' Materializes.'" International headlines announced the "diavoletto di Maxwell diventa realtá," "la diabolica nanomacchina del Signor Maxwell," "Gesetzestreue Dämonen," "Maxwellsmoleculaire duveltje," "El tridente informativo del demonio de Maxwell," and "Демон."[199]

Scientists continued to look for Maxwell's demon in extreme settings. Perhaps he could be found in conditions nearing zero temperature?[200] A large-scale study funded by the European Commission, at more than three million euros, for three-year-long periods starting in January 2013 was aptly titled INFERNOS, an acronym for "Information, Fluctuations, and Energy Control in Small Systems."[201] With these and other initiatives, research on demons continued to permeate future and emerging technologies programs around the globe.

9
Biology's Demons

The idea that Maxwell's demon was a secret agent who kept life going occupied the imagination of some of the most important biologists of the twentieth century. His entropy-defying antics were widely used to explain why certain organisms were capable of producing more energy than they took in, and why they could organize themselves in progressively more complex ways. He was also suspected of having been present at that fateful moment when life itself began. Maxwell's demon did not act alone. Laplace's demon was soon presumed to play a stabilizing role in biological systems.

What is life? In a series of lectures with that title delivered in Dublin in 1943, the physicist Erwin Schrödinger noted that the "chromosome fibers" that carried our hereditary material seemed to behave like the "all-penetrating mind, once conceived by Laplace." Chromosomes were tiny and hard to see. How could they contain such detailed and prescient information? Schrödinger surmised that something in them had to act "causally," so that it "could tell from their structure whether the egg would develop . . . into a black cock or into a speckled hen, into a fly or maize plant, a rhododendron, a beetle, a mouse or a woman." By then, some scientists were calling this hereditary unit a "gene," and the molecule that carried the "genetic" material would soon be referred to as DNA. This was only a beginning. As more and more biologists sought to elucidate the mysteries of inheritance and survival, they studied physicists' demons, hoping to understand living systems.

After World War II, biologists were particularly taken by a "physically respectable cousin" of Maxwell's demon whom they termed "Darwin's demon." This demon controlled the inheritance of good or bad genes. Studies of him led to key breakthroughs in our understanding of genetic transmission and inheritance. Brain scientists soon became interested in these demons, using them to develop new models of intelligence and consciousness.

Renowned anthropologists, philosophers, and sociologists—including Bruno Latour, Michel Serres, and Pierre Bourdieu—studied the actions of demons in laboratories, schools, and offices. At the dawn of the next millennium, some of the most important aspects of our lives were found to be vulnerable to, and at times helpless against, these demons.

At the turn of the century the editor of *Nature*, the renowned astronomer Sir Norman Lockyer, urged researchers to look for Maxwell's demons in living processes: "It has been suggested that vital processes afford the most likely region in which to seek for the existence of Maxwell's 'demons,' and should their non-existence be established, information as to the relative efficiency of the human individual compared with a perfectly reversible thermodynamic engine is much to be desired."[1]

Maxwell's demon could be responsible for breathing life into life. Scientists such as Lockyer started suspecting that he was the thingamajig powering living beings who thrived in a universe marked by entropic descent. Something about life seemed eerily demonic, as if powered by something quite unnatural. In some senses, it seemed to have the upper hand over engines or other physical systems: living systems adapted themselves to changing conditions, they reproduced, they created, and they could be highly efficient systems that at times seemed to create more energy than they took in. For Lockyer, determining their essential features was important curatorially as well, as living systems and their demons could affect decisions about the topics that should be included in the journal *Nature*.

The idea that a vital force powered living organisms was closely associated with the work of the French philosopher Henri Bergson, whose book *Creative Evolution* (1907) took Europe by storm. Bergson fought against a mechanistic understanding of living and evolutionary processes associated with Cartesian philosophy and with certain popularizers of Darwinian evolution. The philosopher introduced his argument with the hypothesis that "Laplace formulated with the greatest precision," citing the relevant passages of Laplace's oeuvre. He challenged the claim that the "past, present and future would be open at a glance to a superhuman intellect capable of making the calculation."[2] From there, Bergson went on to criticize Emil du Bois-Reymond and Thomas Huxley. Without using the term "demon," he boldly denied the possible existence of the *intelligence surhumaine* proposed by Laplace, opening up for his numerous readers the alternative possibility that other demons might be in charge.

Bergson was not the only one contesting Cartesian materialism and Laplace's thesis during those years. The German biologist Hans Driesch, an expert on the complex process of cell division in embryos, clinched the association between the powers of Maxwell's being and those of living beings. In his celebrated Gifford Lectures at the University of Aberdeen, the embryologist lectured about "the 'Demons' of Maxwell" as a "famous instance" of violations of the second law of thermodynamics. Order could arise from disorder in the molecular world as much as in ours. "The 'demon' deals with these molecules," he said, "as our workman deals with his bricks."[3]

Driesch's conclusions represented the culmination of a long search for the forces that organized molecules into living forms. Did this mean that Maxwell's creature really existed? In the German version of his text, Driesch closed with a dramatic assertion: "His [Maxwell's] demons exist: we are them."[4] In the English version, this sentence was missing; the text pointed to the same conclusion, but added a note of caution: "Let it be well understood: Maxwell's argument rests upon a fiction," yet even so, "it may really be applied to life as to a natural autonomous reality."[5]

Under the towering influence of Bergson and Driesch, researchers across the fields—including Alan Archibald Campbell-Swinton (engineer), James Johnstone (biologist and oceanographer), and Ralph Lillie (biologist and physiologist)—were among the many who started to think that living systems and Maxwell's demons had much in common. "Now lately, in London," there was a renewed interest in thinking about life as an entropy-resisting force, "I suppose as the result of Bergson's remarkable writings," explained Campbell-Swinton. "There has been raised quite seriously the question as to whether living organisms are subject to these laws of thermodynamics or not." He surmised that "it is probable that there exist living things which in some fashion or other do very much what was the business of Maxwell's demons to do, and in this manner extract the energy that they require from the general stock." Harnessing energy from these entropy-fighting bio-beings might be the answer to many problems, including fuel scarcity and pollution: "We should only have to cultivate the right kind of organisms in sufficient masses, and they would do all this for us."[6] The possibility of breeding organisms with these enhanced powers soon started to fascinate more researchers.

Similarly, the Scottish James Johnstone focused intently on "Maxwell's famous fiction" to understand entropy-resisting living systems. He concluded that "the irreversibility of physical phenomena, the fact that energy tends to dissipate itself, the second law of thermodynamics, depend on the assumption that Maxwell's demons exist only in the imagination." But what appeared to be true in the realm of "experimental physics" failed when life in general was taken into consideration. "The processes of terrestrial life as a whole are reversible, or tend to reversibility," Johnstone conceded, but "we must not introduce demonology into science." He assured the reader that, "lest this fiction of Maxwell's should savour of mysticism, or something equally repugnant, we shall state the idea in quite unexceptionable terms." This could be done by considering only the "helter-skelter" molecules that sometimes moved as if pushed around by a physically respectable demon.[7]

Lillie, a Canadian physiologist interested in how orderliness could arise spontaneously, weighed the evidence for the claim that certain biological organisms violated the second law and thus might resemble Maxwell's demon. In a *Science* article, he took his colleagues to task for uncritically espousing the idea "that life may play the part of the Maxwellian demon," and he associated this view with an uncritical vitalism. During those years, Lillie had noted a growing agonistic conflict that he described as "vitalism versus mechanism," in which vitalists argued that living systems could never be completely reduced to purely physical and material components. Using that phrase as a subtitle to his article, Lillie at first acknowledged that both sides had valid arguments.[8] But for the most part, the confrontation proved to be a huge boon for the physical sciences, to the detriment of the life sciences. Critics of the latter discipline argued that it had never been able to get rid of mysterious "vital forces" that served as a cover for the belief in intelligent design long associated with God's intervention in natural processes. In the midst of such debates, demons thrived.

How consequential were Maxwell's demons? Where would their influence stop? Could a sorting demon not only give shape to organisms but also be affecting our instincts, actions, and behavior? Could it lead us to act virtuously or viciously? The American zoologist and eugenicist Samuel Jackson Holmes looked at the sorting demon as an enforcer of behavioral traits that could be inherited in the process of evolution: "Apparently we have to do with a selective agency which preserves and intensifies certain kinds of behavior and rejects others on the basis of their results—a kind of 'sorting demon' in the realm of behavior," he wrote in a *Science* article titled "The Beginnings of Intelligence."[9] Noting that "pleasure and pain apparently function as agents for the reinforcement of certain reactions and the stamping out of others," he would soon be seduced into playing the role of this demon to better the human race by means of eugenics. Just as his colleagues in England, such as the eugenicist Francis Galton and his protégé Karl Pearson, worried about the effects of the immigration of Polish and Russian Jews on the British, Holmes became concerned about the dangerous repercussions for Americans of the influx of Mexican immigrants across the southern border.

The relation of Maxwell's creation to Darwin's selecting being started to intrigue many more thinkers during those years. One of New England's most influential Christian preachers, Reverend Samuel Phillips Newman Smyth, followed the debates pertaining to the nature of the "sorting" intelligence behind evolution. He had a clear idea of who that being might be like. Physics provided Smyth with the model of Maxwell's demon: "We may refer for this purpose to an ingenious hypothesis which some speculative trouble with the laws of heat led the great physicist, Clerk Maxwell, to put forth, and which has become known as Maxwell's hypothesis of the sorting demon." "As matter of fact," explained the reverend, "nature throughout its age-long process seems to have been very intelligently sifted."[10] Experts on evolution attributed a similar sorting or sifting action to nature. Physics had its demon, and evolutionary theory would have one too.

WHAT IS LIFE?

Without using the term "demon," the physicist Erwin Schrödinger rocked the imagination of biologists by arguing that living systems were characterized by the interplay between "enigmatic" entropy-defying tricks associated with Maxwell and Brownian demons and the stability associated with Laplace's mind. Immediately after it first appeared, Schrödinger's book *What Is Life?*—based on his series of lectures at Trinity College (1943)—became one of the most influential texts in modern biology.

If defying entropy was one clear characteristic of life, another, according to Schrödinger, was maintaining order. Although "the all-penetrating mind, once conceived by Laplace," was amply questioned in physics, it appeared safe in biology. Schrödinger surmised that something inside our cells seemed to contain powers akin to those of the mind imagined by Laplace. "The structure of the chromosome fibres," he wrote, functioned as a "code-script."

Starting within a fertilized egg, it could tell how an organism would develop as it grew:

> In calling the structure of the chromosome fibres a code-script we mean that the all-penetrating mind, once conceived by Laplace, to which every causal connection lay immediately open, could tell from their structure whether the egg would develop, under suitable conditions, into a black cock or into a speckled hen, into a fly or maize plant, a rhododendron, a beetle, a mouse or a woman.[11]

Living organisms could reproduce in ways that maintained exceptional order. The "code-script" mentioned by Schrödinger was like Laplace's intelligence in that it housed *in nuce* the entire past and future of an organism.

"It is by avoiding the rapid decay into the inert state of 'equilibrium' that an organism appears so enigmatic," explained Schrödinger. "So much so, that from the earliest times of human thought some special non-physical or supernatural force (vis viva, entelechy) was claimed to be operative in the organism, and in some quarters is still claimed." When he investigated the nature of this apparently "supernatural force," Schrödinger discovered that life was alive for one main reason: "It feeds upon negative entropy," he stated.[12]

Not all living organisms were equally successful at creating order from disorder and surviving decay. Schrödinger asked readers to consider the work of Raphael. The difference between a "masterpiece of embroidery, say a Raphael tapestry, which shows no dull repetition, but an elaborate, coherent, meaningful design traced by the great master," and "an ordinary wallpaper in which the same pattern is repeated again and again in regular periodicity" was stark. Both created order, but according to Schrödinger, only the first reduced the entropy of a small section of the universe.[13] Masterpieces in literature and the fine arts were to be admired as much as beautiful living forms for reducing entropy. Could these systems be bred, systematized, or maximized?

Research based partly in physics, partly in chemistry, and partly in biology boomed after the war. Numerous historians have written about the greater interaction between biologists and physicists during these years. When the Washington Conference on Theoretical Physics resumed after World War II, the physicist George Gamow, a Ukrainian-born physicist who had fled the Soviet Union, chose the crossover topic "Physics of Living

Matter." John von Neumann, Leó Szilard, and Niels Bohr all attended and soon thereafter started collaborating with biologists.

Gamow had been unable to obtain a security clearance and was left out of war work. He used his time off productively, writing a book series based on a fictional character named Mr. Tompkins. One of Mr. Tompkins's adventures included witnessing the mischief of Maxwell's demon. Gamow's short story "Maxwell's Demon," published in *Mr. Tompkins Explores the Atom*, described the demon as a "fast fellow" working "the way a good sheep dog rounds up and steers a flock of sheep," but with such "speed and accuracy" that he made "champion tennis players" look "like hopeless duffers." The demon was pictured "as a tall, elegantly dressed butler." He seemed like "Mephistopheles himself, straight out of grand opera." Not being wicked at all, he liked "to play practical jokes." One lazy afternoon, he surprised Mr. Tompkins by playing with the molecules in his evening cocktail. "Holy entropy!" exclaimed Tompkins's father-in-law, who was sitting close by and staring, bewildered, at the highball glass as it acted up. "It's boiling!" he exclaimed.[14] Gamow's story would have been only mildly amusing if not for the fact that it described in entertaining terms what the new knowledge of atomic physics could bring to the world.

"I am supposed to divide my time between finding out what life is and trying to preserve it by saving the world," wrote Szilard to Niels Bohr, years after the war ended. "At present the world seems to be beyond saving and that leaves me more time free for biology."[15] After being released from the atomic bomb project, the physicist shifted his interest to biology, taking up a professorship at the new Institute of Radiobiology and Biophysics at the University of Chicago.

Szilard soon started collaborating with the French biologist Jacques Monod, known among his peers as the architect of molecular biology. Monod was part of a group of prominent scientists who would go on to revolutionize molecular and evolutionary biology, the best known today being James Watson and Francis Crick, who determined the helical structure of DNA. Determining the shape of DNA was only a beginning. For that information to be useful, biologists needed to understand how DNA functioned, and they now faced an entirely new set of questions. Watson and Crick had answered a *what* question. It was now time to ask *how* DNA did its job and *why*. How was DNA copied and replicated? How was DNA selected when two parents were involved in reproduction? What mechanisms blocked or allowed certain traits to be passed on? Soon, physics' demons would enter into the life sciences as Watson and Crick, among many other biologists, avidly read Schrödinger's published lectures and sank their teeth into the science of thermodynamics as they set about answering these questions.

During the war, Monod joined the resistance and would later move to the United States, where he worked at the California Institute of Technology. His work on DNA transcription processes stands out among his many accomplishments. In work done with his protégé George Cohen, Monod identified particular enzymes in *E. coli* as "the Maxwell demons which channel metabolites and chemical potential into synthesis, growth and eventually cellular multiplication."[16] It was the beginning of revolutionary work on genetic regulation (that is, on the relation of enzymes and genes) that would earn him a Nobel Prize. Monod was grateful. In his Nobel Lecture, he credited Szilard for sharing with him the penetrating intuition that would lead to his discovery of the process of DNA.

For hereditary material to work as a form of Laplacian code-script, the genetic code had to be copied over and over again in cells. A single human cell chromosome contained approximately 140 million nucleotides. The average human genome was made up of approximately 6 billion nucleotides. For this number of nucleotides to be copied within a few hours, replication had to occur at ungodly speeds. Scientists soon hypothesized that a tiny being, analogous to Maxwell's demon, operated inside cells, working away at tremendous velocity and reducing the expenditure of energy during this complicated process. Reproduction came at a cost. Copies of copies grew progressively worse, and fidelity was hard to come by. Reproduction and replication consumed energy. The demons found in the gene reproduction and transcription processes were masters of that game.

Before the copying even started, some of it had to be selected to be passed down to offspring. Who did the selecting? To understand these processes, evolutionary and molecular biologists soon posited new demons, named after James Clerk Maxwell, Charles Darwin, Léo Szilard, and Léon Brillouin or given a hyphenated name combining some of these last names. Just as Maxwell's being shuffled atoms to and fro, so Darwin's being shuffled genetic material. The historian of science Evelyn Fox Keller, in her account of how the discipline of biology became centered on the concept of the gene in the context of information theory, explained that "Darwin's

Being resembles Maxwell's Demon in one respect: it, too, acts like a selecting agent."[17]

EVOLUTIONARY BIOLOGY

Evolutionary biology, which is concerned with entire organisms rather than the cells that constitute them, faced some of the same challenges as molecular biology. Some specimens survived while some did not. Viability was hardly random. The continual production of thriving life forms required more than passive filtering. What mechanism or logic lay behind such complex processes? As molecular biologists strove to understand the process of cell reproduction, evolutionary biologists focused on understanding the development of individual organisms and species.

The sole mention of Maxwell's demon caused a stir during the Darwin Centennial Celebrations at the University of Chicago in October 1959; filled with pomp and circumstance, this event attracted thousands of participants and garnered the attention of the international press.[18] Toward the end of the celebrations, the British biologist Conrad Hal Waddington delivered an influential presentation that would be cited over and over through the next decade. Waddington was among a growing number of biologists who had grown increasingly worried that Darwin's concept of natural selection played a role similar to that of Maxwell's demon in physics. "We had to rely on a Maxwell demon," he explained, "and persuade ourselves not merely that natural selection could show some of the properties of such a useful *deus ex machina* but that it had them so fully developed that we needed nothing further. This was a rather uncomfortable position, and we can now escape from it."[19] Like Waddington, a growing number of scientists who analyzed what exactly was selected and what was not had started to realize that Darwin's natural selection concept contained a paradox. "In the recent past we were working with a theory in which the obvious organization of the living world had to be engendered *ab initio* out of non-organized basic components—'random' mutation, on the one hand, and an essentially unconnected natural selection, on the other," Waddington explained.[20] Biologists sometimes stressed the role of randomness and other times that of heredity, without having a fixed criterion for understanding the relation between the two.

The introduction of Maxwell's demon into biology was problematic for evolutionary biologists. In light of the Darwin Centennial discussions, they tried to find an escape from some of the most "uncomfortable" paradoxes of natural selection. For some, such as Waddington when he spoke that day, this meant no longer being able "to think of natural selection and variation as being no more essentially connected with each other than would be a heap of pebbles and the gravel-sorter onto which it is thrown." A fuller understanding of the process would require "a more inclusive point of view" that could not be limited to genetics. The conference opened the doors of the discipline to the idea that hereditary mechanisms were not wholly contained in the genes and could be a product of nature as much as of nurture. "We have, in fact, found evidence for what may be regarded as a 'feedback' between the conditions of the environment and the phenotypic effects of gene mutations," explained Waddington.[21] According to some, the answer to the riddle of fitness should no longer be sought just in genetics or within the cell, but in the larger context. Soon scientists would coin the term "epigenetics" to explain the outside factors affecting heredity—including the environment and even the particular behavior of a species or individual—that might influence gene expression.

The provocation was exciting to many, but it also threatened to turn investigations about evolution into studies about nearly everything, including society, religion, politics, and history. Where should biologists draw the line? How did their discipline relate to the social sciences, the humanities, and even the arts? Waddington tried to answer some of these questions by advancing the field of theoretical biology, an approach that had to contend with the antics of Maxwell's demons inside living cells as much as outside of them.[22]

DARWIN'S DEMON

To escape some of the problems that the concept of Maxwell's demon introduced into biology, the doyen of biology at Princeton University, Colin Pittendrigh, came up with "a physically respectable cousin of Maxwell's." Pitt, as he was commonly known, would reach fame as "the father of the biological clock." He had been discussing Maxwell's demon with physicists for years and had found that a similar demon, which he decided to call Darwin's, could be useful for understanding evolutionary biology. If Laplace's demon ensured continuity in the universe throughout time, and Maxwell's demon could enable the occasional appearance of organized beings from random ones, Darwin's demon guaranteed that certain organisms would prevail over others by working directly with genes, its new favorite *materia operandi*.

Pittendrigh later recalled that the idea of Darwin's demon first arose in conversation over a "bottle of good claret" imbibed during an evening with the physicist and Nobel Prize–winner Wolfgang Pauli.[23] In that conversation, Pittendrigh described Darwin's demon as "a physically respectable cousin of Maxwell's" because, like the latter, it fully obeyed the laws of physics. Pauli, according to Pittendrigh, had profound reservations about how most biologists understood the concept of life: their discipline was built, he argued, around concepts that seemed to violate the second law. Pauli used an argument common among physicists to buttress the status of their discipline. Biology had long been the target of accusations that it relied on mysterious "vital forces" and was therefore weaker scientifically. Pittendrigh refused to accept such criticisms, even coming from such a prominent physicist. If physicists had come to accept Maxwell's demon, why would it not be possible for biologists to have one of their own? "Darwin's Demon," argued Pittendrigh, "is as physically respectable as Szilard's version of Maxwell's."[24]

Pittendrigh, Monod, and many others of their generation hoped to set biology on a new theoretical footing, one no longer beset by some of the problems associated with vitalism and intelligent design that had haunted the discipline since the nineteenth century.[25] Pittendrigh and Monod could not have been more different from each other, but the two scientists read and admired each other's works. While the Frenchman's underground wartime activities had led the Gestapo to suspect him, the British-born Pittendrigh was a conscientious objector. Both were concerned with regulation and environmental determinants of genetic processes. Pittendrigh tended to look to factors outside of organisms (such as the effects of light on organisms' circadian rhythms), while Monod looked inside cells. Pittendrigh wanted to learn more about the relation of internal biological characteristics to the environment, asking how biological clocks reacted to the changes in light and darkness associated with daytime and nighttime. Monod, in turn, studied the relation of gene expression to particular enzymes.

Pittendrigh popularized the term "Darwin's demon" in a lecture delivered to the prestigious Harvey Society in New York on January 19, 1961, and in the subsequent publication in the society's yearly volume titled "On Temporal Organization in Living Systems." Therein he explained that living systems did not violate the second law because the action of Darwin's demon "was limited to that small enclave of nature we say is living." Just like Maxwell's, Darwin's demon could create free work "locally, but only by increasing entropy in a larger context."[26]

In these texts, Darwin's long-buried "selecting being" who could direct evolution and even create a new race reemerged in a "physically respectful" form. Pittendrigh explained that Darwin's genius was based on his tacit reliance on this being: "The real stroke of genius in the 'Origin of Species' is its implicit substitution of natural selection for some Maxwellian Demon as the architect of living organization."[27]

Pittendrigh had been thinking intently about "the concept of information as negative entropy," which had "been developed in recent years by engineering physicists concerned with communications networks and the design of automata."[28] This exercise had led him to defend a particular insight that would revolutionize biology for the next decades: that evolutionary success should be studied at the level of genes rather than at the level of individuals, groups, species, societies, nations, or cultures. Characteristics that were evident handicaps to survival, like "the plumage of birds of paradise," could be understood as by-products of the struggle for survival of particular genes.[29] A chicken, Pittendrigh reminded his audience, could be just the means for an egg to reproduce itself; an egg could just be the means for genes within it to reproduce themselves.[30] "In a very real sense," explained Pittendrigh, "the developed organism is no more than a vehicle for its genotype."[31] "Darwin's demon" reappeared in the context of genetics and in the "selfish gene" idea, a term that would later be popularized by the biologist Richard Dawkins in his book of that title. That concept considered many organisms, including humans, mere tools for advancing the master plan of highly efficient genes.

Darwin's demon led scientists to the idea that biologists had initially gotten their understanding of the fittest all wrong by relying on traditional human-level categories to judge others and ourselves. Genes, argued some evolutionary biologists, dominated those who possessed them. Unbeknownst to themselves, individuals were instructed to carry out the replication and survival plans of their genes, even if it came at the cost of self-sacrifice. Instead of being guided by moral ideals, they might only be conduits—cogs in their genes' master plan. What appeared to us from a distance to be an unadaptive trait could simply be due to something else flourishing because of it. The evolutionary success of a certain group could be guaranteed at the expense of some of the individuals, typically the "nice guys" of that group. Altruism or love might appear selfless at the level of the organism, but it was not necessarily so at the level of gene expression. The discovery of Darwin's demon by biologists changed the scale at which they evaluated success and failure in nature. According to some biologists, the famous struggle of the fittest should no longer be understood in terms of individuals, such as brave soldiers who sacrificed themselves for their platoon or nation, or through any other moralistic or theistic lens. Our traditional way of attributing vice or virtue could be entirely misguided. A growing number of biologists questioned the worth of traditional values such as honor, mercy, and compassion once they began to see Darwin's demon at work.

Pittendrigh's published lecture opened with a section enigmatically titled "The Handiwork of Darwin's Demon." "The Demon is economic, or unimaginative, on the whole," he explained.[32] The handiwork of this frugal being was none other than the process that created life by passing the genetic code from cell to cell and, with slight modifications, from organism to organism. "The Darwinian Demon has major merits Maxwell's lacks," he explained. For one thing, it was physically real and not hypothetical. Second, "it works repeatedly with no new information." Before going into the technicalities of how the demon functioned, Pittendrigh explained what he was able to do. He was the gatekeeper to life, a bouncer in the coveted club of the living, "standing on the threshold, as it were, of each new generation in the world of life and granting favored entrée to those systems with the more *appropriate* store of information." Who did he let in, and who did he keep out? Did he favor the smartest, the strongest, the fastest? Neither, as it turned out. The only thing the demon cared about was "the perpetuation of the most efficient reproducer." Darwin's demon's only mission was to ensure the survival and reproduction of genes *à toute force*, even with no regard for the reproducer. It determined who was to be the one to survive the previously most successful survivor; he was the winner who won at all costs and who cared only about winning, the ultimate master of the game of mate selection and one-upmanship. What criterion did he use? Pittendrigh admitted that the demon needed to make a certain "value judgment" to operate, yet that judgment was based only on "trivial information." He could make decisions based on a single bit of information. "The single bit of information he exploits," he explained, "is this one criterion of reproductive success." Pittendrigh's demon followed one rule (perpetuating reproductive success) and needed only one piece of information (who was the most efficient reproducer), yet his actions gave rise to the entire living world throughout all of history: "The architecture of living systems is thus the Demon's handiwork, and it reflects both his limited criterion and limited raw materials," Pittendrigh concluded.[33]

With opposition to the Vietnam War growing during the late 1960s, it was a good time for rethinking altruism, civil rights, group identity, and social allegiances. The use of the term "Darwin's demon" increased with the publication in 1966 of an influential book by the evolutionary biologist George Williams, who proposed a radically new understanding of virtue, sex, and survival—all by reference to this creature. Unbeknownst to themselves, argued Williams, all biological creatures were under the control of a creature they did not know or understand. "I believe," Williams concluded, "that the sterility of the workers is entirely attributable to the unrelenting efforts of Darwin's demon to maximize a mere abstraction, the mean."[34] In the coming years, Williams's book would become required reading for defenders of the "selfish gene" theory. In Williams's view, social insects, such as bees, were clearly not selfish, but neither were they selfless drones sacrificing themselves for their queen. Rather, they were driven by their genes, which in turn were driven by Darwin's demon, a creature who cared only about maximizing the chances of survival of certain genes.

Biologists' scrutiny of physicists' knowledge of demons soon included quantum mechanics. In trying to answer the general question of whether nature was probable or capricious, the evolutionary biologist Richard Lewontin reviewed the current knowledge of demons. He started with Laplace's demon and his ability to predict where a die would land on the casino gaming table. "If there were a demon capable of comprehending all the important information about a die," he wrote, "including the exact forces involved in throwing it, its shape, the properties of the gaming table, and so on, then that demon could foretell exactly the outcome of the cast." And although Maxwell's demon showed the need to think probabilistically, even he could not shoo away determinism entirely: "There is a great similarity between Maxwell's demon and Laplace's," Lewontin concluded, as they both considered chance a measure of our ignorance. For Laplace's demon, that was a problem, but it was an opportunity for Maxwell's.[35] The research of quantum physicists invalidated the lessons of both: "Laplace and Maxwell postulate demons that do not exist in fact; Schroediner [sic] and Heisenberg deny the possibility of their existence," wrote Lewontin.[36] By comparing the rate of gene changes across organisms and throughout generations, the biologist concluded that gene variation in particular populations was extremely capricious, even when their environment proved to be comparatively stable. Genetics too had to contend with the effects of quantum indeterminacy and uncertainty.

Environmental changes could also take on the role of selecting agent affecting who survived. The effects of the pesticide DDT on the environment convinced a molecular biologist that the chemical was a "sort of reversible Maxwell's demon."[37] During those years, other biologists started to notice that certain body parts were designed to act like Maxwell's demons: eyelashes swept and propelled particles in a one-way direction while bushy patches of hair, like eyebrows, kept debris away from entering select areas. Intercellular cilia and other membranes at less respectable orifices functioned much like Maxwell's demon trapdoors. The success of these mechanisms keeping things out or letting them in explained much about the evolution of the natural world.[38]

MODERN DEMONOLOGY

The idea that Darwin's demon was central to evolution soon reached a broader audience with the publication of "The Modern Demonology" by the science writer and professor of biochemistry Isaac Asimov, first written for lay readers of his "Science" column in the *Magazine of Fantasy and Science Fiction*. The piece would later be reprinted in one of Asimov's most famous essay collections.[39] Asimov had been thinking about demons while writing his textbook *Life and Energy: An Exploration of the Physical and Chemical Basis of Modern Biology*, which covered in detail the science of Maxwell's demon. Reviewing the creature's long history, he described situations ruled by chance as those where "it is a drunken Maxwell's demon we are dealing with." His staggering drunken meandering explained chance perfectly: "when no step has any necessary relationship with the previous one is sometimes called a 'drunkard's walk.'"[40] Thus intoxicated, this demon was the ultimate scrambler, not only of information but potentially even of our fate and future.

This demon helped Asimov answer a different question: what makes certain organisms survive and reproduce, while others die off? While Maxwell's demon guaranteed the creation of order (while his intoxicated alter ego led to the opposite outcome), this demon led to evolutionary success.

The long-term effects of "natural selection," in Asimov's view, were due to the actions of someone "capable of picking and choosing among mutations, allowing some to pass and others not." Asimov bestowed a famous surname on this picker-and-chooser: "The English naturalist Charles Robert Darwin discovered the demon, so we can call it 'Darwin's demon' even though Darwin himself called it 'natural selection.'"[41]

"To be sure, the demon does not exist," explained Asimov. But could he be summoned into existence? "Mankind's scientific ability is constantly increasing, and the day may come when he will be able, by some device, to duplicate the demon's function." The engineer in us, *homo technikos*, labored heartily to imitate demons. "In the case of Darwin's demon," he wrote, "it is not a matter of imitating the demon, but, rather, of stultifying it." Our efforts could allow those with "bad eyes to get along by means of glasses, diabetics to get along by means of insulin injections, the feeble-minded to get along by means of welfare agencies." But beating demons was not that easy. Our success at imitating one demon could present another demon with opportunities. "The destruction of our technological society in a fit of nuclear peevishness would become disastrous even if there were many millions of immediate survivors," worried Asimov. Those who had adjusted well to city life would be endangered: "The environment toward which they were fitted would be gone, and Darwin's demon would wipe them out remorselessly and without a backward glance."[42]

Could thermodynamics, by itself, be used to explain the emergence of genius? Asimov asked readers to consider the arrangement of words in Shakespeare's plays. Most English-language dictionaries contained each and every word of *Hamlet*. Shakespeare's genius consisted in placing them in a particular and peculiar order. By shuffling them so, he created a work of art and a work of genius. Did Shakespeare decrease the entropy of the universe? "He ate no more, expended no more energy, than if he had spent the entire interval boozing at the Mermaid Tavern," explained Asimov. "Obviously, the words in the plays represent a much higher and more significant degree of order than do the words of the dictionary." There is no denying that in his "cast of twenty and in the space of three hours," Shakespeare managed to portray humanity in a way that "any group of twenty real people could possibly manage in the interval of three real hours."[43]

How did he do it? The only explanation Asimov could adduce was that Darwin's demon, working behind the scenes favoring some over others, at some point let the right genes pass. Some nine months later, Shakespeare was born. "As far as I know, I am the only one who has called it that and drawn the analogy with Maxwell's demon," wrote Asimov.[44] He was wrong. The term was spreading quickly among physicists and biologists alike.

PREDATORS, NICE GUYS, AND SELFISH GENES

Chance and Necessity, written by Monod for a general public, assured Maxwell's demon of a seat at the table. The tome was hailed by Francis Crick for being "written with force and clarity." He also lauded it for explaining a "central vision of life," which, although considered "strange, somber, arid and austere" by most readers, was "shared by the great majority of working scientists of any distinction."[45] Monod expressed the utter futility of returning to "Laplace's world, from which chance is excluded."[46] He also combated the idea that if "certain initial conditions could be formulated, [a universal theory] would also contain a cosmology which would forecast the general evolution of the universe." Instead, the biologist dedicated an entire chapter to Maxwell's demon.[47] Monod conveyed to the public what he had learned from him. "Destiny is written concurrently with the event, not prior to it," Monod explained.[48]

Monod explained that he had found a demon working within cells that was "far more clever" than Maxwell's unwitting creation. Needing a new name for it, he decided to name him after his friend Szilard and his colleague Brillouin. The "Maxwell-Szilard-Brillouin demon" could be found within the proteins of living systems that acted in intelligent ways, following instructions and reproducing them in ways that seemed to violate the second law.[49] Monod focused on how "proteins exercise their 'demoniacal' functions," which permitted them to replicate.[50] "It is the primary structure of proteins," he explained, "that we shall consult for the 'secret' to those cognitive properties thanks to which, like Maxwell's demons, they animate and build living systems."[51]

Living protein demons (Maxwell-Szilard-Brillouin demons) rivaled electronic circuits in their sorting abilities. Yet they were orders of magnitude smaller and much lighter. Enzyme molecules weighed about a "million billion times less than an electronic relay," yet they acted in similar ways, "receiving and integrating inputs from three or four sources, and responding with threshold effect." Monod concluded that "polypeptide fibers," which carried genetic information, "play the role Maxwell assigned to his demons a hundred years ago."[52]

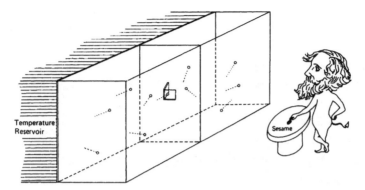

Figure 14. Harold J. Morowitz, *Entropy for Biologists* (New York: Academic Press, 1970), 109.

How did life first emerge in the universe? The question had fascinated thinkers since biblical times. Origin-of-life scientists soon started to suspect demons. How did the earliest life forms that arose from the primeval soup attain characteristics that led to life? The Nobel Prize–winning chemist Ilya Prigogine, known for developing chaos theory, marveled that, even though life's "chances of occurring were next to nil," during the first years of the universe it did indeed occur.[53] Prigogine was among the many scientists in those years who were enthralled by the possibility of exploring knowledge across the boundary between biology and physics. He cited approvingly Monod's *Chance and Necessity*, one of the French scientist's "many authoritative books," for claiming that the existence of life in the universe was likely due to chance.[54] Prigogine reprinted Monod's explanation: "Our number came up in the Monte Carlo game. Is it any wonder if, like the person who has just made a million at the casino, we feel strange and a little unreal?" he asked.[55] Incredible luck had resulted in the emergence of life as we know it. Prigogine added another element to scientists' understanding of this fateful low-probability occurrence.

What "would appear as a miracle"—"like an army of Maxwell's demons working against the laws of physics to produce life"—had a natural explanation that explained the origin of life in the universe.[56] According to Prigogine, the "army of Maxwell demons" that biologists saw as necessary for powering living systems could arise spontaneously in far-from-equilibrium environments.[57] This demon army flourished amid chaos and created order out of it. Prigogine's *La Nouvelle alliance*, coauthored with the philosopher Isabelle Stengers, investigated "those demons that populate the expositions of classical science," including "Laplace's demon, Maxwell's and the God of Einstein," and then proceeded to "exorcise" one of them. Laplace's demon "haunts our imagination, resurges repeatedly, and with it, the nightmare of the insignificance of all things, the solitude hallucinated by those who for a longtime believed they lived in a world built according to their own dimensions."[58]

These questions motivated John Hopfield, a theoretical biologist from Princeton University and Bell Labs, to delve deeper into the study of the Maxwell's demon–like characteristics of cell reproduction processes. His now-famous "kinetic proofreading" article revolutionized the current understanding of error correction processes in biochemical reactions.[59] Even the lightning-fast and super-efficient protein reproduction processes, he argued, should not violate the second law. They were subjected to a delay, which was responsible for the "sense of the direction of flow of time" and which consumed energy. "The mechanism which is to generate the delay," Hopfield concluded, "must consume free energy in order not to be a Maxwell demon."[60] Although some energy was lost, the consumption was exceedingly small. These systems' reproductive abilities were as efficient as anything could get in the universe.

GAME THEORY

During these years, Manfred Eigen, a Nobel Prize–winning expert on cell biology and origin-of-life biochemistry, named one of the new demons that emerged from studies of cell reproduction after Monod. Eigen had embarked on his career with the goal of "uniting Darwin's evolution principle with classical information theory and—after applying this concept to molecular self-organization—providing a quantitative basis for molecular biology."[61] His ambitious coauthored book *The Laws of the Game: How the Principles of Nature Govern Chance* explained to the general public the laws that brought all these disciplines together. The book's goal was to help readers better understand the complex play of lawfulness and lawlessness in nature. When exactly was ascribing the emergence of certain effects to pure chance warranted? When should we instead see them as resulting from fixed laws? Although the universe had evolved and changed, Eigen proposed that certain rules limited the manner in which it could transform itself. "Our world resembles a vast game in which nothing is determined in advance but the rules, and only the rules are open to objective understanding," he explained.[62]

Demons illustrated these rules perfectly. Little did it matter for Eigen whether his readers believed in their actual existence or not; what was important was advising them about what these demons were capable of, how they were limited, and how they were free to rumble. These demons, he explained, were "postulated only to explain certain paradoxes and not because their 'inventors' really believed in them."[63]

Three demons explained the three most essential aspects of nature. The first one, Maxwell's demon, explained the dissipative one-way direction of nature. The second one, Loschmidt's demon, showed its reversible aspects. The third one, Monod's demon, created the seemingly irreversible effects often attributed to living beings. Eigen described these creatures as "chained demons" (*Gefesselte Dämonen*) who could play lots of tricks but were not free to play just any card. The first creature in Eigen's unholy trinity, Maxwell's, illustrated what "every refrigerator knows," that pockets of entropy reversal could be achieved by raising entropy levels outside of an enclosure. Loschmidt's demon had other time-twisting abilities and elflike qualities. To picture him, Eigen urged readers to "imagine an elf that can activate a time switch that suddenly reverses the flow of time." To describe Monod's demon, Eigen quoted passages directly from *Chance and Necessity*, where Monod considered enzymes demonic because of how they could "slide back down the 'slope' of entropy, i.e., somehow move backwards in time." Monod's demon could beat randomness by choosing "those few and valuable mutations that turn up among innumerable others that macroscopic chance produces." This demon's entropy-defying qualities gave rise to "selective evolution" as "a kind of machine that permits us to move backwards in time."[64]

While biologists and chemists paid the most attention to Maxwell's demon (often via Monod's Maxwell-Szilard-Brillouin demon and Eigen's Monod's demon), evolutionary biologists continued to use the term "Darwin's demon," or "Darwinian demon," to explain particularly successful seducers, seductresses, survivors, and predators.[65] Darwin's demon came back strong in this context, growing in significance alongside those from physics and computer science.

Donn Eric Rosen, the chairman of the Department of Ichthyology at the American Museum of Natural History, focused on Darwin's demon in a book review in *Systematic Zoology*. He cited Waddington's claim that biologists tacitly relied on a Maxwell demon to explain how mutations were selected and passed on to others. Problems with trying to understand the logic behind natural selection had not gone away. The problematic position described by Waddington two decades earlier perfectly captured "the real problem with Darwinian selection theory," which "is that it can explain everything, and therefore, nothing." Fitness was defined by reference to survival, and survival by reference to fitness. The argument was circular and thus theoretically defective. "This was a rather uncomfortable position, and we can now escape from it," wrote Rosen. The escape from circularity was not that easy. The concept of Darwin's demon proved to be dangerously divisive amongst biologists. In Rosen's view, it sent "natural selection off to join the ether, phlogiston, and noxious vapors."[66]

Eigen offered a widely popular solution to mend the rifts in his field. In his *Steps towards Life*, the renowned chemist took another look at Monod's "fascinating and ambitious" work and his assumption that enzymes "take on the part of the Maxwell-Szilard-Brillouin demon and convert random fluctuations into information." His colleague, Eigen concluded, was "clearly on the right path." He argued that "selection is more like a particularly subtle demon that has operated on the different steps up to life, and operates today at the different levels of life, with a set of highly original tricks." This demon acted on the fly, and "above all," wrote Eigen, "it is highly active, driven by an internal feedback mechanism that searches in a very discriminating manner for the best route to optimal performance." But if it seemed to be advancing toward a predetermined goal, this was a false impression: in fact, there was no real goal at the end of the process. "It possesses an inherent drive towards any predestined goal, but simply by virtue of its inherent non-linear mechanism, which gives the appearance of goal-directedness," he explained.[67]

What was most exciting about such demons was how they could be used to answer some of the deepest questions in science and philosophy, particularly those pertaining to the relations between biology and physics, life and matter, and the living and the dead. "What is the nature of the physical forces that oppose the tendency towards equilibrium, and thus sustain an extremely improbable state?" Eigen asked. "Can biology be reduced to physics at this decisive point?" The answer depended on the nature of the demons under consideration. If these demons were purely physical, then reduction was possible. "If not, then we would have to accept the presence of a demon that operates outside the physical laws."[68] Eigen's "if" was an intriguing conditional on which the central tenets of modern biology hinged.

NEUROSCIENCE

Programmers' "best friends" successfully migrated to popular accounts of cognitive psychology in the 1990s, when their characteristics and proclivities were used to describe general thought processes. As the use of daemons and demons became commonplace in computer culture worldwide, they started to affect the discipline of neuroscience, to change our understanding of the evolution of intelligent life on earth, and even to explain consciousness. Prominent public intellectuals filled auditoriums, lecture halls, and page after page promoting their theories of intelligence, evolution, religion, history, morality, and free will by reference to computer demons. Computer programmers of earlier decades had studied psychologists to determine how to design computers with AI, but by the end of the century the situation had reversed: conceptually savvy psychologists studied how computers were programmed to better understand minds.

The nature of consciousness and how it first appeared in the universe had baffled philosophers for centuries, but Daniel Dennett found a new angle for understanding it. His book *Explaining Consciousness* appeared to great fanfare in 1991. "Homunculi—demons, agents—," he wrote, "are the coin of the realm in Artificial Intelligence, and computer science more generally."[69] Just as these demons could be used to create intelligent systems out of mechanical ones (creating minds out of brains), so too could they be used to create consciousness. "If one can get a team or committee of relatively ignorant, narrow-minded, blind homunculi to produce the intelligent behavior of the whole, this is progress," he explained.[70]

Dennett knew more about AI than most philosophers from working with computer scientists (including Allen Newell, John McCarthy, and John Haugeland). Like them, he found demons useful for thinking about intelligence, minds, consciousness, materialism, and idealism. According to him, Descartes's demon had helped scientists zero in on the brain as an object of study. "The brain in the vat," he wrote, "is a favorite thought experiment in the toolkit of many philosophers." It is "a modern-day version of Descartes's (1641) evil demon, an illusionist bent on tricking Descartes about absolutely everything, including his own existence."[71] Once the principal functions of the brain had been isolated, neuroscientists were able to move to a second, higher goal: understanding how the brain could give rise to thought and consciousness.

The task at hand first required breaking down complex thought processes into simpler ones. For this, Dennett looked at how AI demons were employed by computer programmers. He admired intelligent systems that functioned most efficiently when organized in a "flow chart" and when "a committee of homunculi (investigators, librarians, accountants, executives)" worked together. Dennett's homunculi were fancy AI demons who relied on less fancy ones, who relied on even less fancy ones. "The function of each is accomplished by subdividing it via another flow chart into still smaller, more stupid homunculi."[72] They were allowed to "interact, develop, form coalitions, or hierarchies, and so forth," but for them to collaborate well, programmers needed to keep these lumpen from fighting with each other or with their superiors. "Competition between homunculi is as tightly regulated as major league baseball," explained Dennett.[73] The trick behind this model of intelligence lay, according to him, in "organizing armies of idiots to do the work."[74]

How ignorant and narrow-minded were the demon agents whose abilities could be harnessed to produce intelligence? "All they have to do," explained Dennett, "is remember whether to say yes or no when asked."[75] They needed only to possess "a modicum of quasi-understanding."[76]

"Joke-detecting demons" explained Dennett, could be simply built by equipping a computer with "demons sensitive to the negative connotations of ethnic-joke telling."[77] Before anyone worried about unconscious bias entering into computer programs, scientists worked hard to introduce it in them. Progress in the difficult work of teaching computers to detect jokes, humor, or irony was slow, but progress in AI exploded when it came to image recognition. The *New Scientist* covered recent advances in "parallel-processing systems analogous to simple neural networks" based on "programs, which could recognise simple patterns, were made up of several independent information-processing units, or 'demons,' and a 'master demon.'"[78]

The term "Darwin's demon," which had mainly been used by insiders and practicing scientists, reached an even wider audience in response to new work published by the biologist Richard Dawkins and by Steven Pinker, a leading authority on cognitive psychology and an accomplished author. Dennett, Dawkins, and Pinker appeared frequently onstage in front of packed audiences confronting some of the most intractable puzzles of the natural world, including consciousness, intelligence, evolution, and the origin of life. Dawkins had gained notoriety decades earlier with the publication of his best-selling *The Selfish Gene*. Part of the excitement around that book had stemmed from its explanation of who did or did not survive in this world by reference to an internal mechanism of genes rather than in traditional moral, social, or political terms. When years later Dawkins was accused by his critics of challenging "the social morality of altruism taught by Christianity, and indeed by every religion in one form or another," he responded by referring to one of the classic texts that had first shown the need to think about evolution at the level of genes, George Williams's *Adaptation and Natural Selection*, which had introduced the term "Darwin's demon" back in the 1960s.[79]

Pinker noted that attempts to understand how our minds worked were now easier thanks to "an inventory of stock demons that do handy things" at the disposal of scientists as much as engineers. "A new way of understanding human intelligence has been born," he explained in *How the Mind Works*, where he referred to cognitive psychologists as "mental software engineers" and "reverse-engineers." Their AI programming techniques helped him "reverse-engineer the psyche" to elucidate how it functioned.[80] The task was much easier than it had been in the past because now they had "a good parts catalogue from which they can order smart demons."[81]

Computational demons traveled under many names. "Data structures are read and interpreted and examined and recognized and revised all the time, and the subroutines that do so are unashamedly called 'agents,' 'demons,' 'supervisors,' 'monitors,' 'interpreters,' and 'executives,'" explained Pinker.[82] They were part of "a production system" that contained "memories" and a set of "reflexes, sometimes called 'demons' because they are simple, self-contained entities that sit around waiting to spring into action."[83] For the most part, they looked up information to find matches. "One of the demons," explained Pinker, "is designed to answer these look-up questions by scanning for identical marks in the Goal and Long-Term memory columns." With a "demon that springs into action" and "a demon that does the checking," the computer—or the mind-as-a-computer—could answer specific questions about the data set that were not completely specified beforehand.

"A computational demon" is smart because it sits "at the top of the chain of command," explained Pinker, so that "the intelligence of the system emerges from the not-so-intelligent mechanical demons inside it."[84] A demon always "shunts control to the loudest, fastest or strongest agent one level down."[85] Once a match is made by a demon, another demon down the line is needed to prevent "the demon from mindlessly making copy after copy like the Sorcerer's Apprentice."[86] The way out of idiocy requires exploiting progressively more idiotic demons: "If we seem to need a smarter demon, someone has to figure out how to build him out of stupider ones."[87] Intelligence allegedly arose from the small actions of hierarchically organized chains of stupid demons, modeled after the built-in computer architecture of those years.

Traditional theories of mind, argued Pinker, had to rely on the problematic concept of the "ghost in the machine." The phrase was used to highlight the paradox of thinking of minds as those possessing powers greater than the powers that could be reduced to biology and physics. Pinker's work purportedly drove away this ghost. "It would not be a ghost in the machine," insisted Pinker, because "demons" could show how thought (and computers) emerged from "just another set of if-then rules."[88]

Could scientists use "demons" to fight away these "ghosts"? If successful, the strategy could be employed to further a secularist and materialist agenda. Pinker argued that the demons that gave rise to thought processes were ultimately reducible to mechanical systems: "In figuring out the mind's software, ultimately we may use only demons stupid enough to be replaced by machines."[89] The basis of intelligence, according to this model, was purely mechanical. "Eventually," Pinker explained in a chapter titled "Aladdin's Lamp," "you have to call Ghostbusters and replace the smallest and stupidest demons with machines—in the case of people and animals, machines built from neurons: neural networks."[90] He cited with approval Dennett's model of "nesting boxes" of homunculi where the very last one "lands you with homunculi so stupid" that "they can be, as one says, 'replaced by a machine.'"[91] By analyzing "thought and thinking" in terms of the computational demons created by software engineers, Pinker argued that these processes were allegedly "no longer ghostly enigmas but mechanical processes that can be studied."[92]

But how did computer demons come to represent that which resided in our brains and led us to think? How could software become something more than a set of instructions? "You still might wonder," Pinker wrote, "how the marks being scribbled and erased by demons inside the computer are supposed to represent or stand for things in the world." Pinker argued that these scribbles might just be symbols, but that, as such, they stood for real things as well as any other—just as "*mother* means mother, *uncle* means uncle, and so on."[93] In his view, the scribbles made automatically by demons working inside a computer stood for actual thinking brains in the same way that other symbols stood for something else. To posit a gap between the two—to sow a rift between demons' representations and the things they represented—would hurt his thesis. Pinker, like Dennett before him, had little patience with those who invoked Searle's demon with such an intent. He dismissed Searle's argument as an overly simplistic and commonsense understanding of understanding. Mocking Searle's Chinese room argument, he portrayed it as exclaiming: "Aw, c'mon! You mean to claim that the guy understands Chinese??!!! Geddadahere! He doesn't understand a word!! He's lived in Brooklyn all his life!!"[94] Pinker would alternate between debating and dismissing Searle's argument for decades to come. Neither he nor the philosopher budged from their initial positions. Pinker would continue to argue in favor of understanding intelligence by reference to demon-based computer programs, while Searle would insist that no such programs could be deemed to be in possession of intelligence of their own.

Notwithstanding the acrimony they incited among scientists and philosophers, AI demons became so powerful during those years that they continued to be used to explain evolution. Dennett's next book, *Darwin's Dangerous Idea*, argued that biological systems might also evolve in ways comparable to the actions of AI demons. "The field of Artificial Intelligence is a quite direct descendant of Darwin's idea," he wrote.[95] Dennett was most taken by the checkers-playing program developed by the machine intelligence pioneer Arthur Samuel, who had been able to create a program that played better than the person who programmed it. "Samuel's legendary program is thus not only the progenitor of the intellectual species, AI, but also of its more recent offshoot, AL, Artificial Life," wrote Dennett, who attributed the program's success to its use of demons. Dennett found that the key to its operation lay in the high number of all possible moves allowed by the game being "ruthlessly pruned by semi-intelligent, myopic demons, leading to a risky, chance-ridden exploration of a tiny subportion of the whole space."[96]

In addition to praising demon-based approaches to machine learning, Dennett admired Eigen's account of natural selection as due to demonlike actions, which Eigen had included as part of his system of "chained demons." In this view, both AI systems and natural organisms could evolve into something that had not been predetermined and surpass the limitations of their own creators. "As we shall see," wrote Dennett, "the capacity of computers to run algorithms with tremendous speed and reliability is now permitting theoreticians to explore Darwin's dangerous idea in ways heretofore impossible, with fascinating results."[97] One of the most stunning conclusions of these investigations was that evolution need no longer be thought of as guided by the vision of a supreme being such as God to reach a particular end. Instead, Dennett's understanding of these processes was based on the possible actions of "semi-intelligent" and "myopic" demons powering all intelligent systems, from computers to living cells.

By the last decade of the century, Darwin's demon was alive and well, traveling under different aliases. Pittendrigh, who had used the term for more than three decades, moved west after retiring from Princeton to become director of the Hopkins Marine Station at Stanford. From there, he reviewed what he had learned from the demon after a lifetime of working and teaching in the field and published his "Reflections of a Darwinian Clock-Watcher." The biologist noted that, when communicating with the public, scientists usually offered other "less sinister" labels than "demon" and came up with "several pen-names." According to Pittendrigh, one alias was "the blind watchmaker."[98] The phrase was popularized by Dawkins, who used it as the title for a best-selling book that sought to clip the wings of the long-standing tradition of viewing the universe as created by God and humans in God's image, opting instead for a gene-centric view that saw neither God nor ourselves as in charge.

In other texts, Pittendrigh continued to express ideas similar to those of Williams and Dawkins. Darwin's demon was responsible for selecting genes and deciding who got to pass them on: "And so it is with Darwin's Demon," Pittendrigh explained, "who stands on the threshold of each new generation granting favored entree to the offspring of those members of the previous generation that were the better reproducers." Darwin's demon was simple and dull, but his determination compensated for what he lacked in imagination and wit. "This single, mindless discrimination is an inevitable consequence of the reproductive process that alone assures perpetuation of the vignette, keeps it in print."[99] The biologist argued that the best way to obtain knowledge about life, its origins, and its future was to inquire about the demon's style, work habits, and motivations:

> How did the Demon get started on programming a day within? What were the earliest selection pressures? Were they different from those prevailing today? Is their impact, given the Demon's lack of imagina-

tion, still detectable in the life of contemporary cells? Knowing his style, what can we expect?[100]

At their most basic, his actions were limited to selecting and copying. "The primary purpose of biological organization is self-perpetuation by self-copying, and as such is the handiwork of Darwin's Demon." He copied genes worth copying, ignoring the others, knowing exactly how to distinguish between the signal and the noise in the perpetuation of life. If one could say that "the genome was the program of a Turing machine," then we could equally say that "Darwin's Demon was the programmer." His "ultimate purpose" in life was clear: "keeping his vignettes in print."[101]

The Princeton professor conceived of "the living world as a vast literature comprising millions of volumes, many still available but even more out of print." They were written sequences of nucleotides: "All are vignettes written in the universal language of nucleotide sequences. All have the same happy ending (reproduction) which, when reached, assures the volume stays in print." Darwin's demon was the "author" of this literature, and "unlike its (divine) Publisher," he was "both knowable and known, and some knowledge of his style is useful to the physiologist as well as to the naturalist."[102] The method behind his madness was not easily, and perhaps not possibly, wholly decipherable: "The literature written by the Demon is no more deducible from a complete command of the nucleotide language, let alone physical law, than the works of Shakespeare or Alfred North Whitehead are deducible from a complete command of the English language."[103] Constrained to interpret his work as one would analyze great works of art or philosophy, Pittendrigh resigned himself to the possibility of never fully comprehending "the Demon."

SOCIAL LIFE

The effects of science's demons were felt in offices, laboratories, and schools as managers, anthropologists, philosophers, and sociologists started studying them. Could their impact on society in general be managed? Starting in the 1960s, corporate managers adopted efficiency models based on Maxwell's demon to run their businesses. Jack Morton, who headed the electronics division of Bell Labs, was one of the most prominent voices to boast about his managerial role being that of a Maxwell's demon. Morton understood his responsibilities as shifting "hot," active, and creative personnel hither, moving those who did repetitive tasks thither, always promoting, demoting, and firing according to demonic rules of efficiency. By acting like a demon, he hoped to manage the scientists under his watch into perpetual innovation.

Morton spelled out how such a feat could be achieved. In various articles, such as "The Manager as Maxwell's Demon," he explained his management principles.[104] Morton portrayed "the manager's job" as "the innovation of innovation." His advice carried a lot of weight in corporate circles, since some of the most influential technological wonders of the century, including the transistor, had been developed during his tenure. The discovery of the transistor had happened at his company, he argued, because "the managers were the 'Maxwell's Demons' who encouraged the acceptance and application of the new, 'hot' ideas generated by creative specialists."[105] Morton's *Organizing for Innovation*, which would become a bible for corporate success, explained that a large research corporation can flourish when it "emphasizes the 'master ecologist' or 'Maxwell Demon' role of the manager" and allows each "master" demon to accumulate the most productive personnel within the enclosure of his domain.[106]

Anthropologists grew curious about science's demons. Could they be used to explain why "scientific" cultures differed from "backward" ones? Perhaps the former employed demons more effectively? Bruno Latour, a young student of theology turned ethnographer before later turning to philosophy, noticed that scientists were acting like Maxwell's demons in their labs. In *Laboratory Life*, published in 1979, he argued that a lab was an "enclosure" similar to the "oven" in which demons operated. Instead of accumulating energetic molecules, the scientist–as–Maxwell's demon accumulated information and knowledge. A well-organized science lab, according to Latour, was a "trapping system for any competent Maxwell's demon wishing to decrease disorder."[107]

When the French anthropologist peeked into one of the most successful and cutting-age laboratories of his time, the Salk Institute in La Jolla, California, he found scientists working away like demons. The Salk Institute had hosted some of the century's most productive scientists, including Szilard and Monod. For two years, from 1975 to 1977, he "followed scientists around" at the prestigious laboratory in the Torrey Pines Mesa, noting with care what they did throughout the day. Latour concluded that the actual job of scientists was essentially one of labeling and sorting, sorting and labeling. "The sorting of facts from the background noise" was generously rewarded, "often heralded by the Nobel Prizes and a flourish of trumpets." Maxwell's demon kinds of activities, he concluded in a separate publication, were responsible for the accumulation of high-tech knowledge in certain places and not others and fostered global inequality as they created rifts in the world between countries with thriving scientific cultures and those without.[108]

Philosophers paid attention. What makes a society prosperous? What makes it learned? What is the relation between scientific practices and wealth creation? One of France's most renowned thinkers, Michel Serres, detailed the powers of Maxwell's demon throughout history in his book *The Parasite*. Serres traced the lineage of Maxwell's creature back to ancient Greece. "Hermes is the god of the crossroads and is the god of whom Maxwell made a demon," he explained.[109] "Maxwell's demon checks the permits, acting as a customs officer," first for molecules, but then affecting the world in general by creating stark imbalances throughout the universe.[110] Like his predecessor in Greek mythology, this disequalizer tended to be positioned at crossroads or gates (including inside semiconductors), creating power imbalances so that "a very few people manage to enslave the greatest number—more or less all of humanity."[111] He held the keys to the gates of heaven on earth. Serres argued for the importance of thinking about the effects of Maxwell's demon actions writ large—whether creating one-way flows, asymmetric balances, or irreversible access points—on the position of everything and everyone from the powerful politician to the lowly ant. A simple reversal could turn predator into prey and king into pauper. "To work is to sort. Maxwell's demon is unavoidable," he concluded.[112] It was an intriguing framework for understanding what society and civilization were all about.

By then, even the eminent French sociologist Pierre Bourdieu used the "metaphor" of Maxwell's demon to explain social inequalities and injustices in society. In his theory of action, the culmination of a lifetime of work, he started by explaining the correlation between educational levels and class differences. No matter how much educators tried to give underprivileged students tools for social advancement, opposite forces seemed to keep them from succeeding. "The school system acts like Maxwell's demon," wrote Bourdieu. "It maintains the pre-existing order, that is, the difference between students with unequal amounts of cultural capital," he concluded. The way the demon operated within institutions was important for understanding how societies functioned. "More specifically, through a series of selection operations, it separates the holders of inherited cultural capital from those who lack them," he explained, and therefore the demon "tends to maintain preexisting social differences."[113] There was an army of these demons. "The more or less coarsely orchestrated actions of thousands of Maxwell's little demons," he noted, "tend to reproduce this order without knowing it or wanting it." Bourdieu cautioned against attributing agency to these demons and considering them as something alien and separate from society, noting that doing so could lead to conspiratorial thinking.

"The metaphor of the demon is dangerous," wrote Bourdieu, "because it favors the phantom of conspiracy, that frequently haunts critical thought, the idea of a malicious will who is responsible for everything that happens, for better or worse, in the social world." His choice of such a term came solely from a desire to "describe a mechanism" for these processes. Those who were caught on demons' bad side experienced life as an "infernal machine." Their effects were felt "as a tragic set of gears, outside and above any agents, because each of the agents is, in a way, forced to participate, to exist, in a game which imposes on him immense efforts and immense sacrifices." For this reason, he concluded, "many speak of the 'hell of success.'"[114]

Recourse to Maxwell's demon to explain all sorts of imbalances continued to be common in specialized circles across fields.[115] More and more sightings occurred inside laboratories, institutions of higher education, computers, brains, and bodies, where demons could be found situated at borders and crossroads, sometimes intercepting individuals, materials, merchandise, or messages, and other times letting them through.

10

Demons in the Global Economy

Can humans innovate in perpetuity? How can they solve the global challenges facing the planet? Will the invention of new demons or the employment of old ones help them?

Twentieth-century growth economists considered "innovation" in terms of the possibility of inventing new technologically advanced demons who could clear the path to an Age of Abundance by unlocking nearly unlimited energy supplies or recycling existing ones in perpetuity. These economists drew from a long tradition based on invoking hidden and invisible forces to understand complex market systems.

An "invisible hand" first appeared in the eighteenth-century writings of the British economist Adam Smith. It ostensibly organized every economic transaction undertaken by each and every individual for private gain into a collective force for the greater good. Laissez-faire economic optimists tended to believe in its existence, while pessimists did not. As global markets grew more complex, many other agents were suspected to be at work in the economy.

In the twentieth century, physicists' demons were the main suspects. If the universe was subject to their whims, the economy was too. From the post–World War II period to the end of the century, the economist Paul Samuelson and a small group of influential colleagues considered their possible actions. The economist Nicholas Georgescu-Roegen stood out among Samuelson's coterie for his concern with depletion, scarcity, inequality, and the health of the planet. These prominent economists studied the latest research from physics to understand the basic elements of finance, such as noise trading and insider trading, and essential economic laws. Financiers looking to line their pockets studied demons as they searched for "Fortune's formula." Maxwell's demon could buy low and sell high, accruing slight differences systematically, slowly but surely. Could he be put to work to transform a pittance into a fortune? Laplace's demon was often cited in attempts to forecast economic trends, take advantage of incoming opportunities, and avert crises and crashes. Critics of the "innovation economy," who read the earth's balance sheet differently, insisted that no such demons could exist and that no such technological wonders would ever come to our rescue. By the time the new digital economy rolled in, some of its most successful companies, such as Microsoft and Amazon, were seen to operate as these demons.

As policymakers sought to better understand the economy, a series of market crashes rocked the twentieth century, leading to the accumulation of wealth into fewer hands and heightening mistrust in experts' ability to steer the planet in the right direction. As a result, experts, technocratic governments, world banks, and monetary policymakers—all central to modern neoliberal societies—found their efficacy and credibility questioned.

Once demons escaped from physics to other disciplines, they never returned. The discovery of nuclear power during World War II led many scientists and economists to believe that other similarly portentous discoveries lay in wait, that future innovations would follow as predictably as previous ones had, and that the economy would grow almost infinitely alongside them. Scientists, economists, technocrats, and policymakers became enthralled by the promises of the innovation economy. Optimistic believers thought that our wits could always come to the rescue so that the economic potential of human capital would always eventually outpace the strain of overpopulation. Even when finding ourselves in a pickle—including one as dramatic as an economic depression—our ability to innovate could help us think ourselves out of it. Pessimists, however, saw a world of limited resources, including intellectual ones.

Growing the economy was on everyone's mind. So was beating the markets. After the war, a small group of economists discovered that knowing about the demons of physics could be useful for learning how to make money, how to break even, and how to lose it. They turned to physics for guidance, sinking their teeth into the discipline's most informative literature and doling out financial advice based on it. The policies they advocated were widely implemented.

Were economic processes fair? Should financial markets be regulated or left alone? Should the demons that were suspected of playing a role in the economy be countered by interventionist financial regulation and strong monetary and tax policies? Or could they be put to good use?

Four principal demons informed the dominant economic theory of those years, mostly associated with the work of the economist John Maynard Keynes. One of the defining features of Keynesian economic theory was the idea (traced back to Szilard, Brillouin, and others) that information had a quantitative value that needed to be taken into consideration to understand how the economy functioned.[1] Laplace's demon was a super-forecaster. It could buy low *and* sell high by having all the information about what would happen next and thus being able to foresee what the stock market would do the next day. It could plan for any booms or busts and bear or bull markets that lay ahead. Maxwell's demon could disrupt economic equilibrium with knowledge or information. By going long on "hot" stocks and short on "cold" ones, it could redistribute wealth by redirecting the flow of money as efficiently as a one-way valve. He was so important that an economic historian recently described the twentieth century as marked by "the proliferation of tiny footprints" of this "unusually tiny *picaro*" who left a "scattering of Demon tracks" in "geographically separate and intellectually diverse territories" far beyond physics.[2] A third demon "far more miraculous than Maxwell's" could recycle resources indefinitely throughout the universe, ridding the world of scarcity and pollution. The old demon of chance was still around. It had a good side: by acting randomly, it could prevent anyone from churning a profit forever and ever. Somebody else might soon get their chance. When it came to financial success, sheer luck played an important role, and perhaps the rich might be little more than *just* lucky.

RANDOM WALKS DOWN WALL STREET

The term "demon of chance" made its debut in economic theory to describe statistical uncertainties in the fluctuations of stock prices. The economist Maurice Kendall introduced the term in a groundbreaking paper detailing this demon's actions on the financial markets. The economist and Nobel Prize laureate Paul Samuelson recalled the furor that arose when it was first presented in 1953: "The economists who served as discussants for Kendall's paper were outraged," he wrote, adding, "Such nihilism seemed to strike at the very heart of economic science."[3]

Kendall tried to predict the pattern that stocks would follow by examining the path they had traveled previously. If only he could figure that out, he would earn a hefty buck. Kendall's "demon of chance" investigations upended the common understanding of wealth and talent distribution in society. "That price changes of common stocks and commodity futures fluctuate somewhat randomly, something like the digits in a table of random numbers or with algebraic sign-patterns like that of heads and tails in tosses of a coin, has commonly been recognized."[4] Many of those making money were not doing so because they were making smart bets; they were just benefiting from positive long-term trends. Kendall used Britain's first programmable computer, the electronic ACE of the National Physical Laboratory, to crunch more numbers for the stock market than anyone had ever done before him. His examination of the price fluctuations of the Chicago wheat market for fifty years proved that most of the movement that stocks exhibited was random.

Kendall concluded that it was impossible to make a profit just by following these fluctuations. "It is impossible to predict the price from week to week from the series itself," he concluded.[5] He insisted that numerous speculators, stockbrokers, and economists who believed that price fluctuations were linked to past performance were dead wrong. Most upward or downward changes appeared to be random, without rhyme or reason:

> The series looks like a "wandering" one, almost as if once a week the Demon of Chance drew a random number from a symmetrical population of fixed dispersion and added it to the current price to determine the next week's price.[6]

Kendall's audience watched in awe as he tried to disentangle long-term trends from short-term random fluctuations. A professor who attended Kendall's presentation at the Royal Statistical Society explained that economists "have been told to be aware of a fearsome devil—serial correlation—and to look to statisticians to cast the devil out from them." But once they learned what statisticians had actually found out, they were in shock: "It seems now that the more statisticians work on the casting out of devils, the more they cast out everything else, including what the economists want."[7] What economists learned from Kendall was eventually incorporated into the "efficient market hypothesis," which would underpin economists' understanding of financial markets for the rest of the century.

Samuelson, whom many of his peers considered to be smarter than almost anyone else, was quite taken by Kendall's "demon of chance."[8] In years to come he would think of this demon in relation to Maxwell's demon, perhaps in part because he himself had suffered greatly from the economic woes of his family. When he turned two years old, his family could no longer afford him, and so he was sent away to live in a foster home. "This whole adventure cost one dollar for food, lodging and love," he recalled later in life.[9] Throughout his life, the economist would be singularly concerned with figuring out how people made it big in life, and how much of their success was due to circumstances, brains, talent, luck, or information. "It is not easy to get rich in Las Vegas, at Churchill Downs, or at the local Merrill Lynch office," he noted.[10] Who did beat the odds? The question obsessed him. To find the answer, he first had "to go off the main turnpike of my discipline of economics" and explore more fully "James Clerk Maxwell's image of a Demon who cheats the second law of thermodynamics."[11]

In the 1960s, "a paper by Paul A. Samuelson on Brownian motion" started circulating among a select circle of economists.[12] Samuelson had figured out that, mathematically, the challenge of predicting stock prices was similar to those faced by physicists trying to determine the trajectory of Brownian motion particles and those examined by Wiener and others who tried to predict the path of airplanes to shoot them down.[13] These paths were so hard to follow that they could only be described using nonlinear mathematical methods (developed in the late nineteenth century by the mathematician Louis Bachelier and others).

SCARCITY VERSUS GROWTH

Physics demons, it was soon discovered, affected the economy beyond fiddling with stock prices. The Harvard-educated, Romanian-born economist Nicholas Georgescu-Roegen made a bold claim in 1966. Most economists, he argued, had gotten "the piquant fable" of Maxwell's demon all wrong. "Time and again," he warned, "we can see the drawback of importing a gospel from physics into economics and interpreting it in a more catholic way than the consistory of physicists."[14] Growth-focused economists were using Maxwell's demon to justify investment in research and development (R&D) in science and technology. Ever since Maxwell first wrote about the demon in 1871, explained Georgescu-Roegen confidently, "the fable has been the object of a controversy which, I submit, is empty." Admitting that, "like many other paradoxes, Maxwell's is still a riddle," the economist stressed that "the fable cannot serve as a basis for any scientific argument."[15]

Samuelson was intrigued. From then on, the demons of physics would never be far from his mind. Samuelson and Georgescu-Roegen had learned from physicists that the demon beating the second law could create a profit out of nothing. They had also learned that the demon could operate only if it had information telling it where molecules were. They had discovered, too, that this information was thermodynamically costly. Applying these insights to an understanding of the markets and the economy led them to conclude that the only way an investor could systematically beat the markets was if he could act with information that was not yet available to or actionable by the rest of the public. That would be unfair. Was the rest a fair game?

After leaving his native Romania, Georgescu-Roegen studied with Karl Pearson in London and became an admirer of *The Grammar of Science*. He then moved to Harvard and eventually took up a professorship at Vanderbilt University, where he referenced demons while loudly protesting against the prevailing neoclassical economic policies of advanced industrial capitalist countries. Throughout his life, Georgescu-Roegen tried to relate economics back to the laws of thermodynamics, arguing that even revolution and class struggle had to be understood in their context. His chapter "From the Struggle for Entropy to Class Conflict" in *Analytical Economics* (1966) explored the links between the two.[16]

According to Georgescu-Roegen, the basic action underpinning all economic activity was "sorting." For this reason, *homo economicus* could be understood as a Maxwell's demon. "The economic process," explained the economist, "depends on the activity of human individuals, who like the demon of Maxwell, sort and direct environmental low entropy according to some definite rules."[17] Georgescu-Roegen considered not one but many "Maxwellian demons," where "not all Maxwellian demons are identical." They ran the gamut of sorting activities, from our cells to our instincts to our minds. "In the case of a single cell, the corresponding Maxwellian activity seems to be determined only by the physico-chemical structure of the cell; in the case of a higher organism, it is a function of its innate instincts as well."[18]

Most important was that "the Maxwellian activity of man depends also on what goes on in his mind, perhaps more on this than on anything else."[19] What were the limits of our intellectual capacity, if any? Could humans always find new ways of creating wealth? "The history of human thought," he admitted, "teaches us that nothing can be extravagant in relation to what thought might discover or where."[20] If so, why focus on scarcity?

The Wall Street crash of 1929 caught almost everyone by surprise. Our intellectual prowess for predicting the future might not have been as strong as once thought. Georgescu-Roegen warned against believing in the "Laplacian demon," or even in a much less powerful demon. "A demon having an ordinary mind deprived of any clairvoyance, but lasting millions of aeons and capable of moving from one galaxy to another, should be able to acquire a complete knowledge of every transient process," he explained. Yet the theory did not match the reality on the ground. "Perhaps," he ventured, "the exceptional properties with which we have endowed our demons violate some (unknown) laws of nature, so that his existence is confined to our paper-and-pencil operations." "The most optimistic expectations do not justify the hope that mankind endowed our demon," he concluded.[21] Unlike the mythical Cassandra, whose oracles were always right but never believed, Laplace's demon was trusted more often than he was correct. Most economists, starting with Léon Walras, had been held in thrall to him.[22] To prove the unlikeliness of his existence, Georgescu-Roegen invited his readers to travel down the familiar pathway of imagining yet another demon: "But let us imagine a new demon, which with the speed of thought can make all the needed observations" of this system "and communicate the solution to everyone concerned." If that were the case, "an individual who comes to experience a new economic situation may alter his preferences," with the result that "*ex post* he may discover the answer he gave to our demon was not right." The changes would then have to be factored in. "Consequently," explained the economist, "our demon will have to keep on recomputing running-away equilibria, unless by chance he possesses a divine mind capable of writing the whole history of the world before it actually happens." The economist ruled that out that possibility. If it could do that, then he would "no longer be a 'scientific' demon," as only a "divine mind" could be clairvoyant in that manner.[23]

Laplace's and Maxwell's demons were not the only demons on his mind. Georgescu-Roegen also considered "Karl Pearson's argument that to an observer traveling away from the earth at a greater speed than light, events on our planet would appear in the reverse order." Yet "since Pearson wrote we have learned that the speed of light cannot be exceeded."[24] In consequence, scientists had learned that time flowed inexorably in one direction. Another famous physics lesson intrigued Georgescu-Roegen even more. He considered the maxim that "no matter how small the probability of an event is, over Time the event must occur infinitely many times." This meant that eternal recurrence, an effect associated with the work of Poincaré and later with Zermelo's demon, would necessarily occur. "Over the limitless eternity, the universe necessarily reaches chaos and then rises again from the ashes an infinite number of times," he wrote. The "thought that the dead will one day rise again from their scattered and pulverized remains to live a life in reverse and find new death in what was their previous birth, is likely to appear highly strange," but it was based on the best science of the day. Georgescu-Roegen was among those who had strong reservations about such a possibility, yet he noted that those "who unreservedly endorse this view do not constitute a rare exception."[25]

The Entropy Law and the Economic Process (1971) appeared in print a few years after *Analytical Economics* (1966). Prefaced by Samuelson, it brought fame to its author. Rising social inequality had made it urgent for economists to understand what made a fair market and a just society. Interventionist regulatory policies faced off against laissez-faire, trickling, invisible hand, and deregulatory philosophies.

Within a few months of his book's publication, Georgescu-Roegen was drawn into the "limits to growth" debate. The debate escalated with the publication of an influential report with that title in 1972 by the Club of Rome, an influential group of academics, politicians, and economists who had come to dire conclusions about the economy based on computer simulations modeling the depletion of natural resources across the planet.[26] For the rest of the century, "ecological economics" and "environmental economics" would compete against "neoclassical economics," dividing the discipline into conflicting camps. In Georgescu-Roegen's new book, many of his earlier points about demons were restated, amplified, and expanded to include the research on demons by Szilard, Lewis, Brillouin, Eddington, and Planck. Taken together, their insights provided ammunition against "standard economics." To these scientists, Laplace's demon was only a servant. "Laplace provided the demon with all the theoretical knowledge necessary to accomplish its task of water boy, so to speak," as Georgescu-Roegen put it. This demon was a simpleton, kowtowing to a higher-up boss. "For this is what Laplace's demon is, a fantastically more efficient water boy than any human, but still a water boy." "Planck's demon," in contrast, was much stronger, more like "the headman who planned the fabulous expedition" and who could "find out . . . how things are related in the universe." His job was bigger: "In contrast with Laplace's, Planck's demon must take over all the duties of the scientific expedition—not only to measure all that is measurable at the atomic level and beyond, but also to frame the right questions about 'motives,' and, above all, to discover the strict causal relationships between a human's thoughts (fleeting or abiding), his motives, and the vibrations of his ganglia (delicate or violent), on the one hand, and the apparent or repressed manifestations of the individual's will, on the other." Georgescu-Roegen emphasized that, "in evaluating Planck's argument, we should first note that it differs fundamentally from Laplace's." He did not believe that this demon would ever succeed in determining the laws of the universe, because perhaps there was no such knowledge to be had. "What if the returning demon tells us that the supposed laws do not exist?" Georgescu-Roegen asked.[27]

In January 1973, the stock market started a perilous decline. It would not halt for almost two years, and nearly half of its value would be wiped away. The Bretton Woods agreement that had been in place after World War II to regulate currencies across nations was in shambles. The bear market was made worse by the oil embargo imposed by many Arab nations against the United States for supporting Israel during the Arab-Israeli conflict. Making money by betting on the stock market was no longer as simple as it had been during the preceding decades.

Samuelson and others' research on the random fluctuations of stock market prices reached a larger audience with the publication of *A Random Walk down Wall Street* by Burton Malkiel. That book famously argued that "a blindfolded monkey throwing darts at a newspaper's financial pages" to pick which stocks to buy and sell would outperform most analysts' portfolios. "The original illustrative analogy of a random walk concerned a drunken man staggering around an empty field," wrote Malkiel as he delved deep into Samuelson's research.[28] He repeated the claim that most stocks fluctuated much like perfectly aimless particles or ambulating drunks, making any mathematical prediction of their next step impossible. "Analysts in pin-striped suits" would fare no better than "bare-assed apes," he concluded.[29] The economist Eugene Fama elaborated on these lessons in his famous article "Random Walks in Stock Market Prices," arguing that the only way analysts could beat the markets was the same way Maxwell's demon could beat entropy: by having new and not-as-yet-discounted information. "If the analyst has neither better insights nor new information, he may as well forget about fundamental analysis and choose securities by some random selection procedure," he concluded.[30] The science of Maxwell's demon now taught economists that noise traders, although necessary for market efficiency, were hopeless at making a profit in the long term. Yet somehow, somewhere, someone beat the odds.

Samuelson studied enough physics to figure out just who could have a consistent advantage in financial markets and why. "Any subset of the market which has a better ex ante knowledge of the stochastic process that stocks will follow in the future," he explained, "is in effect possessed of a 'Maxwell's Demon' who tells him how to make capital gains from his effective peek into tomorrow's financial page reports."[31] After studying the science on the creature, Samuelson concluded that no Maxwell's demon could operate indefinitely without having privileged information. Yet his conclusion, however strong, did not mean that *some* could not have the qualities of Maxwell's demon: "It is not ordained in heaven, or by the second law of thermodynamics, that a small group of intelligent and informed investors cannot systematically achieve higher mean portfolio gains with lower average variabilities." In fact, he pointed out, Morgan Bank and T. Rowe Price were doing just that. "Any Sheik with a billion dollars" could imitate their strategy.[32] Just as the effects of Maxwell's demon were dampened outside of their enclosure, these advantages were exceptional when considering the market as a whole.

From his office at MIT, Samuelson had a front-row view of the latest mathematical research and computational techniques. "Once upon a time there was one world of investment," he wrote. "Now there are two worlds." The "academics with their mathematical stochastic processes" competed against Wall Street's stock and bond salesmen. The two worlds "are still light-years apart: as far apart as the distance from New York to Cambridge; or, exaggerating a bit, as far apart as the vast width of the Charles River between the Harvard Business School and the Harvard Yard."[33]

Experts disagreed on fundamental issues and on the interpretation of physics' demons, and two camps would emerge during the annual meeting of the American Economic Association in May 1974. One focused on the promises of clean energy, renewables, limitless resources, while the other was concerned with pollution, degradation, and scarcity. Did demons justify a pessimistic focus on wear-and-tear or an optimistic orientation toward technological innovation?

The meeting's opening lecture was delivered by Robert Solow, who occupied the office next to Samuelson's at MIT. Chief among technology's optimistic apostles and widely respected for having taught economists how to factor technological innovation into their theories, Solow had received the Nobel Prize for showing that, "in the long run, technological development is the major factor behind economic growth." He would now weigh in on the question of entropy and economic decline and inequality in a talk entitled "The Economics of Resources or the Resources of Economics." He acknowledged the importance of budgeting in "the laws of thermodynamics and life," which "guarantee that we will never recover a whole pound of secondary copper from a pound of primary copper in use, or a whole pound of tertiary copper from a pound of secondary copper in use."[34] "There is leakage at every round," he noted. How could these losses be staved off? Their effects were not as dire, he responded, because "the likelihood of technical progress, especially natural-resource-saving technical progress," would save the world. "Resource-saving inventions" could come to the rescue.[35] "If the future is anything like the past," explained Solow, "there will be prolonged and substantial reductions in natural-resource requirements per unit of real output."[36] But would the future after 1970 be indeed like the past?

A DEMON "FAR MORE MIRACULOUS THAN MAXWELL'S"

Georgescu-Roegen was particularly critical of the "growthmania" preached by most neoclassical economists during those years. It was unsustainable, he protested. Among those he called out was the ecologist Howard Odum, one of the main representatives of systems ecology, for giving the public the illusion that economic growth was inexorable. By using evocative cyclical "energy diagrams" to map paths of energy flows throughout systems, Georgescu-Roegen argued, ecologists were giving readers incorrect ideas about sustainability. Of course, there were no easy answers. Even the possibility that extraterrestrial life forms operated in different ways would have to be fully considered before conclusions could be set in stone. Georgescu-Roegen admitted that "thermodynamics has also a wrap of mystery, for it still does not tell us whether or not its laws are valid for extraterrestrial forms of life." The question was not irrelevant, as "the famous paradox of Maxwell's demon bears on this very issue; hence, the arguments claiming to have resolved it are perforce unavailing."[37] Yet these unresolved issues did not justify what the economist perceived as the unwarranted techno-optimism of his era: "The overpraised and oversold technological developments of our era should not blind us" to the reality of limited natural resources on a strained planet.[38] Solar power and nuclear power were all the rage in the European markets that year. Most economists continued to dream of ever better ways to use and prevent the dissipation of energy and develop technologies that could do what the demon did. Georgescu-Roegen remained unconvinced.

Georgescu-Roegen found evidence of the demon's many failures "all around," including in general "wear-and-tear," "resource depletion," "pollution," and "oxidation, chipping, blowing, and washing away." Limited resources were becoming increasingly hard to ignore. Without the intervention of this being, it was clear to him that "any piece of armament or a two-garage car means less food for the hungry of today and fewer plowshares for some future (however distant) generations like ourselves." He concluded, "What the world needs most [is] a new ethics."[39]

Other scientists weighed in, offering their views about how their work led to innovation and to wealth creation. The nuclear physicist Alvin Weinberg was particularly vocal on this topic. He noted that policy recommendations were increasingly divided between "two conflicting views" that "dominate our perception of man's long-term future." On one side were "the 'catastrophists,'" who "believe that the earth's resources will soon be exhausted and that this will lead to the collapse of society." On the other side were "the 'cornucopians,'" who trusted "that most of the essential raw materials are in infinite supply: that as society exhausts one raw material, it will turn to lower-grade, inexhaustible substitutes." Weinberg sat firmly in the latter camp. In an essay published in the *American Economic Review*, he admitted that what was at stake was much more than the possibility of "recycling" resources. "When an intelligent being sorts used material into separate bins, he diminishes the entropy of the original waste material," he explained. "However, such a macroscopic Maxwell's demon does not change the entropy of mixing appreciably." To do so would require preventing "useful materials from becoming dissolved as individual molecules, which then can be widely dispersed in soils or in the oceans."[40] Weinberg believed that potent alternatives to fossil fuels were right around the corner. Only petty practical difficulties currently kept us from "achieving technological heaven-on-earth." The nuclear physicist optimistically believed there were "no insuperable technical bars to living a decent rather than a brutish life."[41]

Weinberg's sterling credentials (nuclear expert at Oak Ridge National Laboratory, coauthor with Eugene Wigner of the first textbook on nuclear reactors, president of the American Nuclear Society, director of the Office of Energy Research and Development, and, later, director of the Institute for Energy Analysis at Oak Ridge Associated Universities) lent credibility to his pronouncements on the economy. Additionally, he was one of the first scientists to worry about climate change. "At what rate of energy production would the ice caps melt?" he asked. "Will the carbon dioxide or dust thrown into the atmosphere by the burning of fossil fuel threaten the stability of the weather system?" he wondered. Answers did not come easily. "The problem of global effects of energy production, like so many long-range environmental problems," wrote Weinberg, "is everyone's problem, and therefore no one's problem."[42]

Confronting such challenges required knowing more about the demons of physics. Georgescu-Roegen continued to argue that most science advisers and policymakers were mistaken about the role played by Maxwell's demon on the global stage. Even in the face of growing economic crisis, "decision-makers of the most influential economical associations" continued to bet on a future of growth.[43] "Let us remember Maxwell's demon," the economist pleaded, "the demon presupposed to separate the fast-moving from the slow-moving molecules of a gas."[44] The second law limited his powers. "In a world in which that law did not operate, the same energy could be used over and over again at any velocity of circulation one pleased and material objects would never wear out." Yet the law held, and Maxwell's demon illustrated how.[45]

"Without the Entropy Law, you would not dare take a bath," Georgescu-Roegen warned. "One half of the water might become by itself so hot as to scald your neck and the other so cold as to frostbite your toes."[46] He reviewed current research on the demon: "It is now generally believed that this miraculous demon has been 'exorcised' so that, like any other living creature, it must consume a greater amount of available energy than it creates by the separation of 'hot' from 'cold' molecules." Yet some economists and nuclear power optimists such as Weinberg often either forgot their lessons or did not include them fully in their general accounting sheets. As an example, Georgescu-Roegen pointed to the typical accounting table used by economists, whose columns of data for controlled energy, capital goods, consumer goods, recyclable wastes, and population gave the impression that waste and degradation did not matter. "The miraculous features of our demon," he stated, "are tacitly implied in many ideas concerning the unlimited renewability of material resources." That table, wrote the economist, "depicts, in fact, such a demon."[47]

Not only did some economists seem to act as if Maxwell's demon was perfectly capable of solving all energy problems, but they apparently believed in the possible existence of an even more marvelous creature who, unlike a fingered demon, could reverse chemical combustions to recycle energy. "Now, to separate a mixture of, say, nitrogen and oxygen, we need a demon *far more miraculous than Maxwell's*."[48] This being would not be content just to let most fast molecules pass. "Indeed, Maxwell's demon does not have to bring back absolutely every molecule to its initial container. Moreover, it may safely leave some 'hot' or 'cold' molecules in the wrong container; it is only the average speed that counts." The new being could catch each and every one of them: "Our new demon, on the contrary, must not leave even one single molecule mixed with those of the other kind."[49] Could such a demon be found or brought into existence?

When Georgescu-Roegen went through the specific ways in which this demon could be materialized, it evaporated. "To exorcise it, we must not only supply it with enough energy but also endow it with a material existence. And since matter keeps dissipating, the problem now boils down to whether our demon can recycle itself completely while recycling the gas mixture, also completely."[50] The "ghost of friction" would inevitably show up, and the demon would never be able to finish its perfect recycling miracle. "Friction thus appears as a ghost, so to speak, in the backstage of the thermodynamic setup, a ghost that robs us of available energy."[51]

The dream of "unlimited renewability" was simply too seductive: "No wonder that the miraculous features of our demon are tacitly implied in many ideas concerning the unlimited renewability of material resources."[52] Georgescu-Roegen repeated his invitation to change policy to reflect limited natural resources. "My position has been (and still is) that the Entropy Law is the taproot of economic scarcity."[53] Other economists soon joined Georgescu-Roegen's camp. Among them was the prize-winning economist Herman Daly: "Only a Maxwell's Sorting Demon could turn a lukewarm soup of electrons, neutrons, quarks and whatnot, into a resource," he wrote. But "the entropy law tells us that Maxwell's Demon does not exist," insisted Daly, adding that "it is a serious delusion to believe otherwise."[54]

SOCIAL MAXWELL'S DEMONS

While these economists worried about poverty, scarcity, and pollution, some optimistic scientists believed that computers could take on the role of Maxwell's demons and mitigate the problems facing the economy. Such was the claim advanced by Weinberg in a keynote lecture to the annual national conference of the Association of Computing Machinery in Nashville, Tennessee, on October 27, 1980. "A Family of Maxwell's Demons," as his talk was later titled, could be gainfully employed.[55]

In "previewing the Computer Age," Weinberg analyzed a host of demons working away in the world around him. Irrespective of where energy came from, be it from nuclear reactions or fossil fuels, computers could be employed to make more efficient use of it. Audiences that day heard that microprocessors were the prime example of "Macroscopic Maxwell Demons" that could bond together as "System Maxwell Demons" to completely transform contemporary society. "I do not think I am stretching a point too much," he told the audience, "by referring to such microprocessor control systems as macroscopic Maxwell Demons."[56] "System Maxwell demons" could flourish when microchip-based technologies were coupled with older electric and mechanical systems to make the latter more efficient and control larger and larger systems. "Into 1980 fell the 109th birthday of Maxwell's Demon," explained the physicist, who stressed the geopolitical significance of the demon's birthday.

General Motors soon included microprocessors in its cars to adjust the air-to-fuel ratio in carburetors and optimize combustion. What more could be done? One of Weinberg's favorite examples was Trans World Airlines' new flight management computer, which successfully reduced fuel use by a couple of percent points simply by optimizing flight routes and cutting slack in flight paths. At the top of these interrelated demon systems were "Social Maxwell Demons," which acted directly on society by prodding people to act in certain ways. Such demons brought about policy changes and could even limit fuel consumption and curb pollution. Social Maxwell demons of this kind "transmitted signals to innumerable users that it is to their advantage to buy cars that are fuel efficient." They "sent signals to Detroit: design a fuel-efficient car."[57] The hollering of social demons affected the design and use of technologies, the transportation industry, and ultimately the global economy. Weinberg mentioned the importance of the "17 million barrels of oil that flowed each day through the Straits of Hormuz before the Iran-Iraq war" and the consequences of losing that supply.[58] The final lessons Weinberg drew from his analysis of the social Maxwell demon pertained to the need for "replacing oil with nuclear- or coal-generated electricity."[59]

In light of the increased attention to the impact of Maxwell's demon on the economy, ecologists were forced to contend with his limitations. Odum, once a target of Georgescu-Roegen's criticisms, included a section on

Maxwell's demon in his book *Ecological and General Systems*, where he admitted that the creature's deficiencies were apparent "in a system of larger scale" that inevitably led systems away from equilibrium in the long term.[60] His impact was also analyzed in an article in *Science* (1984) on "thermodynamics and economics" that rehearsed the economic lessons of "a demon who can segregate fast and slow molecules." According to the researchers, investment in science and technology would pay off through counterdepletion. The authors proposed R&D as an "information resource" that would create the necessary "negentropy" to "replace the negentropy endowment of coal and oil." "If future increases in accumulated negentropy continue to keep pace with the depletion of resources," they argued, "we can look forward to a steady increase in the standard of living." But "if they do not, the future will be bleak," they concluded.[61] The fate of modern industrial societies was considered to hinge on these questions.

Samuelson's "Microscopic Time Asymmetry of Maxwell's Demon" (1985) offered a humble confession about what economists could not yet explain. By then, Samuelson had studied Maxwell's demon in such detail that he was uniquely attentive to its shape and form; of its appearance, he even noted that "apparently the Demon was male." He described the demon as a being that "could do rather quickly what chance can be expected to accomplish only in aeons of time."[62] One way of understanding Maxwell's demon was to imagine him as a theoretical device for viewing the development of world history as if it were a movie. "So to speak, the Demon speeds up for beginners to see within the lecture hour what nature could only show the students in billions of years." But the ending seemed to change as the movie advanced. "Instead of speeding up nature's film, or fairly paraphrasing its contents," he explained, "the Demon drama substitutes an alternative story."

"Whatever its gender, Maxwell's Demon turns out to be a red herring," concluded Samuelson. For the most part, his effects on the macroscopic world were limited by his dependency on information.[63] To better convey his talents, Samuelson reworked a widely quoted verse from Macbeth's soliloquy. "The Demon is a tale told by a genius," he wrote, full of "sound and fury, and signifying nothing very much."[64] In Samuelson's rephrasing, the word "life" was replaced with "demon," and "idiot" with "genius." But even if he was just a character in a tale, the demon could not be made to disappear from the scientific literature. "Just as you cannot recover a tumblerful of water thrown into the sea, once an imp gets into the literature of science you will never get it out," Samuelson complained.[65]

Nuclear energy enthusiasts and techno-optimistic growth economists were caught off guard the following year when "the word 'Chernobyl' abruptly entered the world's vocabulary." Facts about "an explosion and fire in a Soviet nuclear power plant" were hard to come by, but the extent of the tragedy was becoming clearer by the day.[66] Weinberg, a vocal enthusiast when it came to nuclear technologies, responded with an article titled "A Nuclear Power Advocate Reflects on Chernobyl." Weinberg still believed that nuclear energy was an "essentially inexhaustible energy system, which would confer energy autarky on its user because its requirement for raw uranium was minuscule."[67] In numerous publications, he optimistically described the "benign impact of nuclear power on the environment, at least as compared with fossil fuels." For this reason, nuclear breeders, in his opinion, "will always remain nuclear energy's holy grail."[68]

In light of the Chernobyl disaster, Weinberg became ever more aware of the importance of a lecture he had delivered to the American Academy for the Advancement of Science more than a decade earlier. "We nuclear people," he told his audience that day, "have made a Faustian Bargain with society." The exchange was necessary to obtain a "magical energy source."[69] What had been the consequences? "Fifteen years have passed since I first alluded to nuclear energy as a Faustian bargain," he wrote in light of the new disaster.[70] Five years later, he would repeat the same story. "Twenty years have passed since, at a meeting of the AAAS in Philadelphia, I referred to nuclear energy as a 'Faustian Bargain.'"[71]

While nuclear technology enthusiasts had a moment of reckoning after Chernobyl, economists had theirs. "The efficient markets hypothesis—at least as it has traditionally been formulated—crashed along with the rest of the market on October 19, 1987," recalled Samuelson's nephew, the Harvard economist Lawrence Summers.[72] Right up until that Black Monday, the operation of financial markets based on scientists' knowledge of Maxwell's demon, Brownian motion, random walks, and information theory seemed to be working just fine. But after the crash, Samuelson and his coterie were forced to accept that their assumptions were missing something.

The seventy-four-year-old Samuelson was still following the latest research when major publications headlined the news that a new version of Maxwell's demon could extract work "only by waiting for rare thermal fluctuations." Perhaps new research coming out of physics could explain those little-understood forces at work in the economy that he had missed. As soon as news of the discovery appeared in *Nature*, *New Scientist*, and *Scientific*

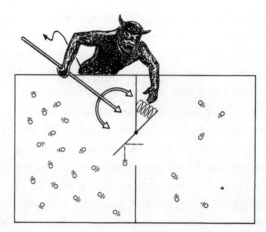

Figure 15. Paul A. Samuelson et al., "Scientific Correspondence: The Law Beats Maxwell's Demon," *Nature* 347, no. 6288 (September 6, 1990): 24. Reprinted by permission of Springer Nature.

American, Samuelson reacted. He penned a letter to the editor of *Nature* explaining that Maxwell's demon, whether in this incarnation or the previous one, was nothing other than a "counterfactual whimsy" fabricated to "get people to think in terms of statistical mechanics."[73] When asked to consider whether the demon could violate the second law, Samuelson responded: "The answer is no." The physics on which Samuelson based his economic theory was still alive, even if hanging by an ever more delicate thread.

ANTISOCIAL DEMONS

The decades after these environmental and economic disasters were sobering for techno-optimists. Gaps between the poor and the rich were widening, and a growing number of economists admitted that the market seemed to have some unfairness built into it, even if they could not exactly explain why or determine exactly who was to blame. While the dispossessed were ending up with little or no money at all, big money appeared to beat marketplace odds. Economists competed to understand why.

The economist Juan Martínez-Alier, president of the International Society for Ecological Economics, summarized the debate between those who used the example of Maxwell's demon to defend growth and those who used it to warn against scarcity. "Growth theory supported, and was in turn supported by, the idea that the situation of the poor could only improve if the rich became richer," he wrote.[74] Why had this logic failed? Martínez-Alier praised Georgescu-Roegen for showing the need for more radical forms of income redistribution "of goods among rich and poor. And of the intergenerational allocation of scarce resources and pollutants—and also, although this was not explicitly included in his analysis, the destructive impact of humans on other species."[75] Deeper lessons could be learned by furthering the demon-based economic analysis inaugurated by these researchers. "Unnatural" demons might be actively interfering with the flow of energy, capital, information, and humans in the world, creating "restrictions to migrations between South and North America, or between North Africa and Europe." "At such borders," explained the economist, "stand a sort of 'unnatural,' political Maxwell's Demons, who successfully maintain (at the cost of many human lives each year), the large differences in the per capita use of energy and materials between adjacent territories."[76]

At the beginning of the new millennium, the economy continued to show evidence of demonic actions, some of which defied available explanations. Shortly before his death, Samuelson returned to the topic of demons. By then, he had lived through the ravages of Black Tuesday, studied the oil energy crises, and witnessed Black Monday, the collapse of Long-Term Capital Management, and the dot-com bubble bust. The economy had grown, but so too had the gap between the monied elites and the rest. He had seen more than enough to accept the idea that some beings in this world acted in ways characteristic of Maxwell's demon.

"Some very rare minds do have the special talent and flair needed for a good long-term batting average," noted Samuelson. "I believe, but cannot prove, that it is largely because big money can (legally!) learn early more future-relevant information." Who were these privileged beings, these financial geniuses who defied the standard laws of economics and the forces of nature tending toward equilibrium? "The large universities, foundations, and millionaire family fortunes" consistently "out-perform the noise traders and the small-college treasurers." Stock pickers who consistently beat marketplace odds, and who thus resembled Maxwell's demons, had an uncanny ability to distill particular "future-relevant" information lurking within the loads of unsophisticated public information available to everyone. Their edge was completely legal, since it did not rely on privileged or insider information. Samuelson compared them to demons. "Those with earlier correct, not-yet-discounted, information do possess a Maxwell Demon," he joked, "that can defy the second law of thermodynamics."[77] Samuelson may have been joking, but he was also serious. By the end of his career, and knowing he was close to the end of his life, he had not been able to come up with a better explanation. The existence of demons who could defy the law of thermodynamics had seemed farcical to him, but the fact that some sectors of the population were consistently getting richer while others were getting poorer was far from comical.

As they tended to do, demons presented opportunities. Jaron Lanier, an interdisciplinary research scientist at Microsoft (the software production house dominating the operating systems market), was most attentive to the powers of Maxwell's demon in the new digital economy. He identified the operating strategy of the banking, tech, and insurance industries as similar. "Finance, and indeed consumer Internet companies and all kinds of other people using giant computers," he wrote, "are trying to become Maxwell's demons in an information network." Who could blame them? Insurance companies were doing it too. "With big computing and the ability to compute huge correlations with big data," the temptation "becomes irresistible," he explained. "And so what you do is you start to say, 'I'm going to . . .'—you're like Maxwell's demon with the little door—'I'm going to let the people who are cheap to insure through the door, and the people who are expensive to insure have to go the other way until I've created this perfect system that's statistically guaranteed to be highly profitable.'" Like many economists had done before him, Lanier turned to thermodynamics to explain the broader consequences of these kinds of business practices. "For yourself you've created this perfect little business, but you've radiated all the risk, basically, to the society at large."[78]

At the world's largest multinational online retailer and e-services business, decisions were consciously made to imitate the masterful tactics of Maxwell's creature. Without mentioning demons, Jeff Bezos, founder and majority owner of Amazon, announced that the company had reached $100 billion in annual sales faster than any other company had ever done it. In his letter to shareholders for 2017, he explained that one reason behind Amazon's success was his decision-making strategy. "Some decisions are consequential and irreversible or nearly irreversible—one-way doors—, and these decisions must be made methodically, carefully, slowly, with great deliberation and consultation." Bezos elaborated on why he found it useful to think in terms of one-way doors. "If you walk through and don't like what you see on the other side," he warned, "you can't get back to where you were before."[79]

A wide range of thinkers—from managers and businessmen to social scientists and philosophers—took scientists' lessons about one demon (or a small set of them) to heart. Yet they have remained oblivious to the wider impact of these creatures on the formation of the modern world and our knowledge of it. They were wise by the demon, but foolish by the lot.

CONCLUSION
The Audacity of Our Imagination

The ink on the demon papers has yet to dry. Scientists in cutting-edge laboratories around the world continue to use the term "demon" when trying to grasp something they do not yet fully understand. To dispel these demons, they create new experiments and technologies, revealing in the process new aspects of the universe. Successful researchers conquer unknowns by chipping away at the borders of the unimaginable. What we currently know about demons and all they can do is practically equivalent to what we know about the universe.

The main function of science's demons is at first exploratory, but it does not remain so. These characters remain active long after their initial utility has passed. As scientists and engineers build new demons and improve on existing ones, the technologies that emerge from their work populate our world, as butterflies who have broken free of their chrysalis take to the sky. Today we need to live not only among technologies that have been modeled after the demons of our imagination—from virtual reality to bombs to artificial intelligences—but with the demons that are still within us, some guiding research and innovation and others dividing us.

Figments of our imagination require our full attention. As scientists discover new laws and particles, they imagine others. As they imagine even newer aspects of nature, they get to see even more. How much farther can we push the envelope of our imagination?

Scientists have passed down knowledge about demons across generations. Researchers dig eagerly into relevant scholarship, citing it copiously, before venturing to add one more rookie demon to a growing list. Knowing the preexisting literature on them is necessary for determining the capabilities of fresh newcomers. Some scientists focus their entire careers on understanding a single creature or a small set of creatures, and for the most part, they trade this information within select groups. When a demon is tagged with a scientist's last name, the research of that valued member of the profession lives on even after they die off. As younger professionals take on the challenge of investigating the demons named after their mentors with a celerity that can no longer be matched by those who first raised those speculations, they extend the research programs of their patronyms. Awards and valuable patents are often bestowed on successful practitioners of this art.

Scientific discovery is often essentialized as magically bursting into the world with the proverbial "eureka!" exclamation. This caricature obscures too much: it hides a rich tradition of inventiveness behind that moment. The role of the imagination in the science, and in determining reality more generally, is much more pervasive.

Most scholars no longer believe that a single characteristic can define in its entirety the scientific method, yet paeans to science often stress one much-vaunted quality: the idea that scientific decisions about the nature of reality are made on the basis of experimental verification or falsification.[1] Yet our traditional understanding of science as hypothesis creation followed by testing leaves us at a loss when we try to understand where investigations are heading. It explains how a necessary part of science's compass works, but it has nothing to say about where it points. It does not illuminate how the possible can and does become actual.

MYTH

"Discoveries need a kind of mythology," wrote the influential philosopher of science Hans Reichenbach.[2] Reichenbach's views about science dominated philosophical discussions on the topic for nearly half a century. They still inform most learned and popular discussions about science and its virtues, which were originally shaped by the philosophical movement he helped establish, known as "logical positivism."

One might think that after highlighting the relation of scientific discovery to mythology, as Reichenbach did, logical positivists would jump at the opportunity to study their connection. Yet their goal was just the opposite. Reichenbach only raised the topic in order to dismiss it. In highly influential texts, he forcefully argued against studying the mythologies needed by discovery and worked hard to exclude them from our understanding of science. Indeed, his distaste for them was such that he preferred to consider science via a "fictive construction" in which it played no role.³ It did not matter to him that this constructed version differed greatly from how science was actually practiced, even by practitioners as prominent as Einstein, with whom he collaborated closely. It did not matter to him that scientists themselves acted differently in their laboratories or that their writing often introduced mythical and magical creatures and themes. For him, this depurated fictive semblance—one contrived to be logical and rational—was a worthy substitute. Under Reichenbach's influence, all those aspects of science that could not fit into a tight empirical and analytical model were discarded, treated as detritus to be swept under the rug and shooed away from view.

Philosophy of science has not yet fully recovered from the distorted view of science that Reichenbach and other logical positivists presented to the public. While some critics of logical positivism have offered several correctives, to this day most of the discussions about science continue to approach it as sundered from the imaginary and the mythical.

The fictive construction of science proposed by logical positivists, which stood in the place of actual scientific practices, was used to draw a wedge between high, elite technical knowledge and lowly, practical everyday ways of knowing. This same fabrication divided moderns from premoderns, a division associated with the differences between reason and irrationality, the civilized and the savage, the West and the rest. In consequence, the idea that something *useful* could be learned from understanding demons was completely lost to history.

Science does not lose its enviable status as a valuable and distinct form of knowledge if its mythical aspects are acknowledged. On the contrary, it is stronger. If certain kinds of narratives have been especially useful to scientists, why should we continue to erase them from our accounts of scientific progress? An understanding of science that considers how scientists themselves describe their work risks only adding new insights to our repertoire of knowledge. To our knowledge of the natural world we can add a higher level of *knowledge about knowledge*.

We have feared demons during some of our darkest moments. Violent exorcisms motivated by political interests, greed, intolerance, ignorance, and superstition have led to bloody executions and persecutions. But demons have also been invoked during the best moments of our civilization. By referring to them, we have learned that causality does not always hold at the subatomic level, that there are limits to the speed of message transmission, that we cannot have perpetual motion machines, that living systems spend energy too, and much more. These beings have helped us understand how to combat decay by reversing entropy locally and shown us how to produce work as effortlessly as possible. They have helped us explore the explosive interior of the atom, the recondite corners of the universe, and the territories in between that are imperiled by invasive species and beset by disasters that may strike when least expected. Demons have helped us learn how to control our future and how to increase our understanding of our past. They have also taught us how to escape from our reality and evade our responsibilities. Many of our complex technological systems are modeled after comparatively simple creatures. We might begin to better understand the former with the latter.

Demons are not just the spurious beings in the feeble minds of the superstitious. They are neither just psychological delusions or simplistic heuristic fictions nor simply auxiliary midwives who help scientists deliver knowledge. Their status within nature is akin to that of many other tools of thought that are considered entirely respectable by philosophers, such as concepts, numbers, classes, and categories. In contrast to these abstract or formal entities, demons are colorful and concrete. They have names, proclivities, and personalities. They structure our fears and desires, tempt us into taking new directions, and lead us to invent new, faster, smarter technologies and better, more comprehensive theories. They are often far-fetched because they are far-fetching.

The myth of "Western rationality" is many things, but it is primarily a myth. The desire to study the world and to understand how we can thrive in it is not unique to modern science. It cannot be confined to any specific historical period or geography. The persistence of traditional narrative tools within the highest levels of elite science does not show that we have never been modern or that our quest to become so is as yet unfinished. On the contrary, it proves that our struggle to distinguish between the true and the false takes radically different forms in different places and eras. In all of these cases, the imagination plays a key role.

Understanding the relation of science to myth can help us broaden our understanding of the latter. It offers an opportunity to rescue myth from the conception to which it has been relegated—as something belonging to "other" cultures. Myths are not "machines for the suppression of time" that offer only antiquated lessons about the world, as the prominent anthropologist Claude Lévi-Strauss once famously claimed.[4] Rather, the myths underlying science are machines that create time for the physical, biological, and social sciences.

Knowledge, even in its newest and most modern forms, relies on larger-than-life tales of old. That is why science is so effective at solving old mysteries and creating new ones.

ETHICS: THE FOLLY AND THE WISE

Modern scientific culture has a sturdy legacy of strengthening the divide between the true and the good, severing the physical from the moral. The basic procedures that characterize much of science and laboratory work have no room for moral deliberations. Measurement practices, such as noting how much time has passed or how much distance one has traveled and determining lengths, weights, and temperature, are now extremely simple and as common as exchanging a greeting with a handshake or a hello. They are so integral and transparent in our culture that they are no longer seen as any different from many of the other phatic and automatic actions that characterize much of our everyday modern lives.

"A scientific philosophy," Reichenbach acknowledged, "cannot supply moral guidance; that is one of its results and cannot be held against it." Reichenbach defended a logical and positivist view of science precisely because it helped differentiate science from messier areas of culture that were marred by judgment calls and debatable dilemmas. By focusing on the binaries of the true and the false, he thought we could bypass thinking about the right and the wrong. "You want the truth and nothing but the truth?" asked Reichenbach. "Then do not ask for moral directives."[5]

There are many benefits to maintaining a divide between the world of science and that of ethics. For one, it uncomplicates things, liberating us from the paralyzing task of having to consider every calculation in terms of good and evil. We do not need to distinguish the fair from the foul when engaged in most scientific activities, just as we do not need to during most of the regular and automatic aspects of our lives. A get-it-right mentality and a materialist, operationalist, or instrumentalist understanding of knowledge are extraordinarily useful and productive in society. It is the perfect lubricant that makes the gears of techno-scientific modernity run smoothly. Because these procedures work so well and are so common, it can be tempting to believe that *all* of science works this way: take a measurement and prove or disprove a fact. It can be tempting to think that science's deepest mysteries can be answered with yeas or nays on the basis of transcendental truths. They cannot.

Creating new knowledge requires much more than fixed answers to canned questions. Science does not advance from black-and-white thinking, it is not moved by known solutions, and it does not budge even an inch from its incontestable truths. Knowledge, scientific and otherwise, arises from direct commerce with the unreal. In the halfway house between the real and the symbolic, imagination and innovation meet, greet, and exchange hats, fantasy hardens and solidifies, and fact and fiction cross-dress.

The separation between scientific verities and ethical ones has had another consequence: it makes ethics secondary (at best) to science. As science and technology give us the means to find solutions based on the true or the false rather than on the benevolent or the evil, the list of ethical challenges around us grows. Why?

Our sovereignty with respect to technology is limited. By the time most discussions around use, regulation, and ethics start, it is already too late. The ship has sailed before it even left dry dock and touched the water. Once wheels are set in motion, all we can do is step on the brakes. The gun is cocked and loaded. New technologies enter the world and affect us immediately, for better *and* for worse. Regulatory policy is limited by a narrow *ex post facto* notion of control that leaves us with too few levers to steer our future in a desired direction. If we accept that responsibility starts with our imagination, accountability can be initiated before the clock strikes midnight.

The persistence of debates about demons in some of the most technical literature of our times reveals that the project of distinguishing between the true and the false continues to involve at a deep and fundamental level a differentiation between the folly and the wise. Science is neither value-free nor demon-free.

Despite the better angels of our nature, the development of science and technology bears the imprint of very basic primal drives, impulses, dreams, fears, and desires. We continue to seek knowledge for glory, for gain, for the benefit of just a few, or for the good of all, just as we have done throughout history. Technologies have been built throughout modernity whose ultimate control baffles us. Benign technologies can be transformed into dangerous ones. How can we obtain the benefits of science and technology without engaging its risks? How can we escape from the vicious cycle of creating harmful technologies even as we lust for better ones?

When considering the dangers of technology, the German philosopher Martin Heidegger was compelled to remind his readers that technology itself was not dangerous; the danger lay in its "essence" (*dämonie*).[6] "There is no demonry of technology," he wrote, "but rather there is the mystery of its essence." "There are many omens of the arising of this demonism," he observed, surveying the brave new world of science and technology erected around him.[7] Like the highfalutin' doctors in Molière's play who accounted for the sleep-inducing qualities of opium through its dormitive qualities, Heidegger explained technology's demonic qualities via its demonic essence. To break from such circularity, a study of demons, literally as a "term of art" in science, can serve as firmer ground.

THEN AS NOW

The list of similarities between ancient, religious, folk, and science demons is long. Throughout all civilizations, whether ancient, non-Western, or modern, key characteristics of demons have remained surprisingly constant. These creatures play havoc with the natural world in characteristic ways. Ancient and modern demons appear in similar kinds of locations, operating in ways that are pretty similar in both past and present. Across a wide range of fields and epochs, numerous thinkers have coincided fairly consistently in their assessment of where and when the world is most likely to bend in unlikely directions.

Demons are good to think with because they challenge us to outwit them. Since they are not believed to possess the absolute powers often attributed to God in monotheistic religions, their abilities are considered contiguous in some senses with our own. (Laplace's intelligence came somewhat close to being godly, but even Laplace avoided describing him in those terms.) Demons have to learn. They have to work. They are better than us in some ways, being often endowed with more acute minds and senses and swifter, defter bodies, but they are much worse than us in other ways. Like us, their agency is circumscribed. To be effective, their interventions need to be minimal, accomplishing only what is necessary to get the job done as cheaply as possible. They are subjected to the same natural order that we are. Like us, they try to subvert that order, but can only do so little by little and within limits. They do what they can. We do what we can. Demons in lore, legend, and science are masters at adroitly manipulating natural causes and excel in the exploitation of the hidden, the occult, and the anomalous. That is why they are found at the cutting edge of science and technology. They baffle expectations and luxuriate in surprise, so discovery comes naturally to them. Their personality is somewhat mischievous, though not entirely malevolent.[8]

Examining the transformation of the figure of the demon in modern times takes us perforce on a global tour of science. From the seventeenth century onwards, demons migrated from the Netherlands to revolutionary France; from there they were imported into Victorian England, before arriving with great difficulty in Germany and settling back down in France. They emigrated from Europe to America in the decades before World War II, arriving first in the Physics Departments of Princeton and Harvard before moving to the pioneering cybernetic and computing initiatives at MIT. From the East Coast they moved west, first to the spacious state-funded laboratories of the Midwest, then to California, and later to privately funded multidisciplinary institutes scattered across the nation. They are now active across all areas of science, all over the globe. Celebrated physicists, cell and evolutionary biologists, neuroscientists and cognitive psychologists, sociologists, and economists all keep up with the most recent research about these creatures. By the end of the millennium, the very synapses of our brains were understood by reference to them. Demons no longer stand outside of our minds, nor outside of knowledge itself. When we read a book, try out the newest virtual reality headset, sit in a state-of-the-art theater, or admire the cosmos from the comfortable seat of a planetarium, we are the willing victims of Descartes's demon. He continues to inspire researchers to create more realistic narratives, spectacles, and seamless entertainment technologies. Recently, a demon related to Descartes's has been charged with responsibility for causing confirmation bias across the internet, trapping its victims in media bubbles and echo chambers by showing them only the facts they want to believe in.[9]

Despite our intention to hold them back with the power of reason, demons have kept mutating. As a growing succession of researchers continue to advance knowledge by removing traces of the dark forces of religion, superstition, and ideology, these creatures have kept reproducing. Sticking around in the age of reason, they have made it their comfortable home.

Understanding the similarities between old and new demons is as important as understanding their differences. Both old and new demons are comfortable with changing the regular order of events, but only old ones are described as using black magic incantations in reverse to cast spells and decode hidden messages. Both may be associated with noise, but only the old demons are described as appearing with the loud boom of thunderstorms, the howling of the wind, or the rustling of leaves. Both types are found in chaotic environments, but only in ancient accounts do demons appear when weather currents clash during a tempest. Both old and new demons interfere with biological reproduction, but only the former appears in the shape of incubi and succubi.

Demons embody opposites and are paradoxical: stupid yet smart, mechanical yet lively. New ones combine the powers of most of their ablest progenitors. They can wreak havoc on basic laws of causality. Effects follow like clockwork, until one day all causal links disappear. Causation becomes correlation. These unnatural born innovators cannot *but* think outside of the box. Just when we believe that our calculations are all set, they add another variable. Or they do just the opposite, discounting our previous experience. When we count on getting another chance, they can suddenly intervene to limit the number of tries we thought we had. We may expect results to be independent of past outcomes, as we do when we toss a coin, but we may suddenly discover that the results were linked in ways we just had not imagined. The law of large numbers no longer applies. Our memory is wiped out, and gone with it is our ability to understand, determine, or predict the future. When we have finally landed on a correct answer, it turns out that the only right choice was "none of the above." Results are suddenly greater than the sum of the parts. Demons can change the rules of the game, the object of the game, and the boundaries of the game, stumping even the cleverest minds. They imperil distinctions between winners and losers, being masters of the art of subverting hierarchies.

While demons' skills in performing these odd jobs do not fulfill our wishes exactly, they often exceed expectations in other ways. We reward them by tagging them with the last name of the scientist they successfully stumped. As some wane into has-beens, others rise to fame. When scientists introduce new demons, these spick-and-span newborns are born with the sagacity of old men. Far from weakening them, age seems only to strengthen them.

Modern scientific culture continues to be based on ancient practices. Basic sorting, gathering, and containing activities are as relevant today as ever. So are the demons associated with those practices. Sorting demons pre-

Figure 16. L. Darling and E. O. Hulburt, "On Maxwell's Demon," *American Journal of Physics* 23, no. 7 (October 1955): 470. Reproduced by permission of the American Association of Physics Teachers.

cede Maxwell's being by nearly half a millennium. Although associated with menial tasks and servitude, these repetitive tasks accumulate to great effect. Since biblical times, these sorting activities were acknowledged to have a moral component, with God described as having the power to separate the righteous from the cursed, just as a shepherd culling the sheep from the goats.[10] These connections persist in the etymology of the word "sorcery," a word that comes from the Latin term to sort. Illustrations of this old demon engaged in the work of selection often included angels and devils as his helpers and sometimes depicted him wielding a trident and other implements.

Sorting occurs naturally or artificially, actively or passively, with membranes, filters, or hinges, in the farthest corners of the universe or near us. Some of the first artifacts of early human civilization, such as stone and clay pots and vessels, were tools devised for the purposes of selection and separation, and insulated chambers were often employed to contain what had been separated. *Jinn* (the Arabic word from which the word "genie" stems) are usually found imprisoned in bottles. (Aladdin's genie lives inside a lamp and is awakened only when it is rubbed or opened.)[11] In illustrations of hell in the Christian tradition, sinners are often portrayed as contained in insulated cauldrons filled with hot liquids, and the entrance to hell or heaven is often marked by a hinged door closely watched by a demon or an angel. Maxwell's demon still works with these doors and insulated containers.

The practice of weighing—central to measurement-based scientific practices—involves sorting, traditionally into left and right trays in scales. Exactitude in physics requires a balanced weighbridge unaltered by a demon. Scientific descriptions of balance and equilibrium featuring Maxwell's demon culling molecules left and right shares the iconographic conventions of the Last Judgment. Christian demons are often portrayed as interfering with the balance used by archangels, mainly Saint Michael; they are the psychopomps tasked with the weighing of souls. Justice has long been represented as blind, while the demons who are known for interfering with it are depicted, by contrast, as observant, in physics and beyond.

Demons lurk in places of opportunity like the fulcrum or the balance—a symbol of equality and justice—where minute actions can lead to greater inequality. Demons then as now are contract enforcers who limit repentance and who bring back the sins of the past. In medieval times, they were frequently portrayed as rent collectors. When payment was refused, they often took a soul or a human life as sacrifice in exchange for the payment owed.[12] In scientific texts, Maxwell's demon and Gabor's demon appear as the rent collectors of a universe conceived entropically—no one can make a profit from nothing. When work without requisite expenditure appears, a demon is often suspected.[13] This is most clearly exemplified in the demon of chance in economic theory as well as in Maxwell's demon.

Demons, devils, and other supernatural creatures have long been associated with extreme velocities and fancy modes of transportation. In the Gemara section of the Talmud, Yosef the *sheida* is presented as having the power of instantaneous transmission while a number of other *shedim* (a Hebrew word for "demons") are known to travel extremely fast, hovering and flying through the air. The capacity for speed is typical in fairy tales and epic poems featuring genies or *divs* (Persian for "demons").[14] This power is also central to depictions of divinity in Christian traditions.[15] Seventeenth-century demonologies often described witches being transported to faraway places in a single night for their Sabbat meetings. Many of the demons covered in this book—most clearly, Maxwell's demon's colleague—are also extremely fast.

Laplace's, Maxwell's, and Maxwell-Szilard-Brillouin's demons, the quantum demons (Einstein's, Compton's, Born's, Planck's), and those in our cells and brains (Monod's and Searle's), in addition to being extremely fast, tend to be really big or really small. Giant demons share some similarities with the giants of Northern Teutonic tales, who personify the brute forces of nature, but unlike these giants, who are often described as stupid and easily outwitted, Christian demons and devils, such as the Behemoth, are usually much harder to fool.[16] With its enlarged brain and near-infinite memory, Laplace's demon shares these mythical dimensions. Extreme size in the other direction is also characteristic. Most tiny creatures are impish and compensate for their small size—or, sometimes, complete lack of mass—by operating at incredible velocities and attaching themselves to fulcrums from which they can almost lift the world. High-brow science and *haute* expert knowledge is shot through with low-brow B-genre fascination with extreme size and speed.

Some demons, primarily Lucifer, are associated with light or darkness. In modern physics, Maxwell's demons, his colleague, and other quantum demons are masters at handling light, electricity, and information. As far back as Plato, good demons were described as transporters whose function was to carry messages from humans to the gods in the sky, sometimes in the form of prayers. They also had the ability to carry the goodwill of the gods back to the human world and announce divine judgments. Later, malicious demons were thought of as differing from angels in their ability to distort, rather than convey, messages and information. The demons that appear in information and communication theory also carry messages and are similarly believed to have the power to listen in on communication networks and intercept messages. Demons thrive with noise and in states of intoxication. Brownian demons, with their random, "drunken," and unpredictable movements, personify noise's essential qualities.[17]

Demons are believed to thrive in chaotic environments where extreme contrasts meet. In the digital era, they are found at the border between a 0 and a 1, where they can be responsible for turning one digit into the other. In black body radiation, they operate at the scale of the minimum amount of energy needed to overcome the radiation of systems in equilibrium. Demons are frequently portrayed as beings that interfere with or distort memories.[18] In computer science, the bottleneck preventing computers from calculating everything lies in the size of their memory and their ability to access, erase, rewrite, and learn from it.

Biology's demons, mainly Monod's and Maxwell-Szilard-Brillouin's, are similar to those that traditionally were thought to possess the power to reanimate dead bodies. In the early twentieth century, biology's demons were closely associated with a vital force capable of bringing the inert to life. Demons often interfere with regular processes of reproduction, with repercussions for our understanding of fidelity, understood as a physical concept and a moral one. In modern biology, demons appear at the center of copying and reproductive processes, acting directly on replicating DNA.

Demons often favor young maidens and children as victims. Darwin's demon, a master predator, continues this tradition, feasting on the feeble and preying on youngsters. Others, such as Loschmidt's and Zermelo's demons, can work in a world in reverse: thriving in an inverted universe, they can work backward as well as forward, and in reverse as well as in regular order. Searle's demon illustrates another common characteristic of demons: their limited agency and the difficulty in determining who is acting, who is thinking, and who is to blame. Related to this ability is their penchant for intervening in cases where it is hardest to tease apart the natural from the artificial and nature from nurture. Darwin's demon also excelled in this capacity.

Demons appear as atomic and subatomic particle manipulators, faster-than-light messengers, feedback experts, chain reactors, efficient triggers, reproductive opportunists, signal attendants, gene selectors, data handlers, code followers, and uncannily adept information jugglers ("certain RNAs—are but Maxwell demons").[19] These shape-shifting protean masters of disguise doff old skins with as much ease as they don new ones. They can hide. They can impersonate. They adapt. When one is eliminated, another one reappears. These expert time-twisters are particularly adept at acting nearly effortlessly, at key moments, directing or changing our fates by the simple push or pull of a lever or switch. Yet others have an uncanny ability to position themselves at strategic places where small actions can have untold consequences. They can be devilishly fast or diabolically slow.

CORDON SANITAIRE

Numerous books have been written about the rise of reason and the decline of magic and superstition that led to the modern world as we know it. Most accounts date the beginning of this shift to the mid-seventeenth century. Since then, researchers have continued to draw a firm divide—a *cordon sanitaire*—between medieval culture, the so-called Dark Ages, and the modern, or Enlightened, world. Prominent historians have shown how the "quiet exit of demons from theology" set science off on the path we recognize today.[20] In consequence, we now typically see them as belonging to the spiritual, artistic, or psychological realms.

Explaining such an abrupt before-and-after change has proved challenging for historians, who have to offer radically original interpretations of past eras and foreign lands to explain beliefs that appear so counterintuitive from their perspective. How could so many cultures around the world for such long periods of time espouse beliefs that we now consider to be at odds with rational experimental and scientific methods? Scholars continue to spill ink trying to explain the supposed backwardness of the people of the past and those in the present who live in far-off places.[21] There is another answer: a line between modern and premodern ways of thinking is impossible to draw because no such firm line can be found separating knowledge from the imagination—then as now.

The persistence of the category of the demon in science is evidence of an interesting anomaly. The term was never dropped from the lexicon of truth-seeking enterprises, even when they became secular and scientific. Instead of a *cordon sanitaire*, modern demons reveal that a *liaison infecté* continues to connect modern culture with past eras in a way that has merged the lessons of old with the lessons of today.

"Always choose the devil you know" goes a common saying. Must we stay the course?

Division lies at the root of the word *diabolo*. When discussing problems pertaining to the global economy and the health of our planet, we often invoke science to argue for contradictory solutions, some considered utopian, others dystopian. Who is right in these debates matters, but the debates themselves can be as productive as agreement and consensus are rare in science. Debate motivates researchers to create new experiments and technologies and drives scientists to sharpen their arguments. Debate forces them to improve their theories, tweak their plans, and revise and resubmit proposals. Disagreement about what is real and what is not is a generous fountain.

Demons divide, but if we take on the role of *advocatus diaboli* vis-à-vis imaginary demons, they can help us learn more about science, technology, nature, and culture. More importantly, by studying their place in history, we can learn essential lessons about the limits of modern knowledge itself. The smartest and most rational hominids among us—the most brilliant scientists and engineers of the last centuries—do not have an absolute monopoly on knowledge, nor on its consequences. *Homo faber* cannot fully command its tools, nor can *homo sapiens* completely control its own mind. *Homo imaginor* can do much more. By shining a light on the relation between the imaginary and the real across history, we can rehearse a different way of thinking about the universe.

To overcome the mistakes of the past and build a better tomorrow, we cannot rely on science without history. By helping us step more firmly into nature's shoes, our imagination can lead us to transcend our most immediate fears and desires and to see that nature can be more than a problem to be solved, a territory to be conquered, or a source from which we can extract power. It is as essential for understanding nature as it is for understanding others.

An account of the world that does not include the underworld is half empty. Science's demons can be tamed if we venture to peer into their glassy eyes and follow their deeds and misdemeanors. We have learned much *from* them. We can learn much more by thinking *about* them, as they offer us the opportunity to uncover a new understanding of nature as both humanistic and scientific. Demons' long sojourn through our history reveals the world of today as the future of those who came before us and as the past of those who will come after us.

If we open our eyes to them, demons might help us carry the weight of a future built on the dreams of yesteryear.

POSTSCRIPT
Philosophical Considerations

It is time to fold, to put my cards on the table and accept that my strongest card is *nonexistent*. In one sense the game has been lost, but in another it is not. Almost nothing matters to us more than that which cannot be, is no longer, or has not yet appeared. Nonexistents capture our imagination.

Once we start thinking seriously about existence, it is easy to realize how quickly we slip from thinking about existents to pondering about nonexistents, and how hard it is to draw a line between the two. Actual existing objects play small roles in our general everyday discourse. Societies are organized around legal and juridical concepts that can never be pinned down to any particular thing or referent. They also have meager roles to play in science. In what sense can numbers, colors, shapes, organizing categories, and other tools of thought be thought to exist? Our minds, as they flash back and foreshadow, rarely consider the world in terms of the limited sphere of the really real.[1] Our emotions, as they oscillate between grief and hope, are similarly oblivious to the real. The transformation of the imaginary into the real is hardly a simple process, but it has a story that can be told.

Numerous philosophers have dedicated themselves to establishing the exact differences between the existent, the nonexistent, the imaginary, the possible, and the actual.[2] Typical examples from the scholarly literature dedicated to this question include the liar who says, "I am now lying," round squares, golden mountains, and Plato's beard. To distinguish varieties of existence, some of the most meticulous and rigorous among these scholars argue in favor of introducing into their discipline an "existential quantifier" with the symbol "$\exists x$" to designate varying degrees of plausibility—to anything and everything. None of their examples, however, come close to those that concern scientists. Nonexistents are *not* equivalent to *everything* that does not exist. Only a select set of them have been most useful in the history of science and technology.

This rabbit hole runs deep. Existents and nonexistents are opposites that resemble each other. As the philosopher Palle Yourgrau has argued, "What distinguishes the actual from the possible, the existent from the nonexistent, is not what they are but *whether* they are."[3] Unstable and ambiguous, they can somersault from one to the other. Their power derives from their variability and instability.

Red-in-tooth-and-claw debates about existence have often been overshadowed by attempts to taxonomize, categorize, and fix it. Since the beginning of the Enlightenment, the desire to legislate the real estate comprising reality was essential for the establishment of modern forms of government.

The work of the eighteenth-century jurist Jeremy Bentham illustrates just how necessary this undertaking was for the Enlightenment. Bentham is now celebrated as a proponent of utilitarianism, a practical-minded approach to philosophy that dominated early British empiricism and inspired progressive legal and social reforms aimed at creating a better and more just society. The British thinker advocated a new approach to prisons and asylums, proposing "panopticon" observation towers from which to survey individuals placed in institutions for "punishing the incorrigible, guarding the insane, reforming the vicious, confining the suspected, employing the idle, maintaining the helpless, curing the sick, instructing the willing in any branch of industry, or training the rising race in the path of education."[4]

Throughout his reformist projects, Bentham was inspired by Descartes's materialism and urged his followers to look at society as a machine that could be understood, fixed, and oiled like one. Given "matter and motion," Descartes had said, he would "make a physical world," Bentham wrote. "Give me the human sensibilities—joy and grief, pain and pleasure—and I will create a moral world," Descartes had added.[5] Bentham's "Theory of Fictions" remains largely unknown. Why would a thinker so concerned with thinking about society in materialistic terms be inspired to write a work about nonexistents in which he concerned himself with a study of what was most immaterial, such as a devil with "a head, body and limbs, like man's, horns like goat's, wing like bat's, and a tail like a monkey's"?[6]

Bentham's "Theory of Fictions" was never finished. Bodies proved easier to manage than hearts and minds. The indefatigable jurist was unable to bring order to all sorts of "non-entities" that proliferated around him. His failure at creating a rulebook for the unreal showed just how hard it was, then as now, to round them up into one coherent bunch, and not from lack of trying. On the contrary: significant resources continue to be dedicated to policing what is allegedly the most real and pressing. Rarely do we think about what and who is included in that assessment and why.

To the practical and common-sense-minded, forays into the unreal might seem like senseless and useless philosophical exercises, as cockamamie a pursuit as listening to the sound of silence. After all, it is difficult enough to work with elements whose existence is not up for debate, such as what impinges on our safety, income, and health. But the process of prioritization that brings certain people, objects, concerns, and ideas under the purview of government and into the sight lines of society should concern us.

In the early nineteenth century, the aristocratic Austrian philosopher Alexius Meinong became convinced that intellectuals around him were in thrall to a "prejudice in favor of the real."[7] To combat this prejudice, Meinong attempted to create an entirely new area of knowledge, called "the Theory of Objects," that went against the general trend dominating most intellectual inquiries of his era. His philosophical project consisted in trying to find a place for "homeless" (*Heimatlos*) objects that did not fit within the realm of the "real" (*Wirklich*). They too, he argued, existed in some strange capacity. Otherwise, we could not even talk about them.

After he uttered these words, the philosophical community rallied against him, united in a common fervent cry of "I am not a Meinongian!"[8] The mathematician Bertrand Russell, who had at first been sympathetic, changed tack. He became one of the first to attack, accusing Meinong of introducing into philosophy an "unduly populous realm of being," as primitives were wont to do.[9] Russell and his acolytes embarked instead on the opposite project: trying to cut many of these non-real objects out of the picture. Under their influence, most philosophers still concur that the generally-agreed-upon list of the "existing" should not be up for modification.[10] In consequence, philosophy became dominated by a school of thought called analytical philosophy, whose implicit aim has been to police the boundary around the real.

A few years later, the German philosopher Hans Vaihinger made a second attempt, but the initiative behind his controversial *The Philosophy of "As If": A System of the Theoretical, Practical, and Religious Fictions of Mankind* (1911) largely backfired.[11] Like Meinong, Vaihinger ruffled the feathers of a generation of scholars who tried to combat him in every possible way. He too galvanized an entire philosophical movement, logical positivism, against him.[12] In light of these difficulties and ugly debates, few scholars have tried to pursue these investigations much further.

The "bête noir of analytical philosophy" opened up by the pioneers Meinong, Vaihinger, and those few followers who sided with them motivates much of my own argument.[13] They barely fathomed the dimensions of the labyrinth before them. Perhaps it was wise to not venture further in. Perhaps it was not.

The most consequential initiative to legislate the boundaries of existence did not come from philosophy but from psychology. In the 1920s, Sigmund Freud found that irrationality and pathology were the source of many beliefs in nonexistents. When his method was used to understand past historical eras, things got confusing rather quickly. How could so many of the witness accounts of past eras that seemed extremely strange to us, including collective ones, be explained? Freud had to develop a radical strategy of historical reinterpretation alongside psychoanalysis.

Freud read one of the most important demonologies of early modern times, Johann Weyer's *De Praestigiis Daemonum et Incantationibus ac Venificiis* (On the tricks of demons and on spells and poisons) (1563) with fascination. He immediately claimed that it was one of "the ten most significant books" of all time. Weyer's demonology helped Freud reach the conclusion that those who claimed to see or hear demons were usually the victims of some kind of disease, some pathology that led them to see the unreal as real.[14] His interest in the text lay not in what it said about demons, but in what it revealed about mental illness and the precarious psychological state of those who believed in them.

Psychological interpretations of such beliefs represented a culmination of a centuries-long effort that succeeded in culling many superstitious beliefs and most demons from the natural world. Even progressives within the Catholic Church, the old bastion of demonology, started to reinterpret cases of demonic action as due to subconscious psychological inferences, à la Freud.[15] Such bold acts of historical reinterpretation proved to be widely satisfying and popular, even as they served to pathologize nearly everyone in the long history of humanity except for a few modernizers.

The wisdom of the ancients, the baroque, and the gothic was excised from our repertoire of advanced knowledge. Yet as more and more philosophers and scientists tried to understand the world by reducing it to its bare material essentials, the more they ended up relying on imaginary creatures, categories, and concepts. The contradiction is increasingly difficult to ignore.

What has prevented scholars from studying the numerous references to the otherworldly in science? Sometime between the fall of 1960 and the spring of 1961, two of the most respected philosophers of science came close to considering modern science as a belief system, one just like any other, comparable (in essence) to the belief systems of ancient civilizations and religious cultures. Thomas Kuhn had just finished a draft for an essay that would transform the field of history and philosophy of science, *The Structure of Scientific Revolutions*. It would become an instant sensation and achieve fame for, among other things, coining the word "paradigm." Its author studied the advances of science by focusing on the process through which widely shared beliefs were supplanted by new ones, starting with the Copernican Revolution. Kuhn insisted that facts and observations alone were not enough to bring a new paradigm into existence. For this reason, he was immediately accused of proposing a view of science that, instead of being progressive, was "through-and-through relativistic"—an accusation he forcefully resisted.[16]

"I believe that people in the 15th century saw ghosts with the same intensity as they saw real things," wrote the philosopher of science Paul Feyerabend to his colleague.[17] Kuhn had just shared his manuscript with Feyerabend, in the hope of getting some positive feedback and constructive criticism. Instead, Feyerabend claimed that Kuhn made it seem throughout his book that science was comparable to witchcraft. One obvious question that arose in the minds of readers, including Feyerabend, pertained to the emergence of science as a distinctly modern way of thinking. "What difference does it make that the modern witchcraft is being called 'science'?" he asked. "After all, a name does not make the difference. Nor does the existence of institutions, of journals, of instruments, of technicalities," he concluded.[18]

Similarities between science and witchcraft were evident. According to Feyerabend, the "phenomenal character" of the beings seen by the populace of the Dark Ages "was the same" as the objects studied by today's science. "The spiritual world—even if it was populated by evil spirits—was much more important and much more secure than the transitory material world of tables, chairs and philosophy books," he argued. Feyerabend had recently read Aldous Huxley's *The Devils of Loudun* (1952), a historical reconstruction of the famous witchcraft case against Urbain Grandier, a priest accused of making a pact with Satan before he was burned at the stake in 1634. Huxley's book showed just how much our assessment of reality was influenced by our beliefs, and this was Kuhn's take too. Feyerabend noted that "the way in which, for example, the theory of witchcraft and demonic influence crept into the most common ways of thinking, and could be preserved for quite a considerable time, offers a vivid illustration" of how scientists themselves resisted changing their models.[19] When confronted with unexplainable facts, scientists often tweaked their theory a bit by adding hypotheses on the fly, in an ad hoc manner. As these aspects of their theories were corrected, the whole conceptual structure on which they stood was rarely discredited.

New paradigms are hard to swallow. Discarding long-held beliefs is just tough. When it comes to understanding the complexity of our belief systems, facts alone are insufficient. Once the existence of folk and religious demons was questioned, old facts required new interpretations, and so they were reconsidered as effects caused by fear, panic, or mental illness. Feyerabend argued that the experience of scientists who did not want to let go of their theories was "not a bit more strange than the denial of angels, demons, or the devil seemed to the faithful who had been brought up in the corresponding beliefs and who in addition had experiences such as hearing voices, partially split personality, fear of corruption, and the like."[20]

But there was another difference between those who believed in the supernatural and those who trusted science, Feyerabend insisted. Science was not the same as any old belief. "The theory of witchcraft brought order into natural phenomena but in such a manner that no conceivable counter argument existed." Feyerabend elaborated: "What distinguishes reason from folly is the kind of procedure adopted: are difficulties taken seriously? Is it admitted that the theory may, after all, be false? Or are they only taken as an incentive to invent ingenious ad hoc hypotheses?"[21] The last point raised by Feyerabend is key. Science's demons are not like any other demons. They remain imaginary until proven real.

"You really are like a witch doctor," Feyerabend scolded Kuhn. "That is, I do not object to your findings, but to the way you present them."[22] Feyerabend's apparent inability to stop thinking about the similarities between science and witchcraft started to affect his assessment of scientific progress more generally. In "Hidden Variables and the Argument of Einstein, Podolsky, and Rosen," he asked: "How was the theory of witchcraft replaced by a reasonable psychology, based as it was upon innumerable direct observations of demons and demonic influence?"[23] When it came to either quantum demons or the devil, facts and observations alone were also not enough to disprove them. "An even more striking example is provided by the phenomena known as the 'appearance of the devil,'" he explained. "These phenomena are accounted for both by the assumption that the devil exists, and by some more recent psychological (and psycho-sociological) theories."[24] To make progress in science required finding concepts that did their explanatory "job equally well, or perhaps even better, because more coherently."[25] The possibility of improving in this way, he maintained, was what set modern science apart from dogmatic thinking.

Feyerabend defended the search for hidden variables in the quantum world. He took the physicist Werner Heisenberg to task for attempting to censor scientists' recourse to imaginary beings in their attempts to find more answers. Heisenberg's injunction that "there is no use discussing what could be done if we were other beings than we are" was, Feyerabend proclaimed, "an astounding argument indeed!" Feyerabend argued instead for the benefits of thinking about—and even trying to become—other beings. "Why should we not try to improve our situation and thereby indeed become 'other beings than we are'?" he asked. "Is it assumed that the physicist has to remain content with the state of human thought and perception as it is given at a certain time and that he cannot (or should not) attempt to change, and to improve upon that state?"[26]

Even as philosophers such as Kuhn and Feyerabend clarified much about the role of observation and facts in science, both failed to see the extent to which the use of fictions—and demons—prevailed throughout science. In his classic study of thought experiments in science, Kuhn simply defenestrated them. "A thought experiment," he concluded, "can teach nothing that was not known before." Without even giving them a chance, he claimed that thought experiments "can teach nothing about the world."[27] They simply did not fit within his paradigm.

Something of an exception to this state of affairs can be seen in the work of Karl Popper. By the time Kuhn published his *Structure of Scientific Revolutions*, Popper had enjoyed decades of notoriety that came from the publication of his influential book *The Logic of Scientific Discovery*, published in German in 1934 and revised for an English-language edition in 1959. Popper had claimed that the best recipe for doing the best science was to create strong hypotheses that could be falsified experimentally. This is still the model that many scientists believe should be followed.

Popper's interests eventually shifted to history and politics, but by the 1950s he was ready to return to thinking about science. He was shocked to see that the best scientists of his generation were not working in the manner described in his book. Compared to the reality of on-the-ground scientific practices, his old views seemed like very distant idealizations. Some of the best of the best seemed to be working in exactly the opposite way from how he had once thought great science should be done. Many of them were using thought experiments in ways that he deemed improper. And many of them were debating demons. Instead of creating hypotheses that could be put to the test and falsified when needed, he found that contemporary scientists were assuming that certain demons could not exist and then developing theories based on that assumption. "The non-existence of a Maxwell demon is simply assumed *ad hoc*," Popper wrote, with exasperation.[28] In years to come, he would become increasingly concerned about these demons.[29] Were there absolute limits to the power of thought itself that come from the natural limits of energy and information in the living and physical universe? He did not believe so.

Emboldened by Popper's criticisms of previous exorcisms, Feyerabend also turned to a study of the concept of information as used by physicists in relation to meaning more broadly, labeling his own contribution an "addendum" to Popper's work.[30] He read John Pierce's *Symbols, Signals, and Noise* textbook, which contained a description of a Maxwell's demon–type machine that showed how a perpetual machine could never be built because even information was subject to the laws of thermodynamics. Reinterpreting the machine described by Pierce, Feyerabend maintained that it "establish[ed] the exact opposite" of what the original author intended. He renamed it the "Hell Machine" (*Höllenmaschine*). For him, as for Popper, the power of our intellect would always remain larger than the sum of its parts.[31]

Toward the end of his life, Popper published *Postscript to the Logic of Scientific Discovery*, the culmination of a lifetime of work. Articles that he had started decades earlier saw the light of day, and his references to demons finally became public. Popper explained that all historical eras were characterized by a belief in a particular demon: "Great determinist thinkers produced their bogies in the eighteenth century, together with Laplace who produced the most famous bogy of all, the 'super-human intelligence' of his *Essay* of 1819, often called 'Laplace's demon.'"[32] It might be tempting to characterize Laplace's being as a deity, yet it was not, Popper warned, "an omniscient God, merely a super-scientist."[33] While trust in this being had waned, the popularity of others had grown. Moreover, "Maxwell's demon is no longer young, but he is still going strong," and "although innumerable attempts have been made on his life . . . almost from the day he was born, and although his non-existence has frequently been proved, he will no doubt soon celebrate his hundredth birthday in perfect health and vigour."[34] Quantum mechanics had not killed Maxwell's creation entirely.[35] "I for one," wrote Popper, "feel confident that he will survive us all, and that all the proofs and explanations of his non-existence, will, as fast as they are produced, be exposed as inconclusive.[36] Explanations of the second law that worked well at a quantum level lost power as scales increased, but they did not disappear.

Popper referred to the dominant demon of his time as "Gabor's Demon." This demon had tried to eliminate Maxwell's demon, only to leave scientists with a greater riddle. Questions pertaining to the validity of the second law of thermodynamics, to the relation between quantum mechanics and classical physics, and to indeterminism and determinism hinged on the possible existence of a demon who slipped back onstage when microscopic events added up sufficiently to create macroscopic effects. Popper's close analysis of the scientific work on particular demons showed "how very open the question still is, after almost a century of intensive study."[37] That effort would lead others to coin the term "Popper's demon" for a creature who could only completely know what had already passed and for whom the future remained unknown and open.

Popper's "Metaphysical Epilogue" was a far cry from his earlier work. He was no longer interested in studying how experiments could be used to falsify hypotheses, or in finding a hard-and-fast recipe for demarcating science from messier knowledge practices. He instead became interested in investigating the utility of metaphysics in the production of knowledge. In his epilogue, Popper dropped the word "demon" modified by the possessive form of "Laplace." He used instead the term "dream," referring to the "classical Laplacean dream of unlimited predictability" as one of the many research programs that drove scientific research.[38] These dreams could also be understood as "dream programmes," or "metaphysical research programmes" if they were systematically advanced. When used more informally, Popper thought of them simply as "pictures." While these pictures were marked by the "vice" of "irrefutability, or lack of testability," Popper argued that they were still extremely useful.[39] "Science needs these pictures," he stressed. "A new picture, a new way of looking at things, a new interpretation may change the situation in science completely." They were more than just tools. "These pictures are not only much-needed tools of scientific discovery, or guides to it," he wrote, "they also help us to decide whether a scientific hypothesis is to be taken seriously; whether it is a potential discovery, and how its acceptance would affect the problem situation in science, and perhaps even the picture itself."[40]

Despite the attention given to science's demons by these philosophers (and despite a growing acknowledgment of the need to think about science beyond a strict hypotheses-and-falsification model), most scholars continue to view the differences between premodern and modern forms of knowledge as directly tied to the rise of a culture of experimentation emerging from a decline in the political authority of the Church in monarchical Europe. Much is lost in such a generalization. One of the greatest losses is the knowledge that what first sets science on its path of discovery—then as now—is the continued and productive use of our imagination. What is key about science's demons is how they *become* real, that is, how our imagination drives discovery and how we can use it to change the world. The might of scientific ways of knowing and the virtuous ways in which they surpass dogmatic or authoritarian ones can be strengthened even more by acknowledging the power of nonexistent things across disciplines and throughout our universe.

An even more difficult challenge for the history and philosophy of science comes from the very nature of this topic. For philosophers of history, Laplace's demon was an "Ideal Chronicler," a being who had access to all of history as a giant data set. Knowing the science of this creature became key for historians concerned with the methodology of their discipline. The philosopher of history Arthur Danto described him as someone who "knows whatever happens the moment it happens, even in other minds. He is also to have the gift of instantaneous transcription: everything that happens across the whole forward rim of the Past is set down by him, as it happens the *way* it happens."[41] The philosopher and intellectual historian Hans Blumenberg understood clearly that historians had to examine their own belief in this creature in order to understand the limits and possibilities of their discipline. Blumenberg tried to learn all he could about him before writing one more thing about history. Particularly important for him was blackhole cosmology: he wondered about the possibility that historical information had already been swallowed by these cosmic giants.[42] Ever attentive to this possibility, he referenced Laplace's demon when thinking about how historians reconstructed narratives of the past. If the choices that determined "the subject" of history for them were implicitly related to beliefs in a certain demon, the historical record available to us was also at their mercy. "The subject of history," Blumenberg wrote, "is determined entirely by its never-before-known object, 'the history.' It is her Laplacian demon, for she knows it by virtue of the logic that it has in common with her."[43]

As other demons competed with Laplace's throughout the nineteenth and twentieth centuries, they showed the fragility of causal understandings of the universe and of history based on one-directional, chronological causation. Once time-twisting demons entered science, historians and philosophers, just as much as scientists, had to contend with the possibility of more complex forms of loopy historical interactions.

It is true, in some respects, that the complex relation between history and science leaves the historian at the mercy of science and its future discoveries. But what is most interesting is how it also leaves the scientist at the mercy of the historian. Scientists and historians are participant-observers of a cosmic drama that is much larger than them. This complication offers an opportunity—to study the sciences and the humanities together—even if it first appears to be much the opposite in disguise.

Notes

All translations from previously untranslated material are mine. Some quotations have been abbreviated without the use of ellipses, and capitalization has sometimes been changed. When available, archival and other sources have been consulted online. Early edition quotations are identified by part, section, scene, or act.

PREFACE

1. My interest in demons in science goes back more than a decade; I started researching them for my course "A Matter of Fact," taught at the Department of the History of Science at Harvard University in the spring of 2008. See Canales, *The Matter of Fact: Exhibition Catalog*, 2; Canales and Krajewski, "Little Helpers."
2. Carroll, *From Eternity to Here*, 400, n. 167.

INTRODUCTION

1. According to the 2014 edition of the *Oxford English Dictionary*, "demon, n. (and adj.)" "apparently aris[es] from Descartes' use of a powerful or omnipotent evil demon which attempts to mislead him as to the nature of reality and existence: see Descartes *Meditationes* (1641) i. §xii."
2. McKelvey, *Internet Daemons*.
3. Leff and Rex, "Overview," in *Maxwell's Demon*, 12; Leff and Rex, *Maxwell's Demon 2*. The physicist Friedel Weinert acknowledges the usefulness of demons as "conceptual models" in Weinert, *Demons of Science*, 3, 5. For a popular account of demons in physics and biology, see Davies, *The Demon in the Machine*.
4. First published as "Le Diable," *Revue du XIXe siècle* (June 1, 1866); republished in Baudelaire, "Le Joueur généreaux," 90.
5. Genesis 3:6.
6. Entrepreneur-inventor Elon Musk used the word to explain those aspects of technology that could not be so easily controlled. "With artificial intelligence we are summoning the demon," he warned. "In all those stories where there's the guy with the pentagram and the holy water, it's like—yeah, he's sure he can control the demon. It doesn't work out." McFarland, "Elon Musk."
7. Holton, "On the Art of Scientific Imagination." In the fury to understand science as an activity transcending the ruses of fiction, the seduction of poetry, the vicissitudes of politics, the imprecision of feelings, and the intolerance of religion, most scholars have neglected to study the role of the imagination in science. In the rare

cases when it is taken into consideration, it is often considered as belonging to the "context of discovery," delimited to those obscure (and largely mythical) eureka moments of occasional inspiration when prepared minds suddenly get the right idea, as if from nowhere. For the "context of discovery" in contrast to the "context of justification," see Reichenbach, *Experience and Prediction*.
8. A good introduction to the topic of thought experiments is Brown, *The Laboratory of the Mind*.
9. Most accounts from science and technology studies (STS) continue to think of imagination as an activity that *precedes* scientific work and is most evident in disciplines *outside* of science, such as in science fiction and literature, from where its impact is felt. An example of this position is represented by the argument that "technological innovation often follows on the heels of science fiction, lagging authorial imagination by decades longer." My focus contrasts with that approach by studying the use of the imagination in science concurrently with it, not before or outside of it, but simultaneously and within it. See Jasanoff, "Future Imperfect," 1.
10. Victor Hugo, *Les Misérables*, vol. V, book 1, chap. 5, 54–55.
11. "The technological-magical fairytale," argues Bloch, "aims at the transformation of things into utility goods which are available at any time." Bloch, *The Principle of Hope*, 367.
12. Sagan, *The Demon-Haunted World*, 201.
13. Ibid., 60.
14. Ayer, "Editor's Introduction," 14.
15. This saying is usually attributed to Louis Pasteur in a lecture at the University of Lille on December 7, 1854: "Dans les champs de l'observation le hasard ne favorise que les esprits préparés."
16. Born, *My Life and My Views*, 202–3.

CHAPTER 1. DESCARTES'S EVIL GENIUS

1. Descartes, *Discours de la méthode* (1637), part 1.
2. Descartes, *Meditationes de prima philosophia* (1641), meditation 1.
3. Descartes listed only one exception to the equivalence between artifacts and natural bodies: "except that the operations of artefacts are for the most part performed by mechanisms which are large enough to be easily perceivable by the senses—as indeed must be the case if they are to be capable of being manufactured by human beings." Descartes, *Les Principes de la philosophie*, part IV, sect. 203.
4. Nadler, "Descartes's Demon and the Madness of Don Quixote."
5. Descartes, *Discours de la méthode*, part 1, p. 7.
6. Cervantes Saavedra, *El ingenioso hidalgo Don Quixote de la Mancha*, chap. XLVII, n.p.
7. Ibid., 284.
8. Ibid., chap. XLVII, n.p.
9. Shakespeare, *The Tragicall Historie of Hamlet*, act III, scene 2. Shakespeare, his characters, and his public had good reasons to doubt the testimony of the senses—including their own—especially when it came to those featuring creatures of common legends. Tortured by not knowing if he should trust the horrible apparition he has just witnessed, the recently orphaned Hamlet thinks up an imaginative way to test its veracity. He commissions a theatrical performance—*The Murder of*

Gonzago—for the enjoyment of the court. His hopes for the evening run high. It is a play about a murder by poison. "The play's the thing, Wherein I'le catch the conscience of the King," he muses. By using fiction, he hopes to be able to get at the actual truth about the ghost and, by inference, learn more details about his father's untimely death. "It is an honest ghost, that let me tell you," he states. For Shakespeare, the play within a play serves as an experiment that transforms the court theater into a kind of laboratory, albeit one that is more humanistic than scientific. It is used not only to test Hamlet's ghost and catch a guilty conscience but to explore how beliefs are formed more generally.

10. Grafton, "The Devil as Automaton."
11. Shakespeare, *The Tempest*, act I, scene 2, in *Mr. William Shakespeares Comedies, Histories, & Tragedies.*
12. Shakespeare, *The Tragedie of Macbeth*, act 5, scene 5, in *Mr. William Shakespeares Comedies, Histories, & Tragedies.*
13. The full quotation is: "Being worldly wise, and taught by continual experience ever since the creation, he judges by the likelihood of things to come *according to the like that has passed before*, and the natural causes in respect of the vicissitude of all things worldly." King James I of England, *Daemonology, in Forme of a Dialogue* (1597), in Tyson, *The Demonology of King James*, 58.
14. "Because demonology is now a dead science," a scholar of Descartes reminds us, "it is easy to miss the fact that it incorporates an important chapter in the history of epistemology." There is no doubt that "demonology drew some of its tenets from theology; but since on the orthodox view demons were part of the natural rather than the supernatural world, able to work wonders but not true miracles by their super-human powers, their study also overlapped that of natural philosophy." Scarre, "Demons, Demonologists, and Descartes," 9.
15. Descartes, *Meditationes de prima philosophia*, meditation 2.
16. Ibid.
17. Descartes to the Illustris Academiae Lugduno-Batavae, May 4, 1647, special collections, University of Leiden library, the Netherlands.
18. Ibid.
19. Descartes, *Meditationes de prima philosophia*, meditation 2.
20. Ibid.
21. Ibid.
22. Descartes, *Meditationes de prima philosophia*, meditation 1.

CHAPTER 2. LAPLACE'S INTELLIGENCE

1. Laplace, "Recherches, sur l'integration des équations différentielles aux différences finies . . . ," sect. XXV (reprinted in Laplace, *Oeuvre complètes*, 144–45).
2. Laplace, *Essai philosophique sur les probabilités* (1814), 2.
3. Gillispie, *Pierre-Simon Laplace, 1749–1827*, 271.
4. Caws, *The Philosophy of Science*, 300.
5. Ehrenberg, *Dice of the Gods*, 26.
6. Georgescu-Roegen, "Energy and Economic Myths," 347.
7. Margenau, "Meaning and Scientific Status of Causality," 136.
8. Livingston, "Introduction," in *Disorder and Order*, 26.
9. Laplace, *Essai philosophique sur les probabilités*, 4.

10. Locke, "Of the Conduct of the Understanding," 9.
11. Locke, "A Discourse of Miracles," 231.
12. Hume, "Sceptical Doubts," in Hume, *Philosophical Essays Concerning Human Understanding*, 65.
13. Hume, "Of Miracles," in Hume, *Philosophical Essays Concerning Human Understanding*, 197.
14. Hume, *Dialogues Concerning Natural Religion*, part 10.
15. Laplace, "Un mémoire sur l'équation séculaire de la lune," 248–49.
16. Laplace, *Essai philosophique sur les probabilités*, 6.
17. Laplace dated the beginning of history to five thousand years earlier, since he was working with a slightly revised biblical scale. He placed the origin of the earth to approximately 1,826,213 days prior. Scientists have since revised their estimate of the age of the solar system to some 4.56 billion years. Using Laplace's logic, the chances that the sun will not rise tomorrow are 1 in 4.56 billion.
18. Arnauld and Nicole, *Logique de Port-Royal*, 389.
19. Arago, "Notice scientifique sur la tonnerre," 475.
20. Napoléon I, "Consuls provisoires," 330.
21. Gillispie, *Pierre-Simon Laplace, 1749–1827*, 177.
22. Fourier, *Théorie analytique de la chaleur*, iv.
23. Ibid., 9, xiv–xv.
24. Cited in Menabrea, "Sketch on the Analytical Engine Invented by Charles Babbage," 667.
25. Babbage, *The Ninth Bridgewater Treatise: A Fragment*, 1st ed., 112.
26. Ibid., 173–74.
27. Ibid., 111.
28. Ibid., 114–15.
29. Ibid., 113.
30. Babbage, *The Ninth Bridgewater Treatise: A Fragment*, 2nd ed., 111.
31. Babbage, *The Ninth Bridgewater Treatise: A Fragment*, 1st ed., 118.
32. Ibid., 116.
33. Somerville, *Personal Recollections*, 140.
34. Ada Lovelace to Charles Babbage, July 10, 1843, British Library, Add MS 37192.
35. Marx and Engels, *The Communist Manifesto*, 8.
36. Marx and Engels, *Das Kapital*, vol. 1, 367.
37. Carlyle, *Sartor Resartus*, 177. Carlyle's text was published in installments in *Fraser's Magazine*, starting in 1833.
38. Ibid., 170.
39. Ibid.
40. Pascal, *Éloge et pensées de Pascal*, article VI, no. XVIII.
41. Voltaire, *Le Siècle de Louis XIV*, 438. Voltaire similarly started to think about small changes that could have changed "the face of Europe." He speculated that the peace of Utrecht, which led to the ascension of the Tories and the beginning of parliamentary democracy in Europe, might not have occurred if Queen Anne had not gotten into a quarrel with Baroness Masham over a pair of gloves and a pail of water.
42. Menabrea, "Notions sur la machine analytique de M. Charles Babbage," 353, 373, 375.

43. Menabrea, "Sketch on the Analytical Engine Invented by Charles Babbage," 670, 675, 89.
44. Ibid., 675.
45. Lovelace, "Notes by the Translator," 697.
46. Charles Robert Darwin to Charles Babbage, [1838], British Library, Add 37191: 81, letter no. DCP-LETT-351.
47. Darwin to Babbage, [February 1839–August 1842], 19, British Library, Add 37191: 299, letter no. DCP-LETT-479.
48. Darwin, "Essay of 1842," sect. ii, p. 6.
49. Darwin, "Essay of 1844," sect. ii, p. 85.
50. Darwin to Asa Gray, from Down Bromley Kent, September 5, [1857], Archives of the Gray Herbarium, Harvard University, letter no. DCP-LETT-2136 (8 pp.).
51. Darwin, "Essay of 1844," sect. ii, p. 85, n. 1.
52. Riskin, *The Restless Clock*, 224.
53. Darwin to Charles Lyell, from Ilkley, Yorkshire, October 11, [1859], American Philosophical Society, letter no. DCP-LETT-2503 (22 pp.).
54. Huxley, "The Genealogy of Animals," 110.
55. Ibid.
56. "The world is to organisms in general," he argued, what "each organism is to the molecules of which it is composed." Huxley, "The Genealogy of Animals," 115.
57. Ibid.
58. Alfred Russel Wallace to Darwin, from Hurstpierpoint, July 2, 1866, Cambridge University Library, DAR 106: B33–8, letter no. DCP-LETT-5140 (12 pp.) (emphasis mine).
59. Wallace referred to the criticisms in Paul Janet's 1866 article, "The Materialism of the Present Day," 618–19.
60. Wallace to Darwin, July 2, 1866.
61. Wallace, *The Action of Natural Selection on Man*, 43 (emphasis mine).
62. Ibid., 53.
63. This phrase was taken up by Karl Pearson in *The Grammar of Science*.
64. Wallace, *Contributions to the Theory of Natural Selection*, 372.

CHAPTER 3. MAXWELL'S DEMON

1. Macduffie, "Irreversible Transformations."
2. Mahon, *The Man Who Changed Everything*.
3. Einstein, "Maxwell's Influence on the Development of the Conception of Physical Reality," 66.
4. Feynman, Leighton, and Sands, *The Feynman Lectures on Physics*, 2:11.
5. Harman, *The Natural Philosophy of James Clerk Maxwell*, 2.
6. Simberloff, "A Succession of Paradigms in Ecology," 10.
7. Watt, "Order and Disorder in the Arts and Sciences," 38.
8. Goldstein and Goldstein, *The Refrigerator and the Universe*, 212.
9. Leff and Rex, "Overview," in *Maxwell's Demon*, 12; Leff and Rex, *Maxwell's Demon 2*.
10. Some examples in literature include Stanislaw Lem's *The Cyberiad* (1965), Thomas Pynchon's *The Crying of Lot 49* (1965), and Ken Kesey's *Demon Box* (1986). For film, see Hollis Frampton's *Maxwell's Demon* (1968). For music, see Richard Einhorn's

"Maxwell's Demon" (1990) and Vintersorg's "A Metaphysical Drama." Other references include Boris and Arkady Strugatsky's science fiction novel *Monday Begins on Saturday* (1965), Arthur C. Parlett's "Maxwell's Demon and Monsieur Ranque" (1950), Max Frisch's novel *Homo Faber* (1957), Kosuke Fujishima's manga comic *Oh My Goddess*, and Christopher Stasheff's *Her Majesty's Wizard*.

11. Leff and Rex, "Overview," in *Maxwell's Demon*, 32; Leff and Rex, *Maxwell's Demon 2*, 39.
12. Add. MSS 7655, V, i/11(a), University Library Cambridge. See the facsimile reproduction in Price, "The Cavendish Laboratory Archives," facing p. 140; reprinted in Harman, ed., *The Scientific Letters and Papers of James Clerk Maxwell*, 3:185–86.
13. Peter Guthrie Tait to James Clerk Maxwell, December 6, 1867, Add. MSS 7655, I, a/4, University Library Cambridge; reprinted in Harman, ed., *The Scientific Letters and Papers of James Clerk Maxwell*, 2:328.
14. Maxwell to Tait, December 11, 1867, reprinted in Maxwell, *The Scientific Letters and Papers of James Clerk Maxwell*, 2:332.
15. Maxwell, *The Scientific Letters and Papers of James Clerk Maxwell*, 3:185–86.
16. Maxwell to John William Strutt, December 6, 1870, reprinted in Maxwell, *The Scientific Letters and Papers of James Clerk Maxwell*, 2:583.
17. Maxwell, *Theory of Heat*, 308–9 (emphasis mine).
18. Ibid., 238.
19. William Thomson, "The Kinetic Theory of the Dissipation of Energy," *Proceedings of the Royal Society of Edinburgh* 8 (1875); William Thomson, "Kinetic Theory of the Dissipation of Energy," *Nature* 9 (1874).
20. Thomson, "Kinetic Theory of the Dissipation of Energy," 442–43; Thomson, "The Kinetic Theory of the Dissipation of Energy," 329.
21. Cited in Gooday, "Sunspots, Weather, and the Unseen Universe."
22. Thomson, "Kinetic Theory of the Dissipation of Energy," 442.
23. Ibid.
24. Ibid.
25. Thomson, "The Sorting Demon of Maxwell," 145.
26. Thomson, "The Kinetic Theory of the Dissipation of Energy," 330; Thomson, "Kinetic Theory of the Dissipation of Energy," 443.
27. Thomson, "The Sorting Demon of Maxwell," 144, 148.
28. Maxwell, *Theory of Heat*, 308.
29. Maxwell, "Molecules," 438.
30. Thomson, "The Sorting Demon of Maxwell," 145 (emphasis in original).
31. Thomson, "Theoretical Considerations on the Effect of Pressure in Lowering the Freezing Point of Water," 575. See also James Thomson's remarks in the *Cambridge and Dublin Mathematical Journal* (November 1850).
32. John R. McClean to James Thomson, Walsall, November 15, 1841, reprinted in Thomson, *Collected Papers in Physics and Engineering*, xx–xxi.
33. Paper read before the Belfast Social Inquiry Society on March 2, 1852, reprinted in Thomson, "On Public Parks in Connexion with Large Towns . . . ," 469.
34. Thomson, "On the Age of the Sun's Heat," 375.
35. "Among other reasons it may be mentioned that the sun radiates out heat from every square foot of his surface at only about 7,000 horse power. Coal, burning at a rate of a little less than a pound per two seconds, would generate the same amount; and it is estimated (Rankine, *Prime Movers*, p. 285, ed. 1852) that, in the

furnaces of locomotive engines, coal burns at from one pound in thirty seconds to one pound in ninety seconds per square foot of grate-bars. Hence heat is radiated from the sun at a rate not more than from fifteen to forty-five times as high as that at which heat is generated on the grate-bars of a locomotive furnace, per equal areas." Thomson, "On the Age of the Sun's Heat," 368–69.
36. Tait, *Lectures on Some Recent Advances in Physical Science with a Special Lecture on Force*, 118.
37. Ibid., 119.
38. Ibid.
39. Ibid., 120.
40. Garnett, "Energy," in *Encyclopedia Britannica*, 210. This description also appears in the 1902 edition of the encyclopedia.
41. Bois-Reymond, *Über die Grenzen des Naturerkennens* (1872). A second edition followed that same year, a third was published in 1873, and then a fourth in 1876. In 1882, the work appeared in a fifth edition together with the "Sieben Welträthseln" in the *Monatsberichten der Berliner Akademie* (1880) and in the *Deutschen Rundschau* 28 (1881). It was republished again in 1886 as part of a volume on Bois-Reymond's lectures. A French translation appeared in the *Revue scientifique de la France et de l'étranger* 14 (1874), and again in 1882. The term "Geist" was translated as "seer expert" in English in *Popular Science Monthly* 5 (1874), where it was republished again in 1882. Bois-Reymond's work was published in Italian in *Giornale internazionale delle Scienze Mediche* 5 (1883).
42. Nernst, *Zum Gültigkeitsbereich der Naturgesetze*, 15.
43. Cassirer, *Determinismus und Indeterminismus in der modernen Physik*, 14.
44. Bois-Reymond, "The Limits of Our Knowledge of Nature," 18.
45. Ibid., 18.
46. Ibid., 18, 20.
47. Ibid., 20, 23, 29.
48. Es regnet, wenn es regnen will,
    Und regnet seinen Lauf;
    Und wenn's genug geregnet hat,
    So hört es wieder auf.
49. Helmholtz, "Wirbelstürme und Gewitter," 139.
50. Ibid.
51. Ibid., 140 (emphasis added).
52. Ibid., 163.
53. Neesen, "W. Thomson. Kinetic theory of the dissipation of energy," 673–74. "Wessen (Maxwell's Dämonen) die Fähigkeit haben, jedes einzelne Molecül in jedem Augenblick zu fassen."
54. Tait, *Lectures on Some Recent Advances in Physical Science with a Special Lecture on Force*, 118–19 (emphasis added).
55. Clausius, "Abschnitt XI: Discussionen über die mechanische Behandlung der Wärme und Electricität," 316, citing Tait, "Preface," xviii.
56. Zöllner, "Thomson's Daemonen und die Schatten Plato's." Zöllner claimed that Thomson and the demon enthusiasts around him were drawing inferences comparable to the ones drawn by Plato's prisoners, who can only see the outside world through the shadows it reflects. Their theories, Zöllner argued, were simplified outlines of a much more complex and imperfectly understood world beyond. Just how

deep was that world beyond our senses and barely under the purview of contemporary scientific theories? For Zöllner, who had an enduring interest in psychic experiments and spiritualism, its depth could not be restricted to the three dimensions of space. Throughout "Thomson's Daemonen und die Schatten Plato's," he refers to the creature under debate by the name of the one who actually named it: Thomson.

57. Tait, "Zöllner's Scientific Papers," 421. Tait discussed the possible existence of "beings (not of course *Thomson's Dämonen* nor *die Schatten Plato's*, for these are unscientific and therefore impossible) in a space of four dimensions."
58. Zöllner, "VI. On Space of Four Dimensions," 231. Republished as the first chapter of Zöllner, *Transcendental Physics*, 31–45.
59. Planck, "Verdampfen, Schmelzen, und Sublimieren," 475, n. 1. Planck cited p. 321 of the German translation of Maxwell's *Theory of Heat* by Felix Auerbach (1877) and p. 373 of Friedrich Neesen's translation (1878).
60. Loschmidt, "Über den Zustand des Wärmegleichgewichts eines Systems von Körpern mit Rücksicht auf die Schwerkraft" (1876), 139; continued in Loschmidt, "Über den Zustand des Wärmegleichgewichts eines Systems von Körpern mit Rücksicht auf die Schwerkraft" (1877), and Loschmidt, "Über den Zustand des Wärmegleichgewichts eines Systems von Körpern mit Rücksicht auf die Schwerkraft" (1878).
61. Loschmidt, "Über den Zustand des Wärmegleichgewichtes eines Systems von Körpern mit Rücksicht auf die Schwerkraft" (1876), 135. See also Loschmidt, "Der zweite Satz der mechanischen Wärmetheorie."
62. Gibbs analyzed the "important question which suggests itself in regard to any case of dynamical motion" of eternal recurrence, or "whether the system considered will return in the course of time to its initial phase, or, if it will not return exactly to that phase, whether it will do so to any required degree of approximation in the course of a sufficiently long time." Without referencing demons, Gibbs concluded that the answer was that it was possible, although extremely unlikely. Yet studying this mere possibility proved to be infinitely alluring. See Gibbs, *Elementary Principles of Statistical Mechanics*, 164.
63. Aphorism 341, in Nietzsche, *Die fröhliche Wissenschaft*, 254. These themes are also explored in his *The Will to Power*.
64. Aphorism 125, in Nietzsche, *Die fröhliche Wissenschaft*, 153–55.
65. Ibid.
66. Particularly relevant during those years were the experiments and calculations of Albert A. Michelson, author of the famous Michelson-Morley.
67. Eddy, "Radiant Heat and Exception to the Second Law of Thermodynamics," 342.
68. Ibid., 339–40.
69. Whiting, "Maxwell's Demons."
70. Dixon and Joly, "On the Ascent of Sap."
71. Pearson, *The Grammar of Science*, 100, n. 2.
72. Adams dedicated chapter 31 of his autobiography to Pearson's *Grammar of Science*, where he wrote: "The kinetic theory of gas is an assertion of ultimate chaos. In plain words, Chaos was the Law of Nature; Order was the dream of man." Adams, *The Education of Henry Adams*, 451, 57.
73. "The rule of Phase Applied to History," originally c. 1909, published in Adams, "The Rule of Phase Applied to History," 279.

74. Henry Adams to Cecil Spring Rice, November 11, 1897, in Adams, *Letters of Henry Adams (1892–1918)*, 135–36.
75. Henry Adams to Brooks Adams, May 2, 1903, in Cater, *Henry Adams and His Friends*, 545–46, 45. The constraints of thermodynamics on democracy and government later caused Henry Adams's pessimism to soar: "As an energy he [man] has but one dominant function:—that of accelerating the operation of the second law of thermodynamics. So far as his reason acts as an energy at all, it is a miraculous invention for this purpose, which inspires wonder and almost worship, but in strictness and reason, does not work,—it is only a mechanism; nature's energy which we have agreed to call Will, that lies behind reason, does the work,—and degrades the energy in doing it!" Adams, *A Letter to American Teachers of History*, 155.
76. I have translated *fingierte* as "invent." See Boltzmann, "Zur Erinnerung an Josef Loschmidt," 231.
77. Boltzmann, *Vorlesungen über Gastheorie*, 254.
78. Einstein to Mileva Marič, [September 13?, 1900], [Milan]. Unless otherwise specified, letters to and from Einstein are cited from Buchwald, *The Collected Papers of Albert Einstein*.
79. Mach, *Die principien der Währmelehre*, 364.

CHAPTER 4. BROWNIAN MOTION DEMONS

1. Arrhenius, "Award Ceremony Speech," 479.
2. Born, "Einstein's Statistical Theories," 166.
3. See, for example, Galison, *Einstein's Clocks, Poincaré's Maps*, 31.
4. Einstein read Poincaré's *Science and Hypothesis* (originally published in 1902), which included a version of Poincaré's text for the Congrès of 1900 with a group of friends who called themselves "the Olympians." The English translation refers to "Maxwell's demon" instead of "the imaginary demon of Maxwell." Poincaré, *Science and Hypothesis*, 179.
5. Gouy, "Note sur le mouvement brownien," 561.
6. Ibid., 564.
7. Jevons, "On the Movement of Microscopic Particles Suspended in Liquids," 174.
8. Ibid., 174.
9. Ibid., 175.
10. The reference is to the British microscopist John B. Dancer. Ibid., 171.
11. Ibid., 180.
12. Ibid., 171.
13. Ibid., 176.
14. Poincaré, "Le Mécanisme et l'expérience," 534. When Poincaré published his lectures on thermodynamics, he chose to avoid entirely the budding molecular program of Maxwell and company, leaving "completely aside the mechanical explanation of Clausius' principle which M. Tait calls 'the true (i.e. the statistical) basis of the second Law of Thermodynamics.'" Poincaré defended himself in the pages of *Nature*. "I desire to stay away completely from all the molecular hypotheses, no matter how ingenious they are," he stated firmly. Poincaré, "Poincaré's 'Thermodynamics.'"
15. Poincaré, "Le Mécanisme et l'expérience," 536–37.

16. Poincaré first started investigating these questions in Poincaré, "Sur les équations de la dynamique et le problème des trois corps."
17. Poincaré, "Le Mécanisme et l'expérience," 536.
18. Ibid.
19. Zermelo's work on this topic is described in Zermelo, "Ueber einen Satz der Dynamik und die mechanische Wärmetheorie." For later research on Loschmidt's demon and Zermelo's demon, see Rothstein, "Loschmidt's and Zermelo's Paradoxes Do Not Exist," and Rhim, Pines, and Waugh, "Time-Reversal Experiments in Dipolar-Coupled Spin Systems."
20. "On croirait voir à la oeuvre le demon de Maxwell." Poincaré, "Relations entre la physique expérimentale et la physique mathématique," 27; Poincaré, *Science and Hypothesis*, 179.
21. Poincaré, "Relations entre la physique expérimentale et la physique mathématique," 27.
22. Ibid.; Poincaré, *Science and Hypothesis*, 179.
23. Lecture delivered at the St. Louis Universal Exposition, St. Louis, MO, September 24, 1904. Reprinted as Poincaré, "The Principles of Mathematical Physics," in *Congress of Arts and Science, Universal Exposition, St. Louis, 1904* (1905); Poincaré, "The Principles of Mathematical Physics," *The Monist* (1905); Poincaré, "La Crise actuelle de la Physique mathématique," in *La Valeur de la science* (1905), 180–99; and, later, Poincaré, *The Value of Science* (1958), 97–98.
24. Poincaré, *Leçons sur les hypotheses cosmogoniques professées à la Sorbonne*, 252–53.
25. Campbell-Swinton, "Scientific Progress and Prospects," 193.
26. Carr, "Life and Logic," 490–91.
27. Ibid., 486.
28. Einstein, "Über die von der molekularkinetischen Theorie der Wärme geförderte Bewegung von in ruhenden Flüssigkeiten suspendierten Tielchen," 549; Einstein, "On the Movement of Small Particles Suspended in Stationary Liquids Required by the Molecular-Kinetic Theory of Heat," 123.
29. For more than a decade after Pearson posed his question, researchers worked hard to understand additional and particular cases (including cases with small n values and additional dimensions) in order to increase our understanding of stochastic processes. Pearson, "The Problem of the Random Walk."
30. Edgeworth, "On the Use of the Theory of Probabilities in Statistics Relating to Society."
31. Perrin, "Mécanisme de l'électrisation de contact et solutions colloïdiales," *Journal de chimie physique* 2 and 3.
32. Oseen, "Award Ceremony Speech," 135–36.
33. Perrin, "Mouvement brownien et realité moléculaire," 11–12.
34. Perrin, *Atoms*, 85.
35. Ibid., 87.
36. Ibid., 87, n. 1.
37. Pais, *"Subtle Is the Lord . . .": The Science and the Life of Albert Einstein*, 100.
38. Smoluchowski, "Essai d'une théoria cinétique du mouvement Brownien et des milieux troubles," 600.
39. Smoluchowski, "Gültigkeitsgrenzen des zweiten Hauptsatzes der Wärmetheorie," 396.
40. Ibid., 390, n. 1.

41. Smoluchowski, "Experimentell nachweisbare, der üblichen Thermodynamik widersprechende Molekularphänomene," 250.
42. Ibid., 247.
43. Ibid., 248.
44. Smoluchowski, "Gültigkeitsgrenzen des zweiten Hauptsatzes der Wärmetheorie," 396.
45. Ibid., 396–97.
46. Ibid., 362. See also Smoluchowski, "Vorträge über die Kinetische Theorie der Materie and Elektrizität."

CHAPTER 5. EINSTEIN'S GHOSTS

1. "Relativity," *Punch* (1923); Buller, "Letters to the Editor: Relativity Limerick."
2. Pearson, *The Grammar of Science*, 2nd ed., 540.
3. It "was suggested to me by Mr. L. N. G. Filon." Ibid.
4. Einstein, "Autobiographical Notes," 15, 53.
5. Ibid., 5.
6. Ibid., 53.
7. Bernstein, *Aus dem reiche der naturwissenschaft*, 151–52; Bernstein, *Naturwissenschaftliche Volksbücher*, supplemental volume, 100.
8. [Eberty], *The Stars and the Earth*, 48–49.
9. Ibid., 36–37, 50.
10. Ibid., 35, 49.
11. Ibid., 25–26.
12. Ibid., 32.
13. Hill, "Recommendatory Letters," 8.
14. Bernstein, *Naturwissenschaftliche Volksbücher*, 11:1.
15. Ibid.; Bernstein, *Naturwissenschaftliche Volksbücher*, 4:106.
16. Bernstein, *Naturwissenschaftliche Volksbücher*, supplement volume, 101. For other historical examples, including Moses, Abraham, Mohammed, Columbus, Luther, and the March Revolution of 1848, see Bernstein, *Aus dem reiche der naturwissenschaft*, 153.
17. Bernstein, *Aus dem reiche der naturwissenschaft*, 151.
18. Poincaré, "Le hasard," as reprinted in Poincaré, *Science and Method*, 71.
19. Poincaré, *Science and Method*, 71.
20. Ibid.
21. Flammarion, *Stories of Infinity*, 15, 47.
22. Ibid., 253 ("bird of the upper regions"); ibid., 240 ("dead leaves").
23. Ibid., 252.
24. Ibid., 160.
25. Ibid., 33.
26. Ibid., 41.
27. Ibid., 43.
28. Ibid., 98.
29. Ibid., 76.
30. Wells, "The Time Machine," 100.
31. Ibid.
32. Proctor, *Other Worlds than Ours*, 312.

33. Ibid., 317–18.
34. Ibid., 318.
35. Ibid., 313.
36. Ibid., 314.
37. Pohle, *Die Sternenwelten und ihre Bewohner*, 187–88.
38. Ibid., 188.
39. Pohle, *God: The Author of Nature and the Supernatural*, 349.
40. Poincaré, "The Principles of Mathematical Physics," 612; reprinted in Poincaré, *The Value of Science*, 100.
41. Einstein, "Zur Elektrodynamik bewegter Körper," 912; reprinted in Einstein, "On the Electrodynamics of Moving Bodies," 162.
42. Einstein, "Zur Elektrodynamik bewegter Körper," 892; reprinted in Einstein, "On the Electrodynamics of Moving Bodies," 141.
43. Einstein, "Zur Elektrodynamik bewegter Körper," 903; reprinted in Einstein, "On the Electrodynamics of Moving Bodies," 152.
44. Einstein, "Zum Relativitätsproblem," 345.
45. "Mais de même qu'on n'a jamais pu me faire croire à revenants, de même je ne crois pas à cette chose gigantesque dont tu me parles et que tu appele l'espace." Einstein, "Sur le problème de la relativité," 147.
46. Einstein to Arnold Sommerfeld, Berlin, November 28, [1915], and Einstein to Michele Besso, [December 10, 1915].
47. Einstein, *Relativity* (translation of *Über die spezielle und allgemeine Relativitätstheorie*), 30.
48. Einstein, *Relativity: The Special and the General Theory*, 61 (translation modified to reflect the original German: "Ein mystischer Schauer ergreift den Nichtmathematiker, enn er von 'vierdimensional' hört, ein Gefühl, das dem von Theatergespenst erzeugten nicht unähnlich ist").
49. Ibid., 75.
50. Eddington, *The Nature of the Physical World*, 121.
51. Einstein, "Die hauptsächlichen Gedanken der Relativitätstheorie."
52. Albert Einstein to Gustav Mie, February 8, 1918, Berlin ("spukt das Gespenst des absoluten Raumes").
53. Einstein, "Marian v. Smoluchowski," 108.
54. Albert Einstein to Sommerfeld, November 28, [1915], Berlin.
55. Johnson, *Modern Times*, 1.
56. Eddington, *The Theory of Relativity and Its Influence on Scientific Thought*, 28; reprinted in Eddington, "The Theory of Relativity and its Influence on Scientific Thought"; Eddington, *The Nature of the Physical World*, 122.
57. Eddington, *The Nature of the Physical World*, 121.
58. Eddington, *The Theory of Relativity and Its Influence on Scientific Thought*, 26; Eddington, *The Nature of the Physical World*, 119.
59. Isaac Newton to Richard Bentley, February 25, 1692.
60. Eddington, *The Theory of Relativity and Its Influence on Scientific Thought*, 27.
61. Ibid., 26.
62. Eddington, *The Nature of the Physical World*, 141.
63. Herschel, *A Treatise on Astronomy*, 222 (emphasis mine).
64. Eddington, *The Nature of the Physical World*, 126.
65. Eddington, *The Theory of Relativity and its Influence on Scientific Thought*, 28.

66. Ibid., 26–27.
67. Ibid.
68. Eddington, *The Nature of the Physical World*, 297.
69. Ibid., 121, 122.
70. Ibid., 121–22.
71. Ibid., 122.
72. Ibid., 118.
73. Ibid., 296–97.
74. Ibid., 297.
75. Langevin, "L'Évolution de l'espace et du temps."
76. Serviss, *The Einstein Theory of Relativity*, 74 (emphasis mine).
77. Einstein, "Geleitwort."

CHAPTER 6. QUANTUM DEMONS

1. Smolin, *Einstein's Unfinished Revolution*, 210.
2. Curie, Curie, and Bémont, "Sur une nouvelle sunstance fortement radio-active, contenue dans la pechblende," 1217.
3. Einstein, "Ist die Trägheit eines Körpers von seinem Energieinhalt abhängig?"
4. Lorentz, "Discourse d'ouverture," 6–7.
5. Albert Einstein to Michele Besso, Prague, October 21, 1911.
6. Born, "Max Karl Ernst Ludwig Planck, 1858–1947," 167.
7. "Discussion du rapport de M. Planck," in Langevin and de Broglie, *La Théorie du rayonnement et les quanta*, 128.
8. "Science on Road to Revolutionize All Existence," *New York Times* (September 28, 1913), M-6.
9. Soddy, *The Interpretation of Radium*, 10, 28.
10. Ibid., 21, 27.
11. Rutherford and Soddy, "LX. Radioactive Change."
12. Could he "predict the behavior of silver or of chlorine or the properties of silver-chloride without having observed samples of those substances than we can at present"? Broad delivered a strong rebuke against mechanical determinism by reference to the abilities and limitations of otherworldly beings. It was doubtful, in Broad's opinion, that nature would decide to act exclusively in a mechanical and predictable way across all its scales. "I cannot see the least trace of self-evidence in theories of the mechanistic type, or in the theory of Pure Mechanism which is the ideal towards which they strive," he wrote, unconvinced. Broad, *The Mind and Its Place in Nature*, 72–73.
13. Spengler, *Der Untergang des Abendlandes*, 165. The first sentence is modified and the second and third sentences omitted in the revised edition and English translation.
14. Ibid., 209.
15. Einstein to Born, January 27, 1920.
16. Einstein to Max and Hedwig Born, [Berlin], April 29, 1924.
17. Ibid.
18. Born, "Quantenmechanik der Stoßvorgänge," 804.
19. Born to Einstein, November 30, 1926.
20. Pais, *"Subtle Is the Lord . . .": The Science and the Life of Albert Einstein*, 443.

21. Paul Ehrenfest to Goudsmit, Uhlenbeck, and Dieke, November 3, 1927, reprinted in Meyenn, Stolzenburg, and Sexl, *Niels Bohr 1885–1962*, 152.
22. Translated as "On the Intuitive Content of Quantum-Theoretical Kinematics and Mechanics."
23. Einstein made a similar comment to Max Born in a December 1926 letter.
24. Bohr, "Discussion with Einstein on Epistemological Problems in Atomic Physics," 218.
25. Ibid., 206.
26. Born, *Physics in My Generation*, 55.
27. Compton, Costa, and Smyth, "A Mechanical Maxwell Demon," 349. Their coauthor Costa worked at the Woonsocket Rayon Company.
28. Unfortunately, they were unable to achieve their "original hope," which was to obtain a more detailed "velocity spectrum" of all the gas molecules. Ibid., 349.
29. Ibid.
30. In light of Oppenheimer's later *mea culpa* lamentations about his role in creating such a weapon, Neumann cynically noted: "Sometimes someone confesses a sin in order to take credit for it." Macrae, *John von Neumann*, 245.
31. Mises, *Wahrsheinlichkeit, Statistik un Wahrheit*, 188, 140.
32. Ibid., 176.
33. Mises, "Über kausale und statistische Gesetzmäßigkeit in der Physik," 147.
34. Eddington, *The Nature of the Physical World*, 297.
35. US Surgeon-General Hugh S. Cummings, "Transcript of the National Radium Conference," December 20, 1928, cited in Moore, *The Radium Girls*, 17.
36. Popper, *The World of Parmenides*, 196.
37. Szilard, "Part II: Leó Szilard: His Version of the Facts," 55.
38. Szilard's article was often cited as having solved the mystery of Maxwell's demon by reference to information. "One may therefore praise or blame Szilard for having opened the path leading to information theory and its mysteries." Ehrenberg, "Maxwell's Demon," 109. "Szilard published in 1929 a very remarkable paper on the problem of Maxwell's demon, and discovered for the first time the connection between information and entropy," recalled the physicist Léon Brillouin. "This was really pioneer work, and the importance of this paper was overlooked until recent developments of the theory brought it to the foreground." Brillouin, *Science and Information Theory*, 176.
39. Szilard, "Über die Entropieverminderung in einem thermodynamischen System bei Eingriffen intelligenter Wesen," 841.
40. Ibid., 851.
41. Ibid., 855.
42. Lewis, *The Anatomy of Science*, 148.
43. "The Anatomy of Science by G. N. Lewis," *Mathematical Gazette* (1927), 430.
44. Lewis, "The Conservation of Photons."
45. Lewis, "The Symmetry of Time in Physics," 573.
46. Ibid., 573.
47. Ibid., 572.
48. Ibid.
49. Ibid., 571.
50. Dannen, "The Einstein-Szilard Refrigerators," 92.
51. Compton, "The Uncertainty Principle and Free Will."

52. Ibid. Compton was citing Lillie, "Physical Indeterminism and Vital Action."
53. Lillie, "The Directive Influence in Living Organisms."
54. "Verachte nur Vernunft und Wissenschaft,/Des Menschen allerhöchste Kraft!/Du lässest doch von Blend- und Zauberwerken/Dich in dem Quantengeist bestärken!/Pass auf, wie jetzt die Schwierigkeiten schwinden,/Und wunderbar wirst du das *Neutron* finden!" Bohr and et. al., "Faust: Eine Histoire," 323.
55. Planck, "Die Kausalität in der Natur," in Planck, *Wege zur physikalischen Erkenntnis*, 237. The lecture was reprinted in Planck, *Vorträge und Erinnerungen* (1970) and left out of the English translation of Planck, *Where Is Science Going?*, for which Einstein wrote the preface. It was translated as "The Concept of Causality in Physics" in Planck, *The Philosophy of Physics*, and in Planck, *Scientific Autobiography and Other Papers*, 148.
56. Ibid.
57. Neumann, *Mathematical Foundations of Quantum Mechanics*, 400.
58. Ibid., 370.
59. Ibid., 359.
60. Ibid., 401.
61. Ibid., 369.
62. Neumann, *Mathematical Foundations of Quantum Mechanics*, 328.
63. Ibid., 369.
64. Ibid., 402.
65. Margenau, "Meaning and Scientific Status of Causality," 138, citing the work of the philosopher Philipp Frank, *The Law of Causality and Its Limits* (1931).
66. Margenau reviewed the work of scientists, philosophers, and mathematicians, including Richard von Mises, who had used the term "Laplace's demon" in his 1928 book on quantum mechanics and statistics. Margenau, "Causality and Modern Physics," 3.
67. Margenau, "Meaning and Scientific Status of Causality," 137.
68. Ibid., 138.
69. Margenau, *The Nature of Physical Reality*, 397.
70. Millikan, *Time, Matter, and Values*, 30.
71. Lichtenstein and Metz, "La Philosophie des mathématiques selon M. Émile Meyerson," 201.
72. Otto Neurath, "Protokollsätze," 204; reprinted as Neurath, "Protocol Sentences," 199.
73. A decade after these first warnings, Neurath would still complain to his colleague Rudolph Carnap that Carnap and some of his colleagues wanted to replace Laplace's ideal with a faulty alternative. He urged Carnap to deny the view "that in principle the Laplace's Demon is the picture of the scientist in action." Carnap accepted that current science could no longer be defined in terms of such an ideal, but still did not give up hope of finding another equally firm empirical and antimetaphysical foundation for it. Neurath to Carnap, April 1, 1944, cited in Reisch, *How the Cold War Transformed Philosophy of Science*, 196; original letter in Rudolf Carnap Collection, Archive of Scientific Philosophy, Hillman Library, University of Pittsburgh, Pittsburgh, PA.
74. Trotsky, *Soviet Economy in Danger*, 29–30.
75. Fedorov, *What Was Man Created For?*, 102.
76. Trotsky, *Soviet Economy in Danger*, 29–30.

340 – NOTES TO CHAPTER 6

77. Laurence, "New 'Gun' Speeds Break-up of Atom."
78. "That is to say, when I set up an experiment I assume that no god, angel, or devil is going to interfere with its course and this assumption has been justified by such success as I have achieved in my professional career." Haldane, *Fact and Faith*, vi.
79. Eddington, *New Pathways in Science*, 64.
80. Ibid., 62.
81. Ibid.
82. Ibid., 69.
83. Ibid., 70.
84. "Holds Roosevelt, Turns Back Chaos: Karl Compton Says That His 'Intelligent Planning' Is Staying Nature's Trend," *New York Times* (June 23, 1934) (reporting on Karl Compton, address before the AAAS, University of California at Berkeley, June 21, 1934).
85. Ibid.
86. Compton, "Science and Prosperity," 388.
87. "Holds Roosevelt, Turns Back Chaos."
88. Compton, "Science and Prosperity," 391.
89. Ibid., 387.
90. Ibid., 391.
91. Einstein to Born, December 3, 1947, cited in Born, *Natural Philosophy of Cause and Chance*, 122.
92. Compton, *The Freedom of Man*, 45.
93. Ibid., 46.
94. Ibid., 7.
95. Ibid., 32–36.
96. Ibid., 60.
97. Ibid., 31.
98. Ibid., 38–39.
99. Ibid., 39–40.
100. Ibid., 46.
101. Ibid., 39.
102. Ibid., 61, n. 11.
103. Ibid., 62.
104. Ibid., 6, citing a passage from Barnes, *Scientific Theory and Religion*, 578.
105. Compton, *The Freedom of Man*, 6.
106. Ibid., 50.
107. "When we view men's actions in the light of science we are thus presented with a new hope," he wrote. "Loyalty to our Maker, who has given us the ability, opportunity and responsibility to mold our lives and our world according to a more perfect pattern cannot but aspire us to work with him heart and soul toward this great end," he concluded. Ibid., 119.
108. Schrödinger, "Die gegenwärtige Situation in der Quantenmechanik," 812.
109. Albert Einstein to Erwin Schrödinger, August 8, 1835, cited in Fine, *The Shaky Game*, 78.
110. Schrödinger, "Die gegenwärtige Situation in der Quantenmechanik," 812.
111. Ibid., 849.
112. Ibid., 812.

113. Erwin Schrödinger to Albert Einstein, August 19, 1935, and Einstein to Schrödinger, September 4, 1935, cited in Fine, *The Shaky Game*, 82–84.
114. Heisenberg, *Physics and Beyond*, 117. For "cleverwoman," see Gustav Heckmann to Grete Hermann, December 17, 1933, reprinted in Crull and Bacciagaluppi, *Grete Hermann*, 221–22.
115. Hermann, "Die Naturphilosophischen Grundlagen der Quantenmechanik," 718. A longer version of this article appeared the same year (1935) under the same title in *Abhandlutlgen der Fries'schen Schule*. For an English translation, see Hermann, "Natural-Philosophical Foundations of Quantum Mechanics," in Crull and Bacciagaluppi, *Grete Hermann*.
116. Hermann, "Natural-Philosophical Foundations of Quantum Mechanics," 260.
117. Ibid., 246.
118. Ibid., 260.
119. Ibid., 241.
120. Ibid.
121. Flint, "Letter to the Editor: A Limit to the Quantum Theory and the Avoidance of Negative Energy Transitions," 313. The letter writer was referring to lectures by the physicist William Wilson.
122. Cassirer, *Determinismus und Indeterminismus in der modernen Physik*, 7.
123. Watson, *Scientists Are Human*, 147.
124. Seaborg, *Adventures in the Atomic Age: From Watts to Washington*, 33.
125. Szilard, "Creative Intelligence and Society." Two drafts of this typescript, dated July 31, 1946, are kept in Leo Szilard Papers, MSS 32, Special Collections and Archives, University of California–San Diego Library.
126. Compton, *The Human Meaning of Science*, 52–53.
127. Ibid., 48–49.
128. Ibid., 35.
129. Albert Einstein et al. to Franklin D. Roosevelt, August 2, 1939, Peconic, Long Island, Franklin D. Roosevelt Presidential Library and Museum.
130. Bridgman, *The Nature of Thermodynamics*, 158.
131. Ibid., 5–6.
132. Ibid., 158.
133. Ibid., x.
134. Quoted in Monk, *Robert Oppenheimer*, 316–17.
135. Ibid., 317.
136. Henry L. Stimson, diary entry, May 31, 1945, Sterling Memorial Library, Yale University, New Haven, CT, cited in Schmitz, *Henry L. Stimson*, 182.
137. Smyth, *Atomic Energy for Military Purposes*, 223.
138. Ibid., 1.
139. Ibid., 2.
140. Farm Hall transcripts, last page, Operation "Epsilon" (August 6–7, 1945), RG 77, entry 22, box 164, National Archives and Records Administration, College Park, MD.
141. Szilard, "Creative Intelligence and Society," 20.
142. Quoted in Frisch, *What Little I Remember*, 159.
143. Albert Einstein to Michele Besso, April 21, 1946, Princeton, NJ. In the letter, Einstein, hoping to be excused for his long period of silence, explained that "the

problem-demon hardly leaves me a minute free, so that I am wearing out my last remaining teeth on mathematical difficulties."
144. Compton, "The Scattering of X-ray Photons," 84.
145. Ibid., 83.
146. Ibid.
147. Ibid., 84.
148. Beyler, "The Demon of Technology, Mass Society, and Atomic Physics in West Germany."
149. Born, *My Life and My Views*, 88.

CHAPTER 7. CYBERNETIC METASTABLE DEMONS

1. Although the coinage of the term is usually credited to Norbert Wiener, an important precedent can be found in André-Marie Ampère's use of the term "cybernetique" to refer to the science of governing.
2. Quoted in Dyson, "The Usefulness of Useless Knowledge," 92.
3. Wiener, "Time, Communication and the Nervous System," 207.
4. Ibid., 208.
5. Platt, "Books That Make a Year's Reading and a Lifetime's Enrichment," BR6.
6. Thurston, "Devaluing the Human Brain," 24.
7. Wiener, *Cybernetics*, 71.
8. Ibid., 12.
9. Ibid., 72.
10. Ibid., 155.
11. Shannon, "*Cybernetics, or Control and Communication in the Animal and the Machine*, by Norbert Wiener."
12. Shannon, "A Mathematical Theory of Communication," pp. 626–27, n. 4.
13. Reichenbach, *Elements of Symbolic Logic*, 390. For another reference to Laplace's "superman," see Reichenbach, *The Rise of Scientific Philosophy*, 162–63.
14. Henry-Hermann, "Die Kausalität in der Physik," 382.
15. Ibid., 380, 382.
16. Brillouin, "Life, Thermodynamics, and Cybernetics," 566.
17. Ibid., 565.
18. Ibid., 565–66.
19. Ibid., 561.
20. Ibid., 560.
21. Ibid., 566.
22. Ibid., 566, n. 11.
23. Ibid.
24. Ibid., 560.
25. Ibid., 561.
26. Ibid., 566.
27. Ibid., 567.
28. The philosopher Rudolph Carnap was among many philosophers who tried to parse out the implications of this research more broadly. He resented the erasure of any sense of *meaning* from information theory. "Prevailing theory of communication (or transmission of information) deliberately neglects the semantic aspects of communication, i.e., the meaning of the messages," he wrote with his coauthor

Yehoshua Bar-Hillel, a mathematician and linguist. The debate between those who sought a purely physical definition of information, such as Shannon, and those who stressed its semantic aspects came to a head at Princeton University in 1952. When Neumann, along with Wolfgang Pauli, represented the former position in a conversation with Carnap, they dissuaded him from publishing his work. Their strong-arming approach was controversial, and Carnap continued to warn scientists to think about the "important relations between the two concepts" of entropy in information theory and in thermodynamics instead of simply equating them uncritically. Carnap and Yehoshua, *An Outline of a Theory of Semantic Information*; Köhler, "Why von Neumann Rejected Carnap's Dualism of Information Concepts"; see also Carnap, *Two Essays on Entropy*. Also relevant are the views of Bertrand Russell, who bristled at the mere suggestion "that communication theory is a branch of physics with the implication, at least, that the former is reducible to the latter." Tillman and Russell, "Language, Information, and Entropy," 127.
29. Brillouin, *Science and Information Theory*, 2nd ed., 168, n. 11.
30. "L. Szilárd took this up [in 1929], and cleared the ground first by showing that a simple observation, which amounts to a selection from n equally likely possibilities, enables the observer to decrease the entropy of the system observed by a maximum of $k \log n$." Gabor, "IV: Light and Information," originally delivered as the Ritchie Lecture on March 2, 1951, at the University of Edinburgh.
31. Gabor, "IV Light and Information," 132.
32. Gabor, *Lectures on Communication Theory*, 4.
33. Brillouin, "Maxwell's Demon Cannot Operate," 335, n. 7.
34. Ibid., 334.
35. Ibid., 336.
36. Ibid., 334.
37. Ibid., 337.
38. Wiener, "Entropy and Information," in *Proceedings of the Symposia of Applied Mathematics of the American Mathematical Society* (89), *Mathematical Review* (305), and Wiener, *Collected Works with Commentaries* (202).
39. Wiener, *The Human Use of Human Beings*, 29.
40. Ibid., 35.
41. Norbert Wiener, lecture delivered October 24, 1950, at New York Academy of Letters, reprinted as Wiener, "Men, Machines, and the World About," in Galdston, *Medicine and Science*. The original recording of the lecture is available at https://www.wnyc.org/story/men-machines-and-the-world-about-them/.
42. Rothstein, "Information, Measurement, and Quantum Mechanics," 174.
43. Rothstein worked at Evans Signal Laboratory; joined Edgerton, Germeshausen and Grier as a senior scientific executive; became the vice president and chief scientist of Maser Optics, Inc.; and then moved to the LFE Electronics Division of Laboratory for Electronics, Inc.
44. Rothstein, *Communication, Organization, and Science*, xciii.
45. Some biographical details on Rothstein appear in *Science* 134 (October 13, 1961): 1060; "Jerome Rothstein: Generalized Life," *Cosmic Search* 1, no. 2 (March 1979): 35; and "Contributors," *IEEE Transactions on Military Electronics* (April/July 1963): 271.
46. Wiener, "Cybernetics," reprinted in Wiener, *Collected Works with Commentaries*, 203.
47. Rothstein, "Information, Measurement, and Quantum Mechanics," 174.

48. Wiener, *Collected Works with Commentaries*, 203.
49. Einstein, "Autobiographical Notes," 57.
50. Albert Einstein to Jerome Rothstein, May 22, 1950, Albert Einstein Archives, Hebrew University of Jerusalem.
51. Pais, *"Subtle Is the Lord . . .": The Science and the Life of Albert Einstein*, 5.
52. Ibid., 9.
53. Bohm, *Quantum Theory*, 609.
54. Bohm, "A Suggested Interpretation of the Quantum Theory in Terms of Hidden Variables."
55. Ghirardi, "Properties and Events in a Relativistic Context," 353; Ghirardi, *Sneaking a Look at God's Cards*, 281, 83. For a recent use of the term "Bohm's demon," see Kiukas and Werner, "Maximal Violation of Bell Inequalities by Position Measurements."
56. "Je dois ressembler à une autruche qui sans cesse cache sa tête dans le sable relativiste pour n'avoir pas à regarder en face ces vilains quanta." Albert Einstein to Louis de Broglie, February 15, 1954, quoted in de Broglie, *Le Dualisme des ondes et des corpuscules dans l'oeuvre de Albert Einstein*, 31.
57. Heisenberg, *Das Naturbild der heutigen Physik*, 25.
58. Heisenberg, *Physics and Philosophy*, 56.
59. Heisenberg, *Physics and Beyond*, 117.
60. Brillouin, *Science and Information Theory*, 1st ed., 163.
61. Rothstein, *Communication, Organization, and Science*, xcii–xciii.
62. Ashby, "Communication, Organization, and Science. By Jerome Rothstein."
63. Muses, "Foreword," in Rothstein, *Communication, Organization, and Science*, lxx-lxxi.
64. Schmeck, "A Scientist Gives Demons Their Due," 2.
65. Ibid..
66. Rothstein, "Thermodynamics and Some Undecidable Physical Questions," 41, n. 1.
67. Rothstein, "Physical Demonology," 99.
68. Ibid., 111.
69. Ibid., 115.
70. Ibid., 116.
71. Rothstein, "Thermodynamics and Some Undecidable Physical Questions," 41, n. 1.
72. Rothstein, "Physical Demonology," 102.
73. Ibid.
74. Foerster, "On Self-Organizing Systems and Their Environments," 40–42.
75. Brillouin, "Information Theory and Its Applications to Fundamental Problems in Physics," 502.
76. Born, *Von der Verantwortung des Naturwissenschaftlers*, 100.
77. Ibid., 97.
78. Ibid., 98.
79. Ibid.
80. Ibid., 99.
81. Ibid., 100.
82. Bell, "On the Einstein Podolsky Rosen Paradox," 195.
83. Ibid., 199.
84. Quoted in Bernstein, "John Stewart Bell: Quantum Engineer," 84.
85. Bastin, *Quantum Theory and Beyond*, ix.

86. Rothstein, "Informational Generalization of Entropy in Physics," in Bastin, *Quantum Theory and Beyond*, 291. For the continuation of his work on Loschmidt's demon and Zermelo's demon, see Rothstein, "Loschmidt's and Zermelo's Paradoxes Do Not Exist."
87. Scientists enlisted the help of even more imaginary friends to understand the world beyond Cartesian categories. In this context, "Wigner's friend" was thought up to explore the bizarre concatenations of effects that could start off from a single quantum effect. Wigner, "Remarks on the Mind-Body Question," in *Symmetries and Reflections: Scientific Essays of Eugene P. Wigner*, 171.

CHAPTER 8. COMPUTER DAEMONS

1. Turing, "Computing Machinery and Intelligence," 459.
2. Project MAC, Progress Report VI.
3. McKelvey, *Internet Daemons*, 56, 85.
4. Turing, "Computing Machinery and Intelligence," 440.
5. Ibid., 460.
6. Ibid., 459.
7. Ibid.
8. Ibid., 460.
9. Sutherland, "Opening Address," x.
10. Rosenblatt, "Two Theorems of Statistical Separability in the Perceptron," 423–24.
11. Ibid., 449.
12. Selfridge, "Pandemonium: A Paradigm for Learning," 513.
13. Ibid., 516.
14. Ibid., 523.
15. Donald M. MacKay cited his articles "Mentality in Machines" and "The Epistemological Problem for Automata" in *Automata Studies*.
16. "Discussion on the Paper by O. G. Selfridge," in *Mechanisation of Thought Processes*, 527.
17. McCarthy, "Programs with Common Sense."
18. McCarthy, "Ascribing Mental Qualities to Machines," 94.
19. "In observing a brain, one should make a distinction between that aspect of behaviour which is available consciously, and those behaviours, no doubt equally important, but which proceed unconsciously." "Discussion on the Paper by O. G. Selfridge," in *Mechanisation of Thought Processes*, 527.
20. Selfridge and Neisser, "Pattern Recognition by Machine," 66.
21. Ibid., 60.
22. Ibid., 66.
23. Ibid.
24. Minsky, "Steps toward Artificial Intelligence," 14.
25. Laird and Rosenbloom, "The Research of Allen Newell," 22. He dated this conversion experience to "a Friday afternoon in mid-November 1954."
26. Newell, "Some Problems of Basic Organization in Problem-Solving Programs," 13, cited in Laird and Rosenbloom, "The Research of Allen Newell," 27.
27. Pierce, *Symbols, Signals, and Noise*, 290. "In order to know" what molecule to act on, a demon needed a modicum of information. "We need one bit of information, specifying which side the molecule is on," Pierce explained. Like his predecessors,

Pierce framed his account of Maxwell's demon in terms of its economic consequences, arguing that "the important thing is that even at the very best we could do more than break even." "The bit which measures amount of information used is the unit in terms of which the entropy of a message source is measured in communication theory. The entropy of thermodynamics determines what part of existing thermal energy can be turned into mechanical work. It seems natural to try to relate the entropy of thermodynamics and statistical mechanics with the entropy of communication theory." Ibid., 201. Pierce's short story titled "The Exorcism" speculated on how far some machine learning improvements might go.
28. Wiener, *Cybernetics*, vii.
29. In an earlier article, Brillouin had concluded that "thought creates negative entropy," citing the genius discoveries of Einstein on relativity and those of others on quantum mechanics. He promoted a view of the scientist as the ultimate intelligent actor who, much like a demon, could save us from the constraints of the natural world. Brillouin, "Thermodynamics, Statistics, and Information," 326.
30. Gleick, *Genius: The Life and Science of Richard Feynman*, 3.
31. Feynman, Leighton, and Sands, *The Feynman Lectures on Physics*, 1:46–3.
32. Ibid.
33. Ibid., 1:46–1.
34. Ibid., 1:46–3.
35. Ibid., 1:46–2.
36. Ibid., 1:46–5.
37. Ibid.
38. "How Will the New 'Golem' Help Israel?," *The Sentinel* (1965), 9.
39. Scholem, "The Golem of Prague and the Golem of Rehovot," in *The Messianic Idea in Judaism and Other Essays on Jewish Spirituality*, 340.
40. Ehrenberg, "Maxwell's Demon," 103.
41. Ehrenberg, "A Note on Entropy and Irreversible Processes."
42. Ehrenberg, "Maxwell's Demon," 107.
43. Ibid., 110.
44. Ibid., 109.
45. Ibid., 107.
46. Ibid., 110.
47. Ibid.
48. Ibid.
49. Ehrenberg, *Dice of the Gods*, 81.
50. Ibid., 90.
51. Ibid., 81; see also Born, "Physics in the Last Fifty Years."
52. Project MAC, Progress Report VI, 3.
53. Project MAC, Progress Report VI, 26.
54. Steinberg, "Mr. Smarty Pants Knows."
55. Project MAC, Progress Report VI, 26.
56. Ibid., 9.
57. Bhushan, "A File Transfer Protocol," 3.
58. Ibid., 1.
59. Seymour Papert, codirector of the Artificial Intelligence Laboratory, would join Charniak's project as thesis adviser, and Terry Winograd, with whom Charniak had worked closely, would offer advice.

60. For the use of the term "Charniak's demon," see Choong Huei Seow, "Of Frames, Scripts, and Stories."
61. Charniak, "Toward a Model of Children's Story Comprehension," 23, 37–38.
62. Ibid., 265.
63. Ibid. For example, the word "kleenex" would summon a set of demons that could move to establish the appropriate context of the story's possibilities: "the person would have a cold, or would be about to sneeze, and we would have a demon which could answer the question 'Why did Janet get the kleenex' in the story." Ibid., 254.
64. Ibid., 123.
65. Ibid., 224.
66. Meyer, "Infants in Children's Stories," 1.
67. Ibid., 1, 30, 7.
68. Ibid., 64.
69. Rieger, "Conceptual Memory," 150. Later Charles "Chuck" Rieger argued in favor of dropping the label "demons" and defining them simply as "watchers." Rieger, "Spontaneous Computation in Cognitive Models."
70. Penrose, *Foundations of Statistical Mechanics*, 225.
71. Ibid., 226.
72. Dotzler, "Demons-Magic-Cybernetics."
73. Foerster, "Responsibilities of Competence," 1–2.
74. Ibid., 1.
75. Ibid.
76. Foerster, "Disorder/Order: Discovery or Invention?," 183–84.
77. Laing, "Maxwell's Demon and Computation," 171.
78. Jauch and Báron, "Entropy, Information and Szilard's Paradox," 222.
79. Feynman, "Cargo Cult Science," 10.
80. Bekenstein, "Black-Hole Thermodynamics," 24.
81. Ibid., 25.
82. "Brillouin admits that the [Maxwell] demon can reduce the *thermal* entropy of the gas. But he points out that the demon requires information to do this," Bekenstein noted. "For example, it must 'see' which molecules are fast and which are slow. Brillouin shows that in acquiring this information (receiving a photon in its eye's retina), the demon inevitably causes an increase in the entropy of the universe." Bekenstein, "Baryon Number, Entropy, and Black Hole Physics," 109.
83. Ibid., 110–11, 17.
84. Ibid., 25, 26.
85. Ibid., 28.
86. Hawking, *A Brief History of Time*, 104.
87. Hawking, "Particle Creation by Black Holes," 201.
88. Hawking, "Black Holes and Unpredictability," 23.
89. Ibid., 24.
90. Hawking, *A Brief History of Time*, 89.
91. Hawking, "Chronology Protection Conjecture."
92. Reuell, "Hawking at Harvard." Hawking first developed these thoughts in his 1997 lecture "Into a Black Hole," delivered in Santiago, Chile.
93. Hawking and Penrose, *The Nature of Space and Time*, 37; see also Hawking and Penrose, "The Nature of Space and Time" (*Scientific American*).

94. Grover Maxwell, in "Open Peer Commentary to 'Minds, Brains, and Programs,'" 437.
95. For a recent survey of the relevant literature and a bibliography, see Mooney, *Searle's Chinese Room and Its Aftermath*.
96. Searle, "Minds, Brains, and Programs," 417.
97. Ibid., 423.
98. "Open Peer Commentary to 'Minds, Brains, and Programs,'" 431.
99. Searle, "Minds, Brains, and Programs," 419.
100. "Open Peer Commentary to 'Minds, Brains, and Programs,'" 428.
101. Ibid., 430.
102. Ibid., 428.
103. Ibid., 445–46.
104. Ibid., 440.
105. Ibid., 433–34.
106. Searle, "Author's Response," 452.
107. "Open Peer Commentary to 'Minds, Brains, and Programs,'" 432.
108. Searle, "Author's Response," 452.
109. Ibid., 456.
110. "Open Peer Commentary to 'Minds, Brains, and Programs,'" 433.
111. Ibid.
112. Ibid., 447.
113. Donald O. Walter, in ibid., 449.
114. These suggestions were those of Haugeland (432), Dennett (439), Hofstadter (434), Roland Puccetti (441), Robert P. Abelson (424), William G. Lycan (435), Robert Wilensky (450), and Howard Ratchlin (444) in "Open Peer Commentary to 'Minds, Brains, and Programs.'"
115. Hofstadter, "Reflections," 378.
116. William G. Lycan, referring to Jerry Fodor, in "Open Peer Commentary to 'Minds, Brains, and Programs,'" 435, n. 5.
117. Hofstadter and Dennett, *The Mind's I*, 474.
118. The authors restated Haugeland's response to the Chinese room argument. "What Haugeland wants us to envision is this: A real woman's brain is, unfortunately, defective. It no longer is able to send neurotransmitters from one neuron to another." "Luckily, however, this brain is inhabited by an incredibly tiny and incredibly speedy Haugeland's demon, who intervenes every single time any neuron would have been about to release neurotransmitters into a neighboring neuron." What could it do? "This demon 'tickles' the appropriate synapse of the next neuron in a way that is functionally indistinguishable, to that neuron, from the arrival of genuine neurotransmitters." Like others of his ilk, "the H-demon is so swift that he can jump around from synapse to synapse in trillionths of a second, never falling behind schedule." The resulting behavior was no different than if the so disabled woman had carried out the thinking: "In this way the operation of the woman's brain proceeds exactly as it would have, if she were healthy." "Now, Haugeland asks Searle, does the woman still think—that is, does she possess intentionality—or . . . does she merely 'artificially signal'?" Hofstadter, "Reflections," 378.
119. Searle, "The Myth of the Computer."
120. Ibid.

121. Searle, "Minds, Brains, and Programs," 423.
122. Weizenbaum, *Computer Power and Human Reason*, 189.
123. See "Rolf Landauer, A Unique Physicist, Dies," *Physics World* (1999).
124. Johnson, "Rolf Landauer, Pioneer in Computer Theory, Dies at 72."
125. Feynman, *Feynman Lectures on Computation*, 150.
126. Ibid., 160.
127. Ibid., 149.
128. Ibid., 213.
129. Ibid., 212.
130. Ibid., 152–53.
131. Ibid., 148.
132. Ibid., 52.
133. Ibid., 146.
134. Landauer, "Irreversibility and Heat Generation in the Computing Process," 181.
135. Bennett, "Logical Reversibility of Computation," 531.
136. Ibid., 525.
137. Bennett, "The Thermodynamics of Computation: A Review," 906.
138. Ibid.
139. This mind, he argued, was exceedingly simple: "The demon's mind has three states: its standard state S before a measurement, and two states L and R denoting the result of a measurement in which the molecule has been found on the left or right, respectively." Ibid., 928.
140. For "Landauer's demon," see Davies, "Universe from Bit," 85.
141. Ibid., 191.
142. Feynman, *Feynman Lectures on Computation*, 184.
143. Ibid., 7.
144. Ibid., 21.
145. Ibid., 166, n. 6.
146. Kornman, "Pattern Matching and Pattern-Directed Invocation in Systems Programming Languages," 96.
147. Ibid.
148. Newell, Laird, and Rosenbloom, "Proposal for Research on SOAR."
149. "Small Wonder," *Los Angeles Times* (July 30, 1986).
150. Feynman, *Feynman Lectures on Computation*, 258.
151. Minsky, *The Society of Mind*, 274.
152. Bennett, "Demons, Engines, and the Second Law," 108.
153. Ibid., 111.
154. Ibid., 116.
155. Ibid.
156. Ibid.
157. Landauer, "Computation: A Fundamental Physical View," 90.
158. Ibid., 88.
159. Herman Weyl, cited in ibid., 90.
160. Ibid., 88.
161. Baeyer, *Maxwell's Demon*, 145, 46, 49, 52.
162. See Rumelhart and McClelland, *Parallel Distributed Processing: Explorations in the Microstructure of Cognition*. The practice of "teaching" computers to "understand" by employing demons became routine by the late 1980s. Charniak's AI textbook

Artificial Intelligence Programming used the basic example of teaching a computer to understand basic family relationships. If a computer was given a file with proper names tagged to kinship categories (sibling, parent, gender, age), it could then be asked to answer more difficult questions that were not already tagged in the database by the programmer. It could determine who was an uncle or aunt merely by having been told that an aunt or uncle was a sibling of a parent of a certain gender. Demons could then be programmed to go through the database and automatically add other details that could then be queried, such as whether someone was older than someone else, after having determined that the person was someone's parent, aunt, or uncle. "So, if we are told that Wotan is Sigmund's father, a demon might be set off that adds the new fact, 'Wotan is older than Sigmund,'" explained Charniak and his coauthors. Charniak, *Artificial Intelligence Programming*, 240.
163. The *modus operandi* of these machines was "similar to that of very young children: successive cycles of strategy reformulation lead Soar through stages of cognitive development much like those studied by the pioneering child psychologist Jean Piaget." Waldrop, "Soar: A Unified Theory of Cognition?," 298.
164. Waldrop, "Toward a Unified Theory of Cognition," 28.
165. Waldrop, "Soar: A Unified Theory of Cognition?," 296.
166. Waldrop, "Toward a Unified Theory of Cognition," 28.
167. Waldrop, "Soar: A Unified Theory of Cognition?," 298.
168. Waldrop, "Toward a Unified Theory of Cognition," 28.
169. Ibid.
170. Ronald Reagan, "The Triumph of Freedom," quoted in Rule, "Reagan Gets a Red Carpet from British."
171. Leffler, *The Design and Implementation of the 4.3BSD UNIX Operating System*, 419. See "Devil Book" entry in Raymond, *The New Hacker's Dictionary*, 125.
172. Nemeth et al., *Unix System Administration Handbook*, 403.
173. Raymond, *The New Hacker's Dictionary*, 118.
174. Ibid., 124.
175. Andrew Leonard, *Bots*, 31.
176. Ibid., 10–11.
177. Ibid., 29.
178. Ibid., 26–27.
179. Ibid., 27.
180. Ibid.
181. McKelvey, *Internet Daemons*, 56, 4.
182. Ibid., 56, 67.
183. Pinker, *How the Mind Works*, 111.
184. Aspden, "Scientific Correspondence: The Law Beats Maxwell's Demon."
185. Maddox, "Maxwell's Demon Flourishes."
186. Caves, "Quantitative Limits on the Ability of a Maxwell Demon to Extract Work From Heat"; Caves, "Maxwell's Demon: From Physics Outlaw to Potent Teachers," 51.
187. Maddox, "Maxwell's Demon Flourishes."
188. Bown, "Science: A Demon Blow to the Second Law of Thermodynamics?"
189. Caves, Unruh, and Zurek, "Comment on 'Quantitative Limits on the Ability of a Maxwell Demon to Extract Work from Heat,'" 1387.

190. Bown, "Science: A Demon Blow to the Second Law of Thermodynamics?"
191. Baeyer, *Maxwell's Demon: Why Warmth Disperses and Time Passes*, 165. The main reference for Zurek's reinterpretation was Smoluchowski's early twentieth-century research. "Smoluchowski, in his discussion of the famous 'trapdoor' (an automated version of Maxwell's demon)," recounted Zurek, concluded that "the second law is safe even from 'intelligent beings.'" But Zurek had one complaint about Smoluchowski's definition of intelligent beings as those whose "abilities to process information are subjected to the same laws of these universal Turing machines." If they were defined differently, then other conclusions might be reached. It was the similarity and difference between intelligent beings and Turing machines that caught Zurek's attention. Zurek, "Thermodynamic Cost of Computation, Algorithmic Complexity and the Information Metric," 124.
192. Caves, Unruh, and Zurek, "Comment on 'Quantitative Limits on the Ability of a Maxwell Demon to Extract Work from Heat.'"
193. Bown, "Science: Sacred Law of Physics Is Safe After All."
194. W. H. Zurek, "Algorithmic Randomness, Physical Entropy, Measurements, and the Demon of Choice," 395.
195. Ibid., 405. Zurek attributed this saying to George Santayana.
196. Lloyd, "Quantum-Mechanical Maxwell's Demon," 3374–75.
197. Ibid., 3381.
198. Shenker, "Maxwell's Demon and Baron Munchausen, 349, n. 6.
199. For these and other references, see University of Manchester, Leigh Group, http://www.catenane.net/pages/pub2010-2006.html.
200. Raizen et al., "Demons, Entropy, and the Quest for Absolute Zero."
201. European Commission grant agreement ID: 308850, part of FP7-ICT Specific Programme "Cooperation": Information and Communication Technologies, topic ICT-2011.9.1 Challenging Current Thinking.

CHAPTER 9. BIOLOGY'S DEMONS

1. "Notes," *Nature* (1899), 40.
2. Bergson, *Creative Evolution*, 37.
3. Driesch, *The Science and Philosophy of the Organism*, 2:199.
4. "Das von ihm Fingierte as wirklich erscheint: Es gibt seine "Dämonen": wir selbst sind sie." Driesch, *Philosophie des organischen*, 2:202.
5. Driesch, *The Science and Philosophy of the Organism*, 2:200.
6. Campbell-Swinton, "Scientific Progress and Prospects," 193.
7. Johnstone, *The Philosophy of Biology*, 116–18.
8. Lillie, "The Philosophy of Biology: Vitalism Versus Mechanism," 842. Lillie noted later that the idea that living systems might not be subject to the laws of thermodynamics was "followed by an active correspondence in the columns of *Nature*, in which Herbert Spencer, Karl Pearson, and other well-known scientific men took part." Lillie, "The Directive Influence in Living Organisms," 483, n. 12.
9. Holmes, "The Beginnings of Intelligence," 478.
10. The Andover-educated Reverend Samuel Phillips Newman Smyth was the pastor at the Centre Congregational Church, the oldest in Connecticut, and a member of the Yale Corporation. Smyth, *Through Science to Faith*, 113, 114.
11. Schrödinger, *What Is Life?*, 20.

12. Ibid., 71, 74.
13. Ibid., 3.
14. Gamow, *Mr. Tompkins Explores the Atom*, 12, 13, 17.
15. Leó Szilard to Niels Bohr, c. 1950, in Lanouette, *Genius in the Shadows*, nos. 315–316.
16. Cohen and Monod, "Bacterial Permeases," 190.
17. Keller, *Refiguring Life*, 54.
18. Smocovitis, "The 1959 Darwin Centennial Celebration in America."
19. Waddington, "Evolutionary Adaptation," 401.
20. Ibid., 399, 401.
21. "The 'feedback' circuit is the simple one, as follows: (1) environmental stresses produce developmental modifications; (2) the same stresses produce a natural selective pressure which tends to accumulate genotypes which respond to the stresses with co-ordinated adaptive modifications from the unstressed course of development; (3) genes newly arising by mutation will operate in an epigenetic system in which the production of such co-ordinated adaptive modifications has been made easy." Ibid., 398.
22. Waddington, "The Basic Ideas of Biology," 2. For Maxwell's demon and self-organizing systems, see Karl, "Towards a Physical Theory of Self-Organization."
23. Pittendrigh remembered the conversation as taking place in the early 1960s, but Pauli died in 1958. Pauli's correspondence includes two letters to Pittendrigh written on March 10 and 12 of 1954. Pittendrigh, "Temporal Organization: Reflections of a Darwinian Clock-Watcher," 19, note.
24. Ibid., 19.
25. Monod found Pittendrigh's concept of "teleonomy" useful for replacing the methodologically suspect idea of "teleology."
26. Pittendrigh, "On Temporal Organization in Living Systems," 19, 94.
27. Ibid., 94.
28. Pittendrigh, "Adaptation, Natural Selection, and Behavior," 395.
29. Ibid., 398.
30. In the nineteenth century, Samuel Butler, a contemporary critic of Darwin, raised the specter of living species not being as in control of their destinies as commonly thought: "It has, I believe, been often remarked, that a hen is only an egg's way of making another egg." Butler, *Life and Habit*, 134. Pittendrigh wrote: "When Samuel Butler, that sharp critic of Darwin, remarked that 'a hen is only an egg's way of making another egg,' he may have spoken more profoundly than he thought." Pittendrigh, "Adaptation, Natural Selection, and Behavior," 398.
31. Pittendrigh, "Adaptation, Natural Selection, and Behavior," 395.
32. Pittendrigh, "On Temporal Organization in Living Systems," 115.
33. Ibid., 94.
34. Williams, *Adaptation and Natural Selection*, 34.
35. Lewontin, "Is Nature Probable or Capricious?," 25.
36. Ibid.
37. Jukes, "DDT: Maxwell's Demon."
38. McClare, "Chemical Machines, Maxwell's Demon, and Living Organisms," 4. McClare cited Karl Popper, who "emphasized that if we define work simply as energy which can lift a weight, then, even without the help of demons, Brownian motion breaks the letter of the second law because it can undoubtedly lift weights (albeit small)." Ibid., 3.

39. Asimov, *View from a Height*.
40. Asimov, *Life and Energy*, 74.
41. Asimov, "Science: The Modern Demonology," 79.
42. Ibid., 75, 81–83.
43. Ibid., 76–78.
44. Ibid., 79.
45. Crick, "Obituary: Jacques Lucien Monod," 430.
46. Monod, *Chance and Necessity*, 115.
47. Ibid., 43.
48. Ibid., 145.
49. Ibid., 69.
50. Ibid., 61.
51. Ibid., 81.
52. Ibid., 94–95.
53. Ibid., 145.
54. Prigogine, "Time, Irreversibility and Structure," 589.
55. Prigogine referred to Monod, *Chance and Necessity*, 146.
56. Cited in "Ilya Prigogine," in Buckley and Peat, *Glimpsing Reality*, 105.
57. Monod, *Chance and Necessity*, 145; Prigogine, "Time, Irreversibility and Structure," 589.
58. Prigogine and Stengers, *La Nouvelle alliance*, 45, 95, 132.
59. Hopfield, "Kinetic Proofreading."
60. Ibid., 4137.
61. Eigen, "Self-organization of Matter and the Evolution of Biological Macromolecules," 516.
62. Eigen and Winkler, *Das Spiel*, 11.
63. Ibid., 182.
64. Ibid., 183.
65. One example of the use of the term "Darwinian demon" as a label for a successful predator, one who more often than not preyed on the young, see Law, "Optimal Life Histories under Age-Specific Predation."
66. Rosen, "Darwin's Demon," 370. Rosen found hope in epigenetics as a complement to genetics: "A renewal of interest in epigenetics has shown that some of the most powerful concepts for understanding the origin of evolutionary novelties have been around for a long time." Ibid., 373.
67. Eigen and Winkler, *Steps towards Life*, 122–23.
68. Ibid., 122. Translation modified to reflect original: "Welcher Art ist die Physik, die uns gestattet, Qualitäten wie 'richtig' oder 'falsch' zu definieren? Lässt sich die Biologie an dieser entscheidenden Stelle auf die Physik reduzieren? Wenn nicht, dann müssten wir das Vorhandensein eines Dämons annehmen, der außerhalb} der physikalischen Gesetze tätig wäre."
69. Dennett, *Consciousness Explained*, 261.
70. Quoted in Pinker, *How the Mind Works*, 79.
71. Dennett, *Consciousness Explained*, 3.
72. Dennett, *Brainstorms*, 123–24.
73. Dennett, *Consciousness Explained*, 261.
74. Quoted in Pinker, *How the Mind Works*, 79.
75. Quoted in ibid.

76. Dennett, *Consciousness Explained*, 439.
77. Ibid., 437–38.
78. Boden, "The Mind of a Very Special Machine?," 35.
79. Sullivan, "Myth, Metaphor, and Hypothesis," 215. Dawkins singled out one sentence that had helped him develop his selfish gene idea: "Thus Williams (1966), returning to Darwin's old problem of the evolution of the sterility in social insects says: 'Yet I believe that the sterility of the workers is entirely attributable to the unrelenting efforts of Darwin's demon to maximize a mere abstraction, the mean.'" Dawkins, "Reply to Lucy Sullivan," 222.
80. Pinker, *How the Mind Works*, 21.
81. Ibid., 111.
82. "The computational theory of mind rehabilitates once and for all the infamous homunculus," where the homunculus is absolutely present and essential. "Talk of homunculi is indispensable in computer science." Ibid., 79.
83. Ibid., 69.
84. Ibid., 79.
85. Ibid., 144.
86. Ibid., 73.
87. Ibid., 111.
88. Ibid., 144.
89. Ibid., 111.
90. Ibid., 99.
91. Dennett cited in ibid., 79.
92. Ibid., 131.
93. Ibid., 80.
94. Ibid., 94.
95. Dennett, *Darwin's Dangerous Idea*, 27.
96. Ibid., 209.
97. Ibid., 51.
98. "According to Dawkins (5) he is 'The Blind Watchmaker,' but for some time (6) I have known him as 'Darwin's Demon.'" Pittendrigh, "Temporal Organization: Reflections of a Darwinian Clock-Watcher," 19.
99. Ibid.
100. Ibid., 49.
101. Ibid., 20.
102. Ibid., 18–19.
103. Ibid., 20.
104. Morton, "The Manager as Maxwell's Demon" (1969).
105. Morton, "The Manager's Changing Role in Technological Innovation," 13, 15.
106. Morton, *Organizing for Innovation*, 122. For more on Maxwell's demon and management science, see Ericson, "The Impact of Cybernetic Information Technology on Management Value Systems."
107. Latour and Woolgar, *Laboratory Life*, 245.
108. Latour concluded that "print plays the same role as Maxwell's demon," by allowing users to sort materials more effectively. Latour, "Drawing Things Together," 34.
109. Serres, *The Parasite*, 43.
110. Ibid., 91.
111. Ibid., 58.

112. Ibid., 86.
113. Bourdieu, *Raisons pratiques*, 40.
114. Ibid., 46–47.
115. Zyga, "Maxwell's Demons May Drive Some Biological Systems"; Tu, "The Nonequilibrium Mechanism for Ultrasensitivity in a Biological Switch"; Dillenschneider and Lutz, "Memory Erasure in Small Systems."

CHAPTER 10. DEMONS IN THE GLOBAL ECONOMY

1. "On the latter Norbert Wiener once commented: 'Here there emerges a very interesting distinction between the physics of our grandfathers and that of the present day. In nineteenth century physics, it seemed to cost nothing to get information' (14, p. 29). In context, the statement refers to Maxwell's Demon—not, of course, to Walras auctioneer. But, mutatis mutandis, it would have served admirable as a motto for Keynes's work." Leijonhufvud, "Keynes and the Keynesians," 410.
2. Mirowski, *Machine Dreams*, 51, 46.
3. Samuelson, "Mathematics of Speculative Price," 18.
4. Ibid., 5.
5. Kendall, "The Analysis of Economic Time-Series-Part I: Prices," 13.
6. Ibid.
7. Comments by Professor R. G. D. Allen, "Discussion on Professor Kendall's paper," in Kendall, "The Analysis of Economic Time-Series: Part I: Prices," 25–26.
8. Krugman, "The Incomparable Economist."
9. Samuelson, "Interview," 43.
10. Samuelson, "Mathematics of Speculative Price," 5.
11. Samuelson, "Economics in My Time" (1986), reprinted in Samuelson, *Collected Economic Papers*.
12. Cootner, "Stock Prices: Random vs. Systematic Changes," 24.
13. Samuelson, "Mathematics of Speculative Price," 5.
14. Georgescu-Roegen, *Analytical Economics*, 119.
15. Ibid., 80–81.
16. Ibid., 92.
17. Ibid., 97.
18. Ibid., 127.
19. Ibid.
20. Ibid., 20.
21. Ibid., 90.
22. "Many still share the idea that the Walrasian system would be an accurate calculating device for a Laplacian demon." Ibid., 118.
23. Ibid., 118–19.
24. Ibid., 85.
25. Ibid., 78, n. 60.
26. Levallois, "Can De-Growth Be Considered a Policy Option?"
27. Georgescu-Roegen, *The Entropy Law and the Economic Process*, 178–79.
28. Ibid., 146.
29. Ibid., 24.
30. Ibid., 59.

31. Samuelson, "Challenge to Judgement," 19.
32. Ibid., 19.
33. Ibid., 17.
34. Solow, "The Economics of Resources or the Resources of Economics," 2.
35. Ibid., 10.
36. Ibid., 10–11.
37. Georgescu-Roegen, "The Steady State and Ecological Salvation," 267, n. 2.
38. Ibid., 270.
39. Ibid., 270.
40. Goeller and Weinberg, "The Age of Substitutability," 7.
41. Ibid., 10.
42. Weinberg, "Global Effects of Man's Production of Energy."
43. Georgescu-Roegen, "Energy Analysis and Economic Valuation," 1023.
44. Ibid., 1037.
45. Ibid., 1041. Georgescu-Roegen did not think that this could be represented completely in unambiguous mathematical terms. "The entropic nature of the economic process notwithstanding," he cautioned, "it would be a great mistake to think that it may be represented by a vast system of thermodynamic equations." Ibid., 1042.
46. Ibid., 1041.
47. Ibid., 1038.
48. Ibid., 1037 (emphasis mine).
49. Ibid., 1037–38.
50. Ibid., 1038.
51. Ibid., 1033.
52. Ibid., 1038.
53. Ibid., 1041.
54. Daly, "Entropy, Growth and the Political Economy," 69.
55. Weinberg, "On the Relation between Information and Energy Systems," 47.
56. Ibid., 49.
57. Ibid., 50.
58. Ibid., 49.
59. Ibid., 52.
60. Odum, *Systems Ecology: An Introduction*, 311.
61. Ayres and Nair, "Thermodynamics and Economics," 69, 71.
62. Samuelson, *Macroscopic Time Asymmetry of Maxwell's Demon*, 1.
63. Ibid., 4.
64. Ibid., 12–13. The original wording from *Macbeth*, act 5, scene 5: "Life's but a walking shadow, a Poore player, / That struts and frets his hour upon the Stage / And then is heard no more. It is a Tale / Told by an idiot, full of sound and fury, / Signifying nothing."
65. Samuelson, *Macroscopic Time Asymmetry of Maxwell's Demon*, 13.
66. Harrison and et. al., "Chernobyl in Context," 2.
67. Weinberg, "Are Breeders Still Necessary?," 345.
68. Ibid., 350.
69. Weinberg, "Social Institutions and Nuclear Energy—II," 33.
70. Weinberg, "A Nuclear Power Advocate Reflects on Chernobyl," 57.
71. Weinberg, "Social Institutions and Nuclear Energy—II," 1071.
72. Shleifer and Summers, "The Noise Trader Approach to Finance," 19.

73. Samuelson, "Scientific Correspondence: The Law Beats Maxwell's Demon," 24, 25.
74. Martínez-Alier, "In Memory of Georgescu-Roengen," 175.
75. Ibid., 176.
76. Ibid., 175.
77. Samuelson, "Foreword," in *Louis Bachelier's Theory of Speculation*, x. Samuelson placed the word "joke" in parentheses.
78. Cherry, "Jaron Lanier: We're Being Enslaved by Free Information."
79. Bezos, "2016 Letter to Shareholders."

CONCLUSION: THE AUDACITY OF OUR IMAGINATION

1. The sociologist and historian Thomas Kuhn aptly described this central characteristic as based on the "prohibition of appeals to heads of state or to the populace at large in matters scientific." Kuhn, *The Structure of Scientific Revolutions*, 168.
2. Reichenbach, *Experience and Prediction*, 403.
3. Ibid., 6.
4. Lévi-Strauss, *Mythologiques: Le Cru et le cuit*, 24.
5. Reichenbach, *The Rise of Scientific Philosophy*, 323.
6. Heidegger, "The Question Concerning Technology," 28.
7. Heidegger, *Einführung in die Metaphysik*, as cited in Heidegger, *Introduction to Metaphysics*, 48–49.
8. I am particularly indebted to Lorraine Daston for sharing with me learned and insightful knowledge on the continuities between early modern demons and modern ones. See Daston, "Marvelous Facts and Miraculous Evidence in Early Modern Europe"; Daston, "Intelligences: Angelic, Animal, Human."
9. Morton's demon was named after the geophysicist and ex-creationist Glenn Morton.
10. Matthew 25:31–46.
11. Consider the Marid *jinn*.
12. See Grimm, Deutsche *Mythologie*; Thiele, *Dänische Volkssagen*; and Carus, *History of the Devil and the Idea of Evil*, 251.
13. In the Grimms' fairy tale, the poor shoemaker is helped by elves, not demons. Gobelin tapestry was so named because goblins were suspected of helping with the incredibly intricate work.
14. The astounding traveling abilities of genies are described in *A Thousand and One Nights*, which features magic carpets, and in the Iranian poem *Shahnameh*, in which genies carry King Jamshid on his throne.
15. The Gospel of Luke relates that the devil took Jesus up into a high mountain and "shewed unto him all the kingdoms of the world in a moment of time." In the Gospel of Mathew, Satan transports Jesus effortlessly, "setteth him on a pinnacle of the temple," and "taketh him up into an exceeding high mountain." After pondering the question of "whether the angel's movement is in time or instantaneous," Saint Thomas Aquinas opted for the former. Aquinas, *Summa Theologiae*, part 1, question 53, article 3.
16. For the history of the transformation of Northern Teutonic giant myths into Christian devils, see Carus, *History of the Devil and the Idea of Evil*, 250. For the enormous size of the devil, see Aquinas, *Commentary on Job*, chap. 40.
17. See Caesarius von Heisterbach, *Dialogus Miraculorum* (c. 1245).

18. An example is Paulus Wann's *Quadragesimale siue tractatulus magistri* (c. 1512).
19. "They are the most cunning ones, as we shall see; their abilities of juggling information are so superlative that it boggles the mind." Loewenstein, *The Touchstone of Life*, 5.
20. "The quiet exit of demons from theology coincides in time and corresponds in structure almost exactly with the disappearance of the preternatural in respectable natural philosophy." Daston and Park, *Wonders and the Order of Nature 1150–1750*, 361.
21. The idea of a *cordon sanitaire* has led many historians to try to reconstitute the underlying logic of the beliefs of the eras marked by the persecution of witches and witchcraft in order to understand nonmodern, outdated modes of thinking. An exemplary account of how demonology made sense to a diverse group of people during those times is Clark, *Thinking with Demons*.

POSTSCRIPT: PHILOSOPHICAL CONSIDERATIONS

1. For the "really real," see Bergson, *Durée et simultanéité*, 95.
2. A classic study on models in science is Hesse, *Models and Analogies in Science*. For more recent scholarship, see Mauricio, *Fictions in Science: Philosophical Essays on Modeling and Idealization*.
3. Yourgrau, *Death and Nonexistence*, 35.
4. Bentham, *Panopticon*, 40.
5. Bentham, *Deontology; Or, The Science of Morality*, 2:10.
6. Bentham described God as an "inferential entity" that was "supreme" and "superhuman." He relegated "heathen Gods, Genii and fairies" to the category of the "fabulous" and considered "angels and devils" to be "superhuman inferential entities" that, in contrast to God, were "subordinate." Ogden, *Bentham's Theory of Fictions*, 9, 137, 16.
7. Meinong, "Über Gegenstandstheorie."
8. Yourgrau, *Death and Nonexistence*, 13.
9. Russell, "My Mental Development," 13.
10. For Russell on Meinong, see Ryle, "Intentionality-Theory and the Nature of Thinking." For revaluations of the thought of these two thinkers, see Griffin and Jacquette, *Russell vs. Meinong: The Legacy of "On Denoting."*
11. Vaihinger, *Die Philosophie des Als ob*.
12. Fine, "Fictionalism."
13. Yourgrau, *Death and Nonexistence*, 13.
14. Freud placed Weyer's *De Praestigiis Daemonum et Incantationibus ac Venificiis* on a par with Copernicus's *De Revolutionibus* and Darwin's *Descent of Man*. He was able to read the book after it had been reissued in response to the interest in diagnosing hysteria, an illness whose symptoms seemed to mirror those that appeared in classic descriptions of demonic possession. Freud produced his famous case study "A Seventeenth-Century Demonological Neurosis" (1923)—which centered on the case of the Bavarian painter Christoph Haizmann, who claimed to have signed a pact with the devil in 1668—after he was intrigued "by the resemblance of this story to the legend of Faust." The psychologist attributed the painter's fear of the devil and his experience of being possessed by him to a psychological neurosis. According to Freud, Haizmann suffered from an extreme form of "projection" that led

him to impute characteristics associated with his father to a constructed demonic figure. "We therefore come back to our hypothesis that the Devil with whom the painter signed the bond was a direct substitute for his father," Freud concluded. Freud, "A Seventeenth-Century Demonological Neurosis," 72.
15. Most demons and apparitions have since been eliminated from religious contexts. "The advances made in the natural sciences," noted the entry for "demon" in the progressive *New Catholic Encyclopedia* (1967), "has forever destroyed the crude concept of a three-storied world in which angels and demons materialize with ingenuous frequency" owing to "the present, more detailed knowledge of the universe." Following Freudian trends, most demons were pushed down deeper into our psyche: "Psychiatry, moreover, has shown that the workings of the subconscious explain many, if not most, of the abnormal conditions that earlier generations had attributed to diabolical activity." The widespread psychologization of the demonic resulted as well in the demotion of demonological knowledge, even within theology. "For these reasons and because the need to reorient theology along more positive lines has been recognized, demonology has not been the object of very much serious study in the 20th century." Elmer, "Demon (Theology of)," 756.
16. Kuhn, "Postscript—1969," 175.
17. Hoyningen-Huene, "Two Letters of Paul Feyerabend to Thomas Kuhn," 364.
18. Ibid., 369.
19. Feyerabend, *Realism, Rationalism, and Scientific Method*, 75.
20. Hoyningen-Huene, "Two Letters of Paul Feyerabend to Thomas Kuhn," 369.
21. Ibid., 361.
22. ibid.
23. In Feyerabend, *Realism, Rationalism, and Scientific Method*, 320, n. 58.
24. Ibid., 323.
25. Ibid., 324.
26. Ibid., 320, n. 58.
27. Kuhn, "A Function for Thought Experiments," 252.
28. Popper, "Irreversibility; Or, Entropy since 1905," 154.
29. Popper commented on the latest scientific research of those years—he considered Szilard's exorcism "completely unsatisfactory"—in Popper, "Irreversibility; Or, Entropy since 1905," 154.
30. Feyerabend, "On the Possibility of a Perpetuum Mobile of the Second Kind."
31. "The Hell Machine I am going to introduce," Feyerabend explained, "is a modification of a machine used by J. R. Pierce." In addition to studying Pierce, he was taken by Smoluchowski's search for "a perpetual source of income," a phrase he cited over and over again. Could it be found? The main mistake of researchers since Smoluchowski lay in their "ambiguous use of the term 'information,'" and the "circular" arguments through which they could prove the second law only by reference to effects derived from it. Feyerabend was optimistic that knowledge could serve to mitigate the otherwise debilitating forces of entropy, siding explicitly with Popper and concluding that the existence of perpetual motion was "in principle possible." Ibid., 412.
32. Karl Popper, "Of Clouds and Clocks," Second Arthur Holly Compton Memorial Lecture, Washington University, April 21, 1965, reprinted in Popper, *Objective Knowledge*, 214, n. 15.

33. Popper, *The Open Universe*, 30.
34. Popper, *The World of Parmenides*, 179.
35. Popper "tried to show that a Maxwell demon exists who is a variant of Szilard's demon, but who does not need to do any work, or to expend any work, or to expend any information." Ibid., 196.
36. Ibid., 179.
37. "One of the most recent and most ingenious proofs and refutations of the nonexistence of Maxwell's demon is due to Professor Dennis Gabor, who constructed a model showing that the demon can exist according to classical physics but not according to quantum physics." "Gabor's demon is sufficiently clever to avoid any classical disproof of his existence; which would show, as Gabor points out, that the second law is false in classical physics and becomes valid only in quantum physics." Ibid.
38. Popper, *Quantum Theory and the Schism in Physics*, 183.
39. Ibid., 189–99.
40. Ibid.
41. Danto, *Analytical Philosophy of History*, 149.
42. Gistau, *Paradigms for a Metaphorology of the Cosmos*, 151–66.
43. Blumenberg, *Lebenszeit und Weltzeit*, 247. "Das Geschichtssubjekt, dessen Funktion sich in retardierenden Störung oder in antreibender Beschleunigung erschöpft, ist ganz durch sein nie zuvor gekanntes Object, 'die Geschichte,' bestimmt. Es ist ihr Laplacescher Dämon, denn es kennt sie kraft der Logik, die es mit ihr gemeinsam hat, im ganzen." Also noteworthy is Fellmann, "Das Ende des Laplaceschen Dämons," and Fellmann, "Wissenschaft als Beschreibung."

Bibliography

Adams, Henry. *The Education of Henry Adams: An Autobiography*. Boston: Riverside Press, 1918.
———. *A Letter to American Teachers of History*. Washington, DC: J. H. Furst, 1910.
———. *Letters of Henry Adams (1892–1918)*, edited by Worthington Chauncey Ford. Boston: Houghton Mifflin Co., 1938.
———. "The Rule of Phase Applied to History." In *The Degradation of the Democratic Dogma*, 267–311. New York: Macmillan Co., 1920.
"The Anatomy of Science by G. N. Lewis." *Mathematical Gazette* 13, no. 190 (1927): 430.
Angrist, Stanley W., and Loren G. Hepler. *Order and Chaos*. New York: Basic Books, 1967.
Arago, François. "Notice scientifique sur la tonnerre." *Annuaire du bureau des longitudes* (1838): 221–618.
Arnauld, Antoine, and Pierre Nicole. *Logique de Port-Royal* (1662). Paris: Hachette, 1874.
Arrhenius, Svante. "Award Ceremony Speech." In *Nobel Lectures, Physics, 1901–1921*, 479–81. Amsterdam: Elsevier, 1967.
Ashby, W. Ross. "Communication, Organization, and Science. By Jerome Rothstein; with a Foreword by C. A. Muses." *Journal of Mental Science* 105, no. 438 (1959): 267.
Asimov, Isaac. *Life and Energy: An Exploration of the Physical and Chemical Basis of Modern Biology*. Garden City, NY: Doubleday, 1962.
———. "Science: The Modern Demonology." *Magazine of Fantasy and Science Fiction* (January 1962): 73–83.
———. *View from a Height*. Garden City, NY: Doubleday, 1963.
Aspden, Harold. "Scientific Correspondence: The Law Beats Maxwell's Demon." *Nature* 347, no. 6288 (September 6, 1990): 25.
Ayer, A. J. "Editor's Introduction." In *Logical Positivism*, 3–28. Glencoe, IL: Free Press, 1959.
Ayres, Robert U., and Indira Nair. "Thermodynamics and Economics." *Physics Today* 37 (November 1984): 62–71.
Babbage, Charles. *The Ninth Bridgewater Treatise: A Fragment*, 1st ed. London: John Murray, 1837.
———. *The Ninth Bridgewater Treatise: A Fragment*, 2nd ed. London: John Murray, 1838.
Baeyer, Hans Christian von. *Maxwell's Demon: Why Warmth Disperses and Time Passes*. New York: Random House, 1998.
Barnes, E. W. *Scientific Theory and Religion*. Cambridge: Cambridge University Press, 1933.
Bastin, Ted, ed. *Quantum Theory and Beyond: Essays and Discussions Arising from a Colloquium*. Cambridge: Cambridge University Press, 1971.
Baudelaire, Charles. "Le Joueur généreaux." In *Petits poèmes en prose: Oeuvres complètes*, 87–91. Paris: Calmann Lévy, 1892.
Bekenstein, Jacob D. "Baryon Number, Entropy, and Black Hole Physics." PhD dissertation, Department of Physics, Princeton University, 1972.

---. "Black-Hole Thermodynamics." *Physics Today* (January 1980): 24–30.
Bell, John S. "On the Einstein Podolsky Rosen Paradox." *Physics* 1, no. 3 (1964): 195–290.
Bennett, Charles. "Demons, Engines, and the Second Law." *Scientific American* 257, no. 5 (1987): 108–16.
---. "Logical Reversibility of Computation." *IBM Journal of Research and Development* 17, no. 6 (1973): 525–32.
---. "The Thermodynamics of Computation: A Review." *International Journal of Theoretical Physics* 21, no. 12 (May 8, 1982): 905–40.
Bentham, Jeremy. *Deontology; Or, The Science of Morality*, edited by John Bowring, 2 vols. London: Longman, Rees, Orme, Browne, Green, and Longman, 1834.
---. *Panopticon; Or, The Inspection-House*. Dublin: Thomas Byrne, 1791.
Bergson, Henri. *Creative Evolution*, translated by Arthur Mitchell. Mineola, NY: Dover Publications, 1998.
---. *Durée et simultanéité: À propos de la théorie d'Einstein*, edited by Élie During, 4th ed. Paris: Quadrige/Presses Universitaires de France, 2009.
Bernstein, Aaron. *Aus dem reiche der naturwissenschaft*, 3rd ed. New York: Charles Schmidt, 1869.
---. *Naturwissenschaftliche Volksbücher*, vol. 4. Berlin: Franz Duncker, 1855.
---. *Naturwissenschaftliche Volksbücher*, vol. 11. Berlin: Franz Duncker, 1856.
---. *Naturwissenschaftliche Volksbücher*, supplemental volume. Berlin: Franz Duncker, 1873.
Bernstein, Jeremy. "John Stewart Bell: Quantum Engineer." In *Quantum Profiles*, 3–89. Princeton, NJ: Princeton University Press, 1991.
Beyler, Richard. "The Demon of Technology, Mass Society, and Atomic Physics in West Germany, 1945–1957." *History and Technology* 19, no. 3 (2003): 227–39.
Bezos, Jeffrey P. "2016 Letter to Shareholders." *Dayone* (The Amazon Blog), April 17, 2017. https://blog.aboutamazon.com/company-news/2016-letter-to-shareholders.
Bhushan, Abhay. "A File Transfer Protocol." *RFC* 114 (April 16, 1971).
Bloch, Ernst. *The Principle of Hope*, translated by Neville Plaice, Stephen Plaice, and Paul Knight, vol. 1. Cambridge, MA: MIT Press, 1986.
Blumenberg, Hans. *Lebenszeit und Weltzeit*. Frankfurt am Main: Suhrkamp, 1986.
Boden, Margaret. "The Mind of a Very Special Machine?" *New Scientist* 1804 (January 18, 1992): 34–37.
Bohm, David. *Quantum Theory* (1951). Mineola, NY: Dover Publications, 1989.
---. "A Suggested Interpretation of the Quantum Theory in Terms of Hidden Variables: I." *Physical Review* 85, no. 2 (1952): 166–79.
Bohr, Niels. "Discussion with Einstein on Epistemological Problems in Atomic Physics." In *Albert Einstein: Philosopher-Scientist*, edited by Paul Arthur Schilpp. La Salle, IL: Open Court, 1949.
Bohr, Niels, et al. "Faust: Eine Histoire." In *Niels Bohr 1885–1962: Der Kopenhagener Geist in der Physik*, edited by K. Stolzenburg, K. V. Meyenn, and R. U. Sexl, 314–42. Braunschweig: F. Vieweg, 1985.
Bois-Reymond, Emil du. *Über die Grenzen des Naturerkennens*. Leipzig: Veit & Comp, 1872.
---. "The Limits of Our Knowledge of Nature." *Popular Science Monthly* 5 (1874): 17–32.
Boltzmann, Ludwig. *Vorlesungen über Gastheorie*, vol. 2. Leipzig: Johann Ambrosius Barth, 1898.
---. "Zur Erinnerung an Josef Loschmidt." In *Populäre Schriften*, 228–52. Leipzig: Johann Ambrosius Barth, 1905.

Born, Max. "Einstein's Statistical Theories." In *Albert Einstein: Philosopher-Scientist*, edited by Paul Arthur Schilpp. La Salle, IL: Open Court, 1949.
———. "Max Karl Ernst Ludwig Planck, 1858–1947." *Obituary Notices of Fellows of the Royal Society* 6, no. 17 (1948): 161–88.
———. *My Life and My Views*. New York: Scribner, 1968.
———. *Natural Philosophy of Cause and Chance*. Oxford: Clarendon Press, 1949.
———. "Physics in the Last Fifty Years." *Nature* 168 (1951): 625–30.
———. *Physics in My Generation*, 2nd ed., rev. Berlin: Springer, 1969.
———. "Quantenmechanik der Stoßvorgänge." *Zeitschrift für Physik* 38 (1926): 803–27.
———. *Von der Verantwortung des Naturwissenschaftlers: Gesammelte Vorträge*. Munich: Nymphenburger Verlagshandlung, 1965.
Bourdieu, Pierre. *Raisons pratiques: Sur la théorie de l'action*. Paris: Seuil, 1994.
Bown, William. "Science: A Demon Blow to the Second Law of Thermodynamics?" *New Scientist* 1725 (July 14, 1990).
———. "Science: Sacred Law of Physics Is Safe After All." *New Scientist* 1734 (September 15, 1990).
Bridgman, Percy W. *The Nature of Thermodynamics*. Cambridge, MA: Harvard University Press, 1941.
Brillouin, Léon. "Information Theory and Its Applications to Fundamental Problems in Physics." *Nature* 183, no. 4660 (February 21, 1959): 501–2.
———. "Life, Thermodynamics, and Cybernetics." *American Scientist* 37, no. 4 (October 1949): 554–68.
———. "Maxwell's Demon Cannot Operate: Information and Entropy: I." *Journal of Applied Physics* 2 (1951): 334–37.
———. *Science and Information Theory*. New York: Academic Press, 1956.
———. *Science and Information Theory*, 2nd ed. Mineola: Dover Publications, 2013.
———. "Thermodynamics, Statistics, and Information." *American Journal of Science* 29 (1961): 318–27.
Broad, C. D. *The Mind and Its Place in Nature*. London: Routledge & Kegan Paul, 1925.
Brown, James R. *The Laboratory of the Mind: Thought Experiments in the Natural Sciences*, 2nd ed. London: Routledge, 2010.
Buchwald, Diana Kormos, ed. *The Collected Papers of Albert Einstein*, 14 vols. Princeton, NJ: Princeton University Press, 1987–.
Buller, A. H. Reginald. "Letters to the Editor: Relativity Limerick." *The Observer* (November 14, 1937): 12.
Butler, Samuel. *Life and Habit*. London: Trübner, 1878.
Campbell-Swinton, Alan Archibald. "Scientific Progress and Prospects." *Nature* 88, no. 2197 (1911): 191–95.
Canales, Jimena. *The Matter of Fact: Exhibition Catalog*. Cambridge, MA: Harvard University Collection of Historical Scientific Instruments, 2008.
Canales, Jimena, and Markus Krajewski. "Little Helpers: About Demons, Angels, and Other Servants." *Interdisciplinary Science Reviews* 37, no. 4 (2012): 314–31.
Carlyle, Thomas. *Sartor Resartus: The Life and Opinions of Herr Teufelsdröckh*. London: Chapman and Hall, 1831.
Carnap, Rudolf. *Two Essays on Entropy*, edited by Abner Shimony. Berkeley: University of California Press, 1977.
Carnap, Rudolf, and Bar-Hillel Yehoshua. *An Outline of a Theory of Semantic Information*, vol. 247. Cambridge, MA: Research Laboratory of Electronics, October 27, 1952.
Carr, Wildon. "Life and Logic." *Mind* 22, no. 88 (1913): 484–92.

Carroll, Sean B. *From Eternity to Here: The Quest for the Ultimate Theory of Time*. New York: Penguin, 2010.

Carus, Paul. *History of the Devil and the Idea of Evil: From the Earliest Times to the Present Day*. Chicago: Open Court, 1900.

Cassirer, Ernst. *Determinismus und Indeterminismus in der Modernen Physik*. In *Göteborgs Högskolas Årsskrift*, vol. 42. Gothenburg: Elanders Boktryckeri Aktiebolag, 1937.

Cater, Harold Dean, ed. *Henry Adams and His Friends*. Boston: Houghton Mifflin Co., 1947.

Caves, Carlton M. "Maxwell's Demon: From Physics Outlaw to Potent Teachers." *Physics World* 4, no. 3 (1991): 51.

———. "Quantitative Limits on the Ability of a Maxwell Demon to Extract Work from Heat." *Physical Review Letters* 64, no. 18 (April 30, 1990): 2111–14.

Caves, Carlton M., William G. Unruh, and Wojciech H. Zurek. "Comment on 'Quantitative Limits on the Ability of a Maxwell Demon to Extract Work from Heat.'" *Physical Review Letters* 65, no. 11 (September 10, 1990).

Caws, Peter. *The Philosophy of Science*. Princeton, NJ: D. Van Nostrand, 1965.

Cervantes Saavedra, Miguel de. *El ingenioso hidalgo Don Quixote de la Mancha*. Madrid: Juan de la Cuesta, 1605.

Charniak, Eugene. *Artificial Intelligence Programming*, 2nd ed. Hillsdale, NJ: Lawrence Erlbaum Associates, 1987.

———. "Toward a Model of Children's Story Comprehension." PhD dissertation, MIT, August 1972.

Cherry, Steven. "Jaron Lanier: We're Being Enslaved by Free Information." *IEEE Spectrum*, July 16, 2013. https://spectrum.ieee.org/podcast/computing/networks/jaron-lanier-were-being-enslaved-by-free-information.

Clark, Stuart. *Thinking with Demons: The Idea of Witchcraft in Early Modern Europe*. Oxford: Clarendon Press, 1997.

Clausius, Rudolf. "Abschnitt XI: Discussionen über die mechanische Behandlung der Wärme und Electricität." In *Die Mechanische Wärmetheorie*, 306–52. Braunschweig: Friedrich Vieweg und Sohn, 1879.

Cohen, George N., and Jacques Monod. "Bacterial Permeases." *Bacteriological Reviews* 21 (1957): 169–94.

Compton, Arthur H. *The Freedom of Man*. New Haven, CT: Yale University Press, 1935.

———. *The Human Meaning of Science*. Chapel Hill: University of North Carolina Press, 1940.

———. "The Scattering of X-ray Photons." *American Journal of Physics* 14, no. 2 (March/April 1946).

———. "The Uncertainty Principle and Free Will." *Science* 74, no. 1911 (August 14, 1931): 172.

Compton, Karl T. "Science and Prosperity." *Science* 80, no. 2079 (1934): 387–94.

Compton, Karl T., J. L. Costa, and Henry De Wolf Smyth. "A Mechanical Maxwell Demon." *Physical Review* 30 (September 1927): 349–53.

Cootner, Paul H. "Stock Prices: Random vs. Systematic Changes." *Industrial Management Review* 3, no. 2 (1962): 24–45.

Crick, Francis. "Obituary: Jacques Lucien Monod." *Nature* 262, no. 5567 (July 29, 1976): 429–30.

Crull, Elise, and Guido Bacciagaluppi, eds. *Grete Hermann: Between Physics and Philosophy*. Dordrecht: Springer, 2016.

Curie, Pierre, Marie Curie, and G. Bémont. "Sur une nouvelle sunstance fortement radio-active, contenue dans la pechblende." *Comptes rendus des séances de l'académie des sciences* 127 (1898): 1215–17.

Daly, Herman E. "Entropy, Growth, and the Political Economy." In *Scarcity and Growth Reconsidered*, edited by V. Kerry Smith, 67–94. Baltimore: Johns Hopkins University Press, 1979.

Dannen, Gene. "The Einstein-Szilard Refrigerators: Two Visionary Theoretical Physicists Joined Forces in the 1920s to Reinvent the Household Refrigerator." *Scientific American* (January 1997): 90–95.

Danto, Arthur C. *Analytical Philosophy of History*. Cambridge: Cambridge University Press, 1965.

Darling, L., and E. O. Hulburt. "On Maxwell's Demon." *American Journal of Physics* 23, no. 7 (October 1955).

Darwin, Charles. "Essay of 1842." In *The Foundations of the Origin of Species: Two Essays Written in 1842 and 1844*, edited by Francis Darwin, 1–55. Cambridge: Cambridge University Press, 1909.

———. "Essay of 1844." In *The Foundations of the Origin of Species: Two Essays Written in 1842 and 1844*, edited by Francis Darwin, 56–255. Cambridge: Cambridge University Press, 1909.

Daston, Lorraine. "Intelligences: Angelic, Animal, Human." In *Thinking with Animals: New Perspectives on Anthropomorphism*, edited by Lorraine Daston and Gregg Mitman, 37–58. Chicago: University of Chicago Press, 2005.

———. "Marvelous Facts and Miraculous Evidence in Early Modern Europe." In *Wonders, Marvels, and Monsters in Early Modern Culture*, 76–105. Newark, DE: University of Delaware Press, 1999.

Daston, Lorraine, and Katharine Park. *Wonders and the Order of Nature, 1150–1750*. New York: Zone Books, 1998.

Davies, Paul. *The Demon in the Machine: How Hidden Webs of Information Are Solving the Mystery of Life*. Chicago: University of Chicago Press, 2019.

———. "Universe from Bit." In *Information and the Nature of Reality: From Physics to Metaphysics*, edited by Paul Davies and Niels Henrik Gregersen, 83–117. Cambridge: Cambridge University Press, 2014.

Dawkins, Richard. "Reply to Lucy Sullivan." *Philosophical Transactions of the Royal Society: Biological Sciences* 349, no. 1328 (August 29, 1995): 219–24.

De Broglie, Louis. *Le Dualisme des ondes et des corpuscules dans l'oeuvre de Albert Einstein: Lecture faite en la séance annuelle des prix du 5 Décembre 1955*. Paris: Institut de France, 1955.

Dennett, Daniel C. *Brainstorms: Philosophical Essays on Mind and Psychology*. Montgomery, VT: Bradford Books, 1978.

———. *Consciousness Explained*. New York: Back Bay Books, 1991.

———. *Darwin's Dangerous Idea: Evolution and the Meanings of Life*. New York: Simon & Schuster, 1995.

Descartes, René. *Discours de la méthode*. Leiden: I. Maire, 1637.

———. *Meditationes de prima philosophia*. Paris: Michaelem Soly, 1641.

———. *Les Principes de la philosophie*. Elzevier, 1644.

Dillenschneider, Raoul, and Eric Lutz. "Memory Erasure in Small Systems." *Physical Review Letters* 102, no. 21 (2009): 210601.

"Discussion du rapport de M. Planck." In *La Théorie du rayonnement et les quanta*, edited by Paul Langevin and Louis de Broglie, 115–32. Paris: Gauthier-Villars, 1912.

"Discussion on the Paper by O. G. Selfridge." In *Mechanisation of Thought Processes: Proceedings of a Symposium Held at the National Physical Laboratory on 24th, 25 th, 26 th, and 27 th November 1958*, 527–31. London: Her Majesty's Stationery Office, 1961.

Dixon, Henry Horatio, and John Joly. "On the Ascent of Sap." *Philosophical Transactions of the Royal Society of London* 186 (1895): 563–76.

Dotzler, Bernhard J. "Demons-Magic-Cybernetics: On the Introduction to Natural Magic as Told by Heinz von Foerster." *Systems Research* 13, no. 3 (1996): 245–50.

Driesch, Hans. *Philosophie Des Organischen*, 2 vols. Leipzig: Von Wilhelm Engelman, 1909.

———. *The Science and Philosophy of the Organism*, 2 vols. London: Adam and Charles Black, 1908.

Dyson, George. "The Usefulness of Useless Knowledge: The Physical Realization of an Electronic Computing Instrument at the Institute for Advanced Study, Princeton, 1930–1958." In *Exceptional Creativity in Science and Technology: Individuals, Institutions, and Innovations*, edited by Andrew Robinson, 83–98. Radnor, PA: Templeton Press, 2013.

[Eberty, Felix]. *The Stars and the Earth, or Thoughts upon Space, Time, and Eternity* (1846). Boston: Lee and Shepard, 1882.

Eddington, Arthur Stanley. *The Nature of the Physical World* (1928), edited by Ernest Rhys. Everyman's Library: Science. London: J. M. Dent & Sons Ltd., 1935.

———. *New Pathways in Science*. New York: Macmillan Co., 1935.

———. "The Theory of Relativity and Its Influence on Scientific Thought." *Scientific Monthly* (January 1923): 34–53.

———. *The Theory of Relativity and Its Influence on Scientific Thought*. Oxford: Clarendon Press, 1922.

Eddy, Henry T. "Radiant Heat and Exception to the Second Law of Thermodynamics." *Proceedings of the American Philosophical Society* 20, no. 112 (1882): 334–43.

Edgeworth, F. Y. "On the Use of the Theory of Probabilities in Statistics Relating to Society." *Journal of the Royal Statistical Society* 76, no. 2 (1913): 165–93.

Ehrenberg, Werner. *Dice of the Gods: Causality, Necessity, and Chance*. London: Birkbeck College, 1977.

———. "Maxwell's Demon." *Scientific American* 217, no. 5 (November 1967): 103–10.

———. "A Note on Entropy and Irreversible Processes." *London, Edinburgh, and Dublin Philosophical Magazine and Journal of Science* 34, no. 233 (1943): 396–409.

Eigen, Manfred. "Self-organization of Matter and the Evolution of Biological Macromolecules." *Naturwissenschaften* 10 (October 1971): 465–523.

Eigen, Manfred, and Ruthild Winkler. *Das Spiel: Naturgesetze steuern den Zufall*. München: R. Piper & Co., 1975.

———. *Steps towards Life*. Oxford: Oxford University Press, 1992.

Einstein, Albert. "Autobiographical Notes." In *Albert Einstein: Philosopher-Scientist*, edited by Paul Arthur Schilpp, 3–94. La Salle, IL: Open Court, 1949.

———. "Die hauptsächlichen Gedanken der Relativitätstheorie." Unpublished manuscript written after December 1916.

———. "Geleitwort." In *Die Gestirne und die Weltgeschichte: Gedanken über Raum, Zeit und Ewigkeit*, edited by Gregory Itelson. Berlin: Gregor Rogoff Verlag, 1924.

———. "Ist die Trägheit eines Körpers von seinem Energieinhalt abhängig?" *Annalen der Physik* 18, no. 13 (1905): 639–41.

———. "Marian v. Smoluchowski." *Naturwissenschaften* 5 (1917): 107–8.

———. "Maxwell's Influence on the Development of the Conception of Physical Reality." In *James Clerk Maxwell: A Commemoration Volume 1831–1931*, 66–73. Cambridge: Cambridge University Press, 1931.

———. "On the Electrodynamics of Moving Bodies." In *The Collected Papers of Albert Einstein*, translated by Anna Beck, 140–71. Princeton, NJ: Princeton University Press, 1989.

———. "On the Movement of Small Particles Suspended in Stationary Liquids Required by the Molecular-Kinetic Theory of Heat" (1905). In *The Swiss Years: Writings, 1900–1909*, vol. 2 of *The Collected Papers of Albert Einstein*, translated by Anna Beck, 123–34. Princeton, NJ: Princeton University Press, 1989.

———. *Relativity: The Special and the General Theory* (1917). New York: Crown, 1961.

———. "Sur le problème de la relativité." *Scientia* 15 (supplement), no. 35–3 (1914): 139–50.

———. *Über die spezielle und allgemeine Relativitätstheorie (gemeinverständlich)*. Braunschweig: Wieveg, 1917.

———. "Über die von der molekularkinetischen Theorie der Wärme geförderte Bewegung von in ruhenden Flüssigkeiten suspendierten Tielchen." *Annalen der Physik* 17 (1905): 549–60.

———. "Zum Relativitätsproblem." *Scientia* 15 (1914): 337.

———. "Zur Elektrodynamik bewegter Körper." *Annalen der Physik* 17 (1905): 891–921.

Elmer, L. J. "Demon (Theology of)." In *The New Catholic Encyclopedia*, 754–56. New York: McGraw-Hill, 1967.

Ericson, Richard F. "The Impact of Cybernetic Information Technology on Management Value Systems." *Management Science* 16, no. 2 (1969): B40–60.

Fedorov, Nikolai Fedorovich. *What Was Man Created For? The Philosophy of the Common Task: Selected Works*, translated by Elisabeth Koutaissoff and Marilyn Minto. London: Honeyglen, 1990.

Fellmann, Ferdinand. "Das Ende des Laplaceschen Dämons." In *Geschichte-Ereignis un Erzählung*, edited by Reinhart Koselleck and Wolf-Dieter Stempel. Munich: Wilhelm Fink, 1973.

———. "Wissenschaft als Beschreibung." *Archiv für Begriffsgeschichte* 18 (1974): 227–61.

Feyerabend, Paul K. "On the Possibility of a Perpetuum Mobile of the Second Kind." In *Mind, Matter, and Method: Essays in Philosophy and Science in Honor of Herbert Feigl*, edited by Paul K. Feyerabend and Grover Maxwell, 409–12. Minneapolis: University of Minnesota Press, 1966.

———. *Realism, Rationalism, and Scientific Method*. Cambridge: Cambridge University Press, 1981.

Feynman, Richard P. "Cargo Cult Science." *Engineering and Science* 37, no. 7 (June 1974): 10–13.

———. *Feynman Lectures on Computation*, edited by Anthony J. G. Hey and Robin W. Allen. Reading, MA: Perseus Books, 1996.

Feynman, Richard P., Robert B. Leighton, and Matthew Sands. *The Feynman Lectures on Physics*, vols. 1 and 2. Reading, MA: Addison-Wesley, 1963 and 1964.

Fine, Arthur. "Fictionalism." In *Fictions in Science: Philosophical Essays on Modeling and Idealization*, edited by Suárez Mauricio, 19–36. Routledge Studies in the Philosophy of Science. New York: Routledge, 2008.

———. *The Shaky Game: Einstein, Realism, and the Quantum Theory*, 2nd ed. Science and Its Conceptual Foundations. Chicago: University of Chicago Press, 1996.

Flammarion, Camille. *Stories of Infinity: Lumen; History of a Comet; In Infinity*, translated by S. R. Crocker. Boston: Roberts Brothers, 1873.

Flint, H. T. "Letter to the Editor: A Limit to the Quantum Theory and the Avoidance of Negative Energy Transitions." *Nature* (February 22, 1936): 313–14.

Foerster, Heinz von. "Disorder/Order: Discovery or Invention?" In *Disorder and Order: Proceedings of the Stanford International Symposium (September 14–16, 1981)*, edited by Paisley Livingston, 177–89. Saratoga, CA: Anma Libri, 1984.

———. "On Self-Organizing Systems and Their Environments." In *Self-Organizing Systems: Proceedings of an Interdisciplinary Conference 5 and 6 May, 1959*, edited by Marshall C. Yovits and Scott Cameron, 31–50. New York: Pergamon Press, 1960.

———. "Responsibilities of Competence." *Journal of Cybernetics* 2, no. 2 (1972): 1–6.

Fourier, Jean-Baptiste-Joseph. *Théorie analytique de la chaleur*. Paris: F. Didot père et fils, 1822.

Freud, Sigmund. "A Seventeenth-Century Demonological Neurosis" (1923). In *The Standard Edition of the Complete Psychological Works*, vol. 19, translated by James Strachley. London: Hogarth Press and Institute of Psychoanalysis, 1953–1974.

Frisch, Otto Robert. *What Little I Remember*. Cambridge: Cambridge University Press, 1979.

Gabor, Dennis. "IV: Light and Information." *Progress in Optics* 1 (1961): 109–53.

———. *Lectures on Communication Theory*. Cambridge, MA: Research Laboratory of Electronics, April 3, 1952.

Galison, Peter. *Einstein's Clocks, Poincaré's Maps: Empires of Time*. New York: W. W. Norton and Co., 2003.

Gamow, George. *Mr. Tompkins Explores the Atom*. New York: Macmillan, 1944.

Garnett, William. "Energy." In *Encyclopedia Britannica*, 205–11. Edinburgh: Adam and Charles Black, 1878.

Georgescu-Roegen, Nicholas. *Analytical Economics: Issues and Problems*. Cambridge, MA: Harvard University Press, 1966.

———. "Energy Analysis and Economic Valuation." *Southern Economic Journal* 45 (1979): 1023–58.

———. "Energy and Economic Myths." *Southern Economic Journal* 41, no. 3 (January 1975): 347–81.

———. *The Entropy Law and the Economic Process*. Cambridge, MA: Harvard University Press, 1971.

———. "The Steady State and Ecological Salvation: A Thermodynamic Analysis." *BioScience* 27, no. 4 (April 1977): 266–70.

Ghirardi, GianCarlo. "Properties and Events in a Relativistic Context: Revisiting the Dynamical Reduction Program." *Foundations of Physics Letters* 9, no. 4 (August 1, 1996): 313–55.

———. *Sneaking a Look at God's Cards: Unraveling the Mysteries of Quantum Mechanics*, translated by Gerald Malsbary, rev. ed. Princeton, NJ: Princeton University Press, 2005. First published in Italian as *Un'occhiata alle carte di Dio* (1997).

Gibbs, Josiah Willard. *Elementary Principles of Statistical Mechanics*. New York: Scribner, 1902.

Gillispie, Charles Couston. *Pierre-Simon Laplace, 1749–1827: A Life in Exact Science*. Princeton, NJ: Princeton University Press, 1997.

Gistau, Alberto Fragio. *Paradigms for a Metaphorology of the Cosmos: Hans Blumenberg and the Contemporary Metaphors of the Universe*. Ariccia, Italy: Aracne, 2015.

Gleick, James. *Genius: The Life and Science of Richard Feynman*. New York: Vintage Books, 1992.

Goeller, H. E., and Alvin M. Weinberg. "The Age of Substitutability." *American Economic Review* 68, no. 6 (December 1978): 1–11.

Goldstein, Martin, and Inge F. Goldstein. *The Refrigerator and the Universe: Understanding the Laws of Energy*. Cambridge, MA: Harvard University Press, 1993.

Gooday, Graeme. "Sunspots, Weather, and the Unseen Universe: Balfour Stewart's Anti-Materialist Representations of 'Energy' in British Periodicals." In *Science Serialized:*

Representations of the Sciences in Nineteenth-Century Periodicals, edited by Geoffrey Cantor and Sally Shuttleworth. Cambridge, MA: MIT Press, 2004.

Gouy, Louis Georges. "Note sur le mouvement brownien." *Journal de physique, theorique, et appliquée* 7 (1888): 561–64.

Grafton, Anthony. "The Devil as Automaton: Giovanni Fontana and the Meanings of a Fifteenth-Century Machine." In *Genesis Redux: Essays in the History and Philosophy of Artificial Life*, edited by Jessica Riskin, 46–62. Chicago: University of Chicago Press, 2007.

Griffin, Nicholas, and Dale Jacquette. *Russell vs. Meinong: The Legacy of "On Denoting."* New York: Routledge, 2009.

Grimm, Jacob and Wilhelm. *Deutsche Mythologie*. Göttingen, 1835.

Haldane, J. B. S. *Fact and Faith*. The Thinker's Library. London: Watts & Co., 1934.

Harman, P. M. *The Natural Philosophy of James Clerk Maxwell*. Cambridge: Cambridge University Press, 1998.

Harrison, Brown, et al. "Chernobyl in Context." *Bulletin of the Atomic Scientists* 43, no. 1 (August/September 1986): 2.

Hawking, Stephen W. "Black Holes and Unpredictability." *Physics Bulletin* 29, no. 1 (January 1978): 23–24.

———. *A Brief History of Time: From the Big Bang to Black Holes*. New York: Bantam Dell, 1988.

———. "Chronology Protection Conjecture." *Physical Review D* 46, no. 2 (July 15, 1992): 603–11.

———. "Particle Creation by Black Holes." *Communications in Mathematical Physics* 43 (1975): 199–220.

Hawking, Stephen W., and Roger Penrose. "The Nature of Space and Time." *Scientific American* 275, no. 1 (1996): 60–65.

———. *The Nature of Space and Time*. Princeton, NJ: Princeton University Press, 1996.

Heidegger, Martin. *Einführung in die Metaphysik*. Tübingen: Max Niemeyer, 1953.

———. *Introduction to Metaphysics*, edited by Gregory Fried and Richard Polt. New Haven, CT: Yale University Press, 2000.

———. "The Question Concerning Technology." In *The Question Concerning Technology and Other Essays* (1954), 3–35. New York: Harper & Row, 1977.

Heisenberg, Werner. *Das Naturbild der heutigen Physik*. Hamburg: Rowohlt, 1955.

———. *Physics and Beyond: Encounters and Conversations*. New York: Harper & Row, 1971.

———. *Physics and Philosophy: The Revolution in Modern Science*. New York: Harper & Row, 1958.

Helmholtz, Hermann von. "Wirbelstürme und Gewitter" (lecture delivered in Hamburg, 1875). *Deutsche Rudschau* 6 (1876): 363–80. Reprinted in *Vorträge Und Reden*, 137–63. Braunschweig: Friedrich Viweg und Sohn, 1896.

Henry-Hermann, Grete. "Die Kausalität in Der Physik." *Studium Generale* 1, no. 6 (1948): 375–83.

Hermann, Grete. "Die Naturphilosophischen Grundlagen der Quantenmechanik." *Naturwissenschaften* 42 (1935): 718–21.

———. "Die Naturphilosophischen Grundlagen der Quantenmechanik." *Abhandlutlgen der Fries'schen Schule* 6, no. 2 (1935): 69–152.

———. "Natural-Philosophical Foundations of Quantum Mechanics." In *Grete Hermann: Between Physics and Philosophy*, edited by Elise Crull and Guido Bacciagaluppi, 239–78. Dordrecht: Springer, 2016.

Herschel, John. *A Treatise on Astronomy*, 3rd ed. Philadelphia: Carey, Lea and Blanchard, 1834.

Hesse, Mary B. *Models and Analogies in Science*. London: Sheed & Ward, 1963.

Hill, Thomas. "Recommendatory Letters." In [Felix Eberty], *The Stars and the Earth, or Thoughts upon Space, Time and Eternity* (1846). Boston: Lee and Shepard, 1882.

Hofstadter, Douglas R. "Reflections." In *The Mind's I: Fantasies and Reflections on Self and Soul*, edited by Douglas R. Hofstadter and Daniel C. Dennett, 373–82. New York: Basic Books, 1981.

"Holds Roosevelt, Turns Back Chaos: Karl Compton Says That His 'Intelligent Planning' Is Staying Nature's Trend." *New York Times*, June 23, 1934, 2.

Holmes, Samuel Jackson. "The Beginnings of Intelligence." *Science* 33, no. 848 (1911): 473–80.

Holton, Gerald. "On the Art of Scientific Imagination." *Daedalus* 125, no. 2 (Spring 1996): 183–208.

Hopfield, John J. "Kinetic Proofreading: A New Mechanism for Reducing Errors in Biosynthetic Processes Requiring High Specificity." *Proceedings of the National Academy of Sciences* 71, no. 10 (October 1974): 4135–39.

"How Will the New 'Golem' Help Israel?" *The Sentinel: Voice of Chicago Jewry*, April 22, 1965, 1.

Hoyningen-Huene, Paul. "Two Letters of Paul Feyerabend to Thomas Kuhn." *Studies in the History and Philosophy of Science* 26, no. 3 (1995): 353–87.

Hugo, Victor. *Les Misérables*. Brussels: A. Lacroix, Verboeckhoven & Co., 1862.

Hume, David. *Dialogues Concerning Natural Religion*. London, 1779.

———. *Philosophical Essays Concerning Human Understanding*. London: A. Millar, 1748.

Huxley, Thomas H. "The Genealogy of Animals" (1869). In *Darwiniana: Essays*, 107–19. New York: D. Appleton, 1896.

"Information, Fluctuations, and Energy Control in Small Systems." *CORDIS* (April 21, 2017). https://cordis.europa.eu/project/id/308850.

"Ilya Prigogine." In *Glimpsing Reality: Ideas in Physics and the Link to Biology*, edited by Paul Buckley and F. David Peat, 98–110. Toronto: University of Toronto Press, 1996.

Janet, Paul. "The Materialism of the Present Day." *Reader* (June 30, 1866).

Jasanoff, Sheila. "Future Imperfect: Science, Technology, and the Imaginations of Modernity." In *Dreamscapes of Modernity: Sociotechnical Imaginaries and the Fabrication of Power*, edited by Jasanoff Sheila and Kim Sang-Hyun, 1–33. Chicago: University of Chicago Press, 2015.

Jauch, J. M., and J. G. Báron. "Entropy, Information, and Szilard's Paradox." *Helvetica Physica Acta* 45 (1972): 220–32.

Jevons, W. Stanley. "On the Movement of Microscopic Particles Suspended in Liquids." *Quarterly Journal of Science* (April 1878): 167–86.

Johnson, George. "Rolf Landauer, Pioneer in Computer Theory, Dies at 72." *New York Times* (April 30, 1999).

Johnson, Paul. *Modern Times: The World from the Twenties to the Nineties*. New York: HarperCollins, 1991.

Johnstone, James. *The Philosophy of Biology*. Cambridge: Cambridge University Press, 1914.

Jukes, Thomas H. "DDT: Maxwell's Demon." *Science* 166, no. 3901 (October 3, 1969): 44.

Karl, Kornacker. "Towards a Physical Theory of Self-Organization." In *Towards a Theoretical Biology*, edited by Conrad Hal Waddington, 94–95. Chicago: Aldine Publishing Co., 1968.

Keller, Evelyn Fox. *Refiguring Life: Metaphors of Twentieth-Century Biology.* New York: Columbia University Press, 1995.

Kendall, Maurice G. "The Analysis of Economic Time-Series: Part I: Prices." *Journal of the Royal Statistical Society: Series A (General)* 116, no. 1 (1953): 11–34.

Kiukas, Jukka, and Reinhard F. Werner. "Maximal Violation of Bell Inequalities by Position Measurements." *Journal of Mathematical Physics* 51, no. 7 (2010).

Köhler, Eckehart. "Why von Neumann Rejected Carnap's Dualism of Information Concepts." In *John von Neumann and the Foundations of Quantum Physics*, edited by Miklos Rédei and Michael Stöltzner, 97–134. Dordrecht: Kuwer, 2001.

Kornman, Brent D. "Pattern Matching and Pattern-Directed Invocation in Systems Programming Languages." *Journal of Systems and Software* 3 (1983): 95–102.

Krugman, Paul. "The Incomparable Economist." *New York Times* (December 15, 2009).

Kuhn, Thomas S. "A Function for Thought Experiments." In *The Essential Tension: Selected Studies in Scientific Tradition and Change*, 240–65. Chicago: University of Chicago Press, 1964. Originally published in *L'Aventure de la science: Mélanges Alexandre Koyré* (1964).

———. "Postscript—1969." In *The Structure of Scientific Revolutions*, 2nd ed., 174–210. Chicago: University of Chicago Press, 1970.

———. *The Structure of Scientific Revolutions*, 2nd ed. Chicago: University of Chicago Press, 1970.

Laing, Richard. "Maxwell's Demon and Computation." *Philosophy of Science* 41, no. 2 (1973): 171–78.

Laird, John E., and Paul S. Rosenbloom. "The Research of Allen Newell." *AI Magazine* 13, no. 4 (1992): 17–45.

Landauer, Rolf. "Computation: A Fundamental Physical View." *Physica Scripta* 35, no. 1 (1987): 88–95.

———. "Irreversibility and Heat Generation in the Computing Process." *IBM Journal of Research and Development* 5 (1961): 183–91.

Langevin, Paul. "L'Évolution de l'espace et du temps." *Scientia* 10 (1911): 31–54.

Lanouette, William. *Genius in the Shadows: A Biography of Leó Szilard, the Man Behind the Bomb.* New York: Skyhorse Publishing, 2013.

Laplace, Pierre-Simon. *Essai philosophique sur les probabilités.* Paris: Courcier, 1814.

———. "Un Mémoire sur l'équation séculaire de la lune" (April 2, 1788). In *Oeuvre complètes*, 241–71. Paris: Gauthier-Villars, 1895.

———. "Recherches, sur l'integration des équations différentielles aux différences finies, et sur leur usage dans la théorie des hasards." *Mémoires de l'académie royale des sciences de Paris (Savants étrangers)* 7 (1776): 37–163. Reprinted in *Oeuvre Complètes*, 69–197 (Paris: Gauthier-Villars, 1891).

Latour, Bruno. "Drawing Things Together." In *Representation in Scientific Practice*, edited by Michael E. Lynch and Steve Woolgar, 19–68. Cambridge, MA: MIT Press, 1990.

Latour, Bruno, and Steve Woolgar. *Laboratory Life: The Construction of Scientific Facts* (1979). Princeton, NJ: Princeton University Press, 1986.

Laurence, William. "New 'Gun' Speeds Break-up of Atom." *New York Times* (June 20, 1933).

Law, Richard. "Optimal Life Histories under Age-Specific Predation." *American Naturalist* 114, no. 3 (September 1979): 399–417.

Leff, Harvey S., and Andrew F. Rex, eds. *Maxwell's Demon: Entropy, Information, Computing.* Princeton Series in Physics, 1–32. Princeton, NJ: Princeton University Press, 1990.

———, eds. *Maxwell's Demon 2: Entropy, Classical and Quantum Information, Computing*. Philadelphia: CRC Press, 2002.
Leffler, Samuel J. *The Design and Implementation of the 4.3bsd Unix Operating System*. Reading, MA: Addison-Wesley, 1989.
Leijonhufvud, Axel. "Keynes and the Keynesians: A Suggested Interpretation." *American Economic Review* 57, no. 2 (1967): 401–10.
Leonard, Andrew. *Bots: The Origin of New Species*. San Francisco: Penguin, 1997.
Lerner, A. Y. *Fundamentals of Cybernetics*. New York: Plenum, 1975.
Levallois, Clément. "Can De-Growth Be Considered a Policy Option? A Historical Note on Nicholas Georgescu-Roegen and the Club of Rome." *Ecological Economics* 69 (2010): 2271–78.
Lévi-Strauss, Claude. *Mythologiques: Le Cru et le cuit*. Paris: Librairie Plon, 1964.
Lewis, Gilbert Newton. *The Anatomy of Science*. New Haven, CT: Yale University Press, 1926.
———. "The Conservation of Photons." *Nature* 118 (1926): 874–75.
———. "The Symmetry of Time in Physics." *Science* 71, no. 1849 (1930): 569–77.
Lewontin, Richard C. "Is Nature Probable or Capricious?" *BioScience* 16, no. 1 (January 1966): 25–27.
Lichtenstein, Léon, and André Metz. "La Philosophie des mathématiques selon M. Émile Meyerson." *Revue philosophique de la France et de l'étranger* 113 (1932): 169–206.
Lillie, Ralph S. "The Directive Influence in Living Organisms." *Journal of Philosophy* 29, no. 18 (September 1, 1932): 477–91.
———. "The Philosophy of Biology: Vitalism versus Mechanism." *Science* 40, no. 1041 (December 11, 1914): 840–46.
———. "Physical Indeterminism and Vital Action." *Science* 66, no. 1702 (August 12, 1927): 139–43.
Lindsay, Peter H., and Donald A. Norman. *Human Information Processing: An Introduction to Psychology*. New York: Academic Press, 1972.
Livingston, Paisley. "Introduction." In *Disorder and Order: Proceedings of the Stanford International Symposium* (September 14–16, 1981), edited by Paisley Livingston, 3–33. Saratoga, CA: Anma Libri, 1984.
Lloyd, Seth. "Quantum-Mechanical Maxwell's Demon." *Physical Review* 56, no. 5 (November 1997): 3374–82.
Locke, John. "A Discourse of Miracles." In *Posthumous Works of Mr. John Locke*. London: Black Swan, 1706.
———. "Of the Conduct of the Understanding." In *Posthumous Works of Mr. John Locke*. London: Black Swan, 1706.
Loewenstein, Werner R. *The Touchstone of Life: Molecular Information, Cell Communication, and the Foundations of Life*. Oxford: Oxford University Press, 1999.
Lorentz, Hendrik A. "Discourse d'ouverture." In *La Théorie du rayonnement et les quanta*, 6–9. Paris: Gauthier-Villars, 1912.
Loschmidt, Josef. "Der Zweite Satz der Mechanischen Wärmetheorie." *Sitzungsberichte der kaiserlichen Akademie der Wissenschaften* 59, no. 2 (1877): 395–418.
———. "Über den Zustand des Wärmegleichgewichtes eines Systems von Körpern mit Rücksicht auf die Schwerkraft." *Sitzungsberichte der kaiserlichen Akademie der Wissenschaften* 73, no. 2 (1876): 128–42, 366–72.
———. "Über den Zustand des Wärmegleichgewichtes eines Systems von Körpern mit Rücksicht auf die Schwerkraft." *Sitzungsberichte der kaiserlichen Akademie der Wissenschaften* 75 (1877): 287–98.

———. "Über den Zustand des Wärmegleichgewichtes eines Systems von Körpern mit Rücksicht auf die Schwerkraft." *Sitzungsberichte der kaiserlichen Akademie der Wissenschaften* 76 (1878): 209–25.
Lovelace, Ada. "Notes by the Translator." In *Scientific Memoires, Selected from the Transactions of Foreign Academies of Science and Learned Societies* 3 (1843): 691–731.
Macduffie, Allen. "Irreversible Transformations: Robert Louis Stevenson's 'Dr. Jekyll and Mr. Hyde' and Scottish Energy Science." *Representations* 96, no. 1 (2006): 1–20.
Mach, Ernst. *Die principien der Währmelehre.* Leipzig: Johann Ambrosius Barth, 1896.
Macrae, Norman. *John von Neumann: The Scientific Genius Who Pioneered the Modern Computer, Game Theory, Nuclear Deterrence, and Much More.* New York: Pantheon, 1992.
Maddox, John. "Maxwell's Demon Flourishes." *Nature* 345, no. 6271 (1990): 109.
Mahon, Basil. *The Man Who Changed Everything: The Life of James Clerk Maxwell.* New York: John Wiley & Sons, 2004.
Margenau, Henry. "Causality and Modern Physics." *The Monist* 41, no. 1 (1931): 1–36.
———. "Meaning and Scientific Status of Causality." *Philosophy of Science* 1, no. 2 (1934): 133–48.
———. *The Nature of Physical Reality: A Philosophy of Modern Physics.* New York: McGraw-Hill, 1950.
Martínez-Alier, Juan. "In Memory of Georgescu-Roengen." In *Varieties of Environmentalism: Essays North and South*, edited by Ramachandra Guha and Juan Martínez-Alier, 169–84. London: Earthscan Publications, 1997.
Marx, Karl, and Friedrich Engels. *The Communist Manifesto* (1848), edited by McLellan David. Oxford: Oxford University Press, 1992.
———. *Das Kapital: Kritik Der Politischen Oekonomie*, vol. 1 of 3 vols. Hamburg: Otto Meissner, 1867.
Mauricio, Suárez. *Fictions in Science: Philosophical Essays on Modeling and Idealization.* Routledge Studies in the Philosophy of Science. New York: Routledge, 2008.
Maxwell, James Clerk. "Molecules." *Nature* (September 1873): 437–41.
———. *The Scientific Letters and Papers of James Clerk Maxwell*, edited by P. M. Harman, vol. 2. Cambridge: Cambridge University Press, 1990.
———. *The Scientific Letters and Papers of James Clerk Maxwell*, edited by P. M. Harman, vol. 3. Cambridge: Cambridge University Press, 1990.
———. *Theory of Heat.* London: Longmans, Green and Co., 1871.
McCarthy, John. "Ascribing Mental Qualities to Machines." In *Formalizing Common Sense: Papers by John McCarthy*, edited by Vladimir Lifschitz, 93–118. Exeter: Intellect, 1998.
———. "Programs with Common Sense." In *Mechanisation of Thought Processes: Proceedings of a Symposium Held at the National Physical Laboratory on 24th, 25th, 26th, and 27th November 1958*, 75–84. London: Her Majesty's Stationery Office, 1961.
McClare, C. W. F. "Chemical Machines, Maxwell's Demon, and Living Organisms." *Journal of Theoretical Biology* 30 (1971): 1–34.
McFarland, Matt. "Elon Musk: 'With Artificial Intelligence We Are Summoning the Demon.'" *Washington Post* (October 24, 2014).
McKelvey, Fenwick. *Internet Daemons: Digital Communications Possessed*, vol. 56 of *Electronic Mediations*. Minneapolis: University of Minnesota Press, 2018.
Meinong, Alexius. "Über Gegenstandstheorie." In *Untersuchungen zur Gegenstandstheorie und Psychologie*, edited by Alexius Meinong, 1–50. Leipzig: Johann Ambrosius Barth, 1904.

Menabrea, Luigi Federico. "Notions sur la machine analytique de M. Charles Babbage." *Bibliothèque universelle de Genève* 41, no. 82 (October 1842): 352–76.

———. "Sketch on the Analytical Engine Invented by Charles Babbage." In *Scientific Memoires, Selected from the Transactions of Foreign Academies of Science and Learned Societies* 3 (1843): 666–90.

Meyenn, K. V., K. Stolzenburg, and R. U. Sexl, eds. *Niels Bohr 1885–1962: Der Kopenhagener Geist in der Physik*. Braunschweig: F. Vieweg, 1985.

Meyer, Garry S. "Infants in Children's Stories: Toward a Model of Natural Language Comprehension." Master's thesis, MIT, August 1972 (memo 265).

Millikan, Robert Andrews. *Time, Matter, and Values*. Chapel Hill: University of North Carolina Press, 1923.

Minsky, Marvin L. *The Society of Mind* (1986). New York: Simon & Schuster, 1988.

———. "Steps towards Artificial Intelligence." *Proceedings of the Institute of Radio Engineers* 49 (January 1961): 8–30.

Mirowski, Philip. *Machine Dreams: Economics Becomes a Cyborg Science*. Cambridge: Cambridge University Press, 2002.

Mises, Richard von. "Über kausale und statistische Gesetzmäßigkeit in der Physik." *Naturwissenschaften* 18 (1930): 145–53. Reprinted in *Erkenntnis* 1: 189–210.

———. *Wahrsheinlichkeit, Statistik un Wahrheit*, vol. 3 in *Schriften zur Wissenschaftlichen Weltauffassung*, edited by Philipp Frank and Moritz Schlick. Wien: Julius Springer, 1928.

Monk, Ray. *Robert Oppenheimer: A Life inside the Center*. New York: Doubleday, 2012.

Monod, Jacques. *Chance and Necessity: An Essay on the Natural Philosophy of Modern Biology* (1970). New York: Vintage Books, 1972.

Mooney, Vincent John, III. "Searle's Chinese Room and Its Aftermath." CSLI 97–202. Stanford, CA: Stanford University, Center for the Study of Language and Information (June 1997).

Moore, Kate. *The Radium Girls: The Dark Story of America's Shining Women*. Naperville, IL: Sourcebooks, 2017.

Morowitz, Harold J. *Entropy for Biologists*. New York: Academic Press, 1970.

Morton, Jack A. "The Manager as Maxwell's Demon." *Innovation* 1 (May 1969): 38–45.

———. "The Manager's Changing Role in Technological Innovation." *Bell Telephone Journal* 47, no. 1 (January/February 1968): 8–15.

———. *Organizing for Innovation: A Systems Approach to Technical Management*. New York: McGraw-Hill, 1971.

Muses, C. A. "Foreword." In *Communication, Organization, and Science* by Jerome Rothstein, vii–lxxxv. Indian Hills, CO: Falcon's Wing Press, 1958.

Nadler, Steven. "Descartes's Demon and the Madness of Don Quixote." *Journal of the History of Ideas* 58, no. 1 (1997): 41–55.

Napoléon I. "Consuls Provisoires." In *Correspondance de Napoléon 1er*, edited by Henri Plon, 324–47. Paris, 1870.

Neesen, Friedrich. "W. Thomson: Kinetic Theory of the Dissipation of Energy." *Fortschritte der Physik Im Jahre* 1874, no. 30 (1879): 673–74.

Nemeth, Evi, et al. *Unix System Administration Handbook*. New York: Prenticc-Hall, 1989.

Nernst, Walther. *Zum Gültigkeitsbereich der Naturgesetze*. Berlin: Norddeutschen Buchdruckerei und Verlagsanstalt, 1921.

Neumann, John von. *Mathematical Foundations of Quantum Mechanics*, translated by Robert T. Beyer. Princeton, NJ: Princeton University Press, 1955.

Neurath, Otto. "Protocol Sentences." In *Logical Positivism*, edited by A. J. Ayer, 199–208. New York: Free Press, 1959.

———. "Protokollsätze." *Erkenntnis* 3 (1932/1933): 204–14.
Newell, Allen. "Some Problems of Basic Organization in Problem-Solving Programs." In *Self-Organizing Systems*, edited by Marshall C. Yovits, G. Jacobi, and Gordon D. Goldstein, 393–423. Washington, DC: Spartan, 1962.
Newell, Allen, John Laird, and Paul Rosenbloom. "Proposal for Research on SOAR: An Architecture for General Intelligence and Learning." Box 32, vol, 22, series 1986–052. Special Collections, Stanford University Libraries.
Nietzsche, Friedrich Wilhelm. *Die fröhliche Wissenschaft*. Chemnitz: Ernst Schmeitzner, 1882.
"Notes." *Nature* 61, no. 1567 (November 9, 1899): 37–40.
Odum, Howard T. *Systems Ecology: An Introduction*. New York: John Wiley and Sons, 1983.
Ogden, C. K. *Bentham's Theory of Fictions*. Paterson, NJ: Littlefield, Adams & Co., 1959.
"Open Peer Commentary to 'Minds, Brains, and Programs.'" *Behavioral and Brain Sciences* 3 (1980): 424–50.
Oseen, C. W. "Award Ceremony Speech." In *Nobel Lectures, Physics 1922–1941*, 135–37. Amsterdam: Elsevier Publishing Co., 1965.
Pais, Abraham. *"Subtle Is the Lord . . .": The Science and the Life of Albert Einstein*. Oxford: Clarendon Press, 1982.
Pascal, Blaise. *Éloge et pensées de Pascal, nouvelle édition, Commentée, corrigée, et augmentée*. Paris: Mr. De ***, 1778.
Pearson, Karl. *The Grammar of Science*. London: Walter Scott, 1892.
———. *The Grammar of Science*, 2nd ed. London: Adam and Charles Black, 1900.
———. "The Problem of the Random Walk." *Nature* 72, no. 1865 (27 July 1905): 294.
Penrose, Oliver. *Foundations of Statistical Mechanics*. Oxford: Pergamon Press, 1970.
Perrin, Jean. *Atoms* (1913). Woodbridge, CT: Ox Bow Press, 1990.
———. "Mécanisme de l'électrisation de contact et solutions colloïdiales." *Journal de chimie physique* 2 (1904–1905): 601–51.
———. "Mécanisme de l'électrisation de contact et solutions colloïdiales." *Journal de chimie physique* 3 (1904–1905): 50–110.
———. "Mouvement brownien et realité moléculaire." *Annales de chimie et de physique* 8, no. 18 (1909): 1–114.
Pierce, John R. "The Exorcism." *Galaxy* (April 1971): 81–91.
———. *Symbols, Signals, and Noise: The Nature and Process of Communication*. Harper Modern Science Series. New York: Harper & Row, 1961.
Pinker, Steven. *How the Mind Works*. New York: W. W. Norton and Co., 1997.
Pittendrigh, Colin S. "Adaptation, Natural Selection, and Behavior." In *Behavior and Evolution*, edited by Anne Roe and George Gaylord Simpson, 390–416. New Haven, CT: Yale University Press, 1958.
———. "On Temporal Organization in Living Systems." *Harvey Lectures* 56 (1960): 93–125.
———. "Temporal Organization: Reflections of a Darwinian Clock-Watcher." *Annual Review of Physiology* 55 (1993): 17–54.
Planck, Max. "The Concept of Causality in Physics." In *Scientific Autobiography and Other Papers*, 121–50. New York: Philosophical Library, 1949.
———. "Verdampfen, Schmelzen, und Sublimieren." *Annalen der Physik* 15 (1882): 446–75.
———. *Wege zur physikalischen Erkenntnis: Reden und Vorträge*. Leipzig: S. Hirzel, 1933.
Platt, John R. "Books That Make a Year's Reading and a Lifetime's Enrichment: A Year's Reading." *New York Times* (February 2, 1964).

Pohle, Joseph. *Die Sternenwelten und ihre Bewohner* (1885). Köln: J. P. Bachem, 1906.

———. *God: The Author of Nature and the Supernatural (De Deo Creante et Elevante)*, translated by Arthur Preuss, 2nd ed. St. Louis, MO: B. Herder, 1916.

Poincaré, Henri. "Le Hasard." *Revue du mois* 3 (1907): 257–76. Republished in Poincaré, *Science and Method*.

———. *Leçons sur les hypotheses cosmogoniques professées à la Sorbonne*. Paris: A. Hermann et fils, 1911.

———. "Le Mécanisme et l'expérience." *Revue de métaphysique et de morale* 1 (1893): 534–37.

———. "Poincaré's 'Thermodynamics.'" *Nature* 45, no. 1169 (March 24, 1892): 485.

———. "The Principles of Mathematical Physics." *The Monist* 15, no. 1 (January 1, 1905): 1–24.

———. "The Principles of Mathematical Physics." In *Congress of Arts and Science, Universal Exposition, St. Louis, 1904* (1904), 604–22. Boston: Houghton Mifflin Co., 1905.

———. "Relations entre la physique expérimentale et la physique mathématique." In *Rapports présentés au Congrès international de physique*, edited by Charles Édouard Guillaume and Lucien Poincaré, 1–29. Paris: Gauthier-Villars, 1900.

———. *Science and Hypothesis* (1902). New York: Dover, 1952.

———. *Science and Method* (1908). New York: Dover, 1952.

———. "Sur les équations de la dynamique et le problème des trois corps." *Acta Mathematica* 13 (1890): 1–270.

———. *La Valeur de la science*. Bibliothèque de philosophie scientifique. Paris: E. Flammarion, 1905.

———. *The Value of Science*. New York: Dover Publications, 1958.

Popper, Karl R. "Irreversibility; Or, Entropy since 1905." *British Journal for the Philosophy of Science* 8, no. 30 (1957): 151–55.

———. *Objective Knowledge: An Evolutionary Approach*. Oxford: Clarendon Press, 1973.

———. *The Open Universe: An Argument for Indeterminism from the Postscript to* The Logic of Scientific Discovery. Totowa, NJ: Rowman and Littlefield, 1982.

———. *Quantum Theory and the Schism in Physics*, edited by William Warren Bartley. Totowa, NJ: Rowman and Littlefield, 1982.

———. *The World of Parmenides: Essays on the Presocratic Enlightenment*. London: Routledge, 1998.

Price, Derek J. "The Cavendish Laboratory Archives," *Notes and Records of the Royal Society of London* 10 (1953): 139–47.

Prigogine, Ilya. "Time, Irreversibility, and Structure." In *The Physicist's Conception of Nature*, edited by Jagdish Mehra, 561–91. Dordrecht: Springer Netherlands, 1973.

Prigogine, Ilya, and Isabelle Stengers. *La Nouvelle alliance: Métamorphose de la science*, 2nd ed. Paris: Gallimard, 1986.

Proctor, Richard A. *Other Worlds than Ours*. New York: D. Appleton and Co., 1896.

Project MAC. Progress Report VI. Cambridge, MA: Massachusetts Institute of Technology, 1969.

Raizen, Mark G., et al. "Demons, Entropy, and the Quest for Absolute Zero." *Scientific American* (March 2011).

Raymond, Eric S. *The New Hacker's Dictionary*. Cambridge, MA: MIT Press, 1991.

Reichenbach, Hans. *Elements of Symbolic Logic*. New York: Macmillan Co., 1947.

———. *Experience and Prediction: An Analysis of the Foundations and Structure of Knowledge*, 4th impression (1952). Chicago: University of Chicago Press, 1938.

———. *The Rise of Scientific Philosophy*. Berkeley: University of California Press, 1951.

Reisch, George A. *How the Cold War Transformed Philosophy of Science: To the Icy Slopes of Logic*. Cambridge: Cambridge University Press, 2005.
"Relativity." *Punch* 165 (December 19, 1923): 591.
Reuell, Peter. "Hawking at Harvard." *Harvard Gazette* (April 18, 2016).
Rhim, Won-Kyu, Alexander Pines, and Justin S. Waugh. "Time-Reversal Experiments in Dipolar-Coupled Spin Systems." *Physics Review* B-3, no. 3 (February 1, 1971): 684–96.
Rieger, Charles J., III. "Conceptual Memory: A Theory and Computer Program for Processing the Meaning of Content of Natural Language Utterances." AIM-233. Stanford, CA: Computer Science Department, Stanford Artificial Intelligence Laboratory, July 1974.
———. "Spontaneous Computation in Cognitive Models." *Cognitive Science* 1 (1977): 315–54.
Riskin, Jessica. *The Restless Clock: A History of the Centuries-Long Argument over What Makes Living Things Tick*. Chicago: Univeristy of Chicago Press, 2018.
"Rolf Landauer, a Unique Physicist, Dies." *Physics World* (April 30, 1999).
Rosen, Donn Eric. "Darwin's Demon." Review of *Introduction to Natural Selection* by Clifford Johnson. *Systematic Zoology* 27, no. 3 (1978): 370–73.
Rosenblatt, Frank. "Two Theorems of Statistical Separability in the Perceptron." In *Mechanisation of Thought Processes: Proceedings of a Symposium Held at the National Physical Laboratory on 24th, 25th, 26th, and 27th November 1958*, 419–50. London: Her Majesty's Stationery Office, 1961.
Rothstein, Jerome. *Communication, Organization, and Science*. Indian Hills, CO: Falcon's Wing Press, 1958.
———. "Informational Generalization of Entropy in Physics." In *Quantum Theory and Beyond: Essays and Discussions Arising from a Colloquium*, edited by Ted Bastin, 291–305. Cambridge: Cambridge University Press, 1971.
———. "Information, Measurement, and Quantum Mechanics." *Science* 114 (1951): 171–75.
———. "Loschmidt's and Zermelo's Paradoxes Do Not Exist." *Foundations of Physics* 4, no. 1 (March 1974): 83–89.
———. "Physical Demonology." *Methodos: Linguaggio e cibernetica* 11, no. 42 (1959): 99–121.
———. "Thermodynamics and Some Undecidable Physical Questions." *Philosophy of Science* 31, no. 1 (1964): 40–48.
Rule, Sheila. "Reagan Gets a Red Carpet from British." *New York Times*, June 14, 1989.
Rumelhart, David E., and James L. McClelland. *Parallel Distributed Processing: Explorations in the Microstructure of Cognition*. 2 vols. Cambridge, MA: MIT Press, 1986.
Russell, Bertrand. "My Mental Development." In *The Philosophy of Bertrand Russell*, edited by Paul Arthur Schilpp, 1–20. Evanston, IL: Northwestern University, 1944.
Rutherford, Ernest, and Frederick Soddy. "LX: Radioactive Change." *London, Edinburgh, and Dublin Philosophical Magazine and Journal of Science* 5, no. 29 (1903): 576–91.
Ryle, Gilbert. "Intentionality-Theory and the Nature of Thinking." *Revue internationale de philosophie* 27 (1973): 104–5.
Sagan, Carl. *The Demon-Haunted World: Science as a Candle in the Dark*. New York: Ballantine Books, 1997.
Samuelson, Paul A. "Challenge to Judgement." *Journal of Portfolio Management* 1, no. 1 (1974): 17–19.
———. *Collected Economic Papers*, vol. 5. Cambridge, MA: MIT Press, 1986.

———. "Economics in My Time." In *Lives of Laureates*, edited by William Breit and Roger W. Spencer, 56–76. Cambridge, MA: MIT Press, 1986.
———. "Foreword." In *Louis Bachelier's Theory of Speculation: The Origins of Modern Finance*, vii–xi. Princeton, NJ: Princeton University Press, 2006.
———. "Interview." In *Roads to Wisdom: Conversations with Ten Nobel Laureates in Economics*, edited by Karen Ilse Horn, 43–56. Cheltenham, UK: Edward Elgar, 2009.
———. *Macroscopic Time Asymmetry of Maxwell's Demon*. Cambridge, MA: MIT Press, 1985.
———. "Mathematics of Speculative Price." *SIAM Review* 15, no. 1 (1973): 1–42.
———. "Scientific Correspondence: The Law Beats Maxwell's Demon." *Nature* 347, no. 6288 (September 6, 1990): 24–25.
Scarre, Geoffrey. "Demons, Demonologists, and Descartes." *Heythrop Journal* 31, no. 1 (1990): 3–22.
Schmeck, Harold M., Jr. "A Scientist Gives Demons Their Due: 'Aladdin's Demon' Added to List." *New York Times* (February 1, 1959).
Schmitz, David F. *Henry L. Stimson: The First Wise Man*. Wilmington, DE: Scholarly Resources, 2001.
Scholem, Gershom. "The Golem of Prague and the Golem of Rehovot." *Commentary* 41, no. 1 (January 1966): 62–65. Reprinted in *The Messianic Idea in Judaism and Other Essays on Jewish Spirituality*, 335–40. New York: Schocken Books, 1995.
Schrödinger, Erwin. "Die gegenwärtige Situation in der Quantenmechanik." *Naturwissenschaften* 48, no. 23 (November 29, 1935): 807–12, 23–28, 44–49.
———. *What Is Life? The Physical Aspect of the Living Cell* (1944). Cambridge: Cambridge University Press, 1955.
"Science on Road to Revolutionize All Existence." *New York Times*, September 28, 1913.
Seaborg, Glenn T. *Adventures in the Atomic Age: From Watts to Washington*, edited by Eric Seaborg. New York: Farrar, Straus and Giroux, 2001.
Searle, John R. "Author's Response." *Behavioral and Brain Sciences* 3 (September 1980): 450–57.
———. "Minds, Brains, and Programs." *Behavioral and Brain Sciences* 3 (September 1980): 417–24.
———. "The Myth of the Computer." *New York Review of Books* 29, no. 7 (April 29, 1982): 3–6.
Selfridge, Oliver G. "Pandemonium: A Paradigm for Learning." In *Mechanisation of Thought Processes: Proceedings of a Symposium Held at the National Physical Laboratory on 24th, 25th, 26th, and 27th November 1958*, 511–26. London: Her Majesty's Stationery Office, 1961.
Selfridge, Oliver G., and Ulric Neisser. "Pattern Recognition by Machine." *Scientific American* 203, no. 2 (August 1960): 60–68.
Seow, Choong Huei. "Of Frames, Scripts, and Stories." Thesis submitted to the Department of Electrical Engineering and Computer Science in partial fulfillment of the requirements for the degree of bachelor of science in computer science and engineering at the Massachusetts Institute of Technology, May 1990.
Serres, Michel. *The Parasite* (1980), translated by Lawrence R. Schehr. Minneapolis: University of Minnesota Press, 2007.
Serviss, Garrett P. *The Einstein Theory of Relativity*. New York: Edwin Miles Fadman, 1923.
Shakespeare, William. *The Tempest*. In *Mr. William Shakespeares Comedies, Histories, & Tragedies*, 1–19. London: Isaac Jaggard and Edward Blount, 1623.

———. *The Tragedie of Macbeth.* In *Mr. William Shakespeares Comedies, Histories, & Tragedies*, 131–51. London: Isaac Jaggard and Edward Blount ,1623.
———. *The Tragicall Historie of Hamlet.* London: N. L. and John Trudell, 1603.
Shannon, Claude E. "*Cybernetics, or Control and Communication in the Animal and the Machine*, by Norbert Wiener." *Proceedings of the Institute of Radio Engineers* 37, no. 11 (November 1949): 1305.
———. "A Mathematical Theory of Communication." *Bell Systems Technical Journal* 27 (July and October 1948): 379–423, 623–56.
Shenker, Orly R. "Maxwell's Demon and Baron Munchausen: Free Will as a Perpetuum Mobile." *Studies in the History and Philosophy of Modern Physics* 30, no. 3 (1999): 347–72.
Shleifer, Andrei, and Lawrence H. Summers. "The Noise Trader Approach to Finance." *Journal of Economic Perspectives* 4, no. 2 (Spring 1990): 19–33.
Simberloff, Daniel. "A Succession of Paradigms in Ecology: Essentialism to Materialism and Probabilism." *Synthese* 43, no. 1 (January 1980): 3–39.
"Small Wonder." *Los Angeles Times* (July 30, 1986).
Smocovitis, Vassiliki Betty. "The 1959 Darwin Centennial Celebration in America." *Osiris* 14 (1999): 274–323.
Smolin, Lee. *Einstein's Unfinished Revolution: The Search for What Lies beyond the Quantum.* New York: Penguin Press, 2019.
Smoluchowski, Marian von. "Essai d'une théoria cinétique du mouvement brownien et des milieux troubles." *Bulletin international de l'académie des sciences de cracovie: Classe des sciences mathématiques et naturelles* 7 (1906): 577–602.
———. "Experimentell nachweisbare, der üblichen Thermodynamik widersprechende Molekularphänomene." *Physikalische Zeitschrift* 13 (1912): 1069–80.
———. "Gültigkeitsgrenzen des zweiten Hauptsatzes der Wärmetheorie." In *Vorträge über die Kinetische Theorie der Materie und Elektrizität*, 89–121. Leipzig: B. G. Teubner, 1914.
———. "Vorträge über die Kinetische Theorie der Materie and Elektrizität." In *Oeuvres de Marie Smoluchowski*, 360–98. Cracow: Jagellonian University Press, 1927.
Smyth, Henry De Wolf. *Atomic Energy for Military Purposes: The Official Report on the Development of the Atomic Bomb under the Auspices of the United States Government, 1940–1945.* Princeton, NJ: Princeton University Press, 1945.
Smyth, Samuel Phillips Newman. *Through Science to Faith.* New York: Charles Scribner, 1902.
Soddy, Frederick. *The Interpretation of Radium*, 4th ed. New York: G. P. Putnam's Sons, 1922.
Solow, Robert M. "The Economics of Resources or the Resources of Economics." *American Economic Review* 64, no. 2 (1974): 1–14.
Somerville, Mary. *Personal Recollections.* London: John Murray, 1873.
Spengler, Oswald. *Der Untergang des Abendlandes: Umrisse einer Morphologie der Weltgeschichte.* Wien and Leipzig: Wilhelm Braumüller, 1918.
Steinberg, R. U. "Mr. Smarty Pants Knows." *Austin Chronicle* (February 8, 2002).
Sullivan, Lucy G. "Myth, Metaphor, and Hypothesis: How Anthropomorphism Defeats Science." *Philosophical Transactions of the Royal Society: Biological Sciences* 349, no. 1348 (August 29, 1995): 215–18.
Sutherland, G.B.B.M. "Opening Address." In *Mechanisation of Thought Processes: Proceedings of a Symposium Held at the National Physical Laboratory on 24th, 25th, 26th, and 27th November 1958*, ix–x. London: Her Majesty's Stationery Office, 1961.

Szilard, Leó. "Creative Intelligence and Society: The Case of Atomic Research, the Background in Fundamental Science." Unpublished manuscript, July 31, 1946.
———. "Part II: Leó Szilard: His Version of the Facts." *Bulletin of the Atomic Scientists* 35, no. 3 (March 1979): 55–59.
———. "Über die Entropieverminderung in einem thermodynamischen System bei Eingriffen intelligenter Wesen." *Zeitschrift für Physik* 53 (1929): 840–56.
Tait, Peter Guthrie. *Lectures on Some Recent Advances in Physical Science with a Special Lecture on Force* (1874). London: MacMillan and Co., 1976.
———. "Preface." In *Sketch of Thermodynamics*, 2nd ed. Edinburgh: David Douglas, 1877.
———. "Zöllner's Scientific Papers." *Nature* 17 (1878): 420–22.
Thiele, Just Mathias. *Danske Folkesagn*. Copenhagen: A. Seidelin, 1819–1823.
Thomson, James. *Collected Papers in Physics and Engineering*. Cambridge: Cambridge University Press, 1912.
———. "On Public Parks in Connexion with Large Towns, with Suggestion for the Formation of a Park in Belfast." In *Collected Papers in Physics and Engineering*, 464–72. Cambridge: Cambridge University Press, 1912.
———. "Theoretical Considerations on the Effect of Pressure in Lowering the Freezing Point of Water." *Transactions of the Royal Society of Edinburgh* 16, no. 5 (1849): 575–80.
Thomson, William. "Kinetic Theory of the Dissipation of Energy." *Nature* 9 (1874): 441–44.
———. "The Kinetic Theory of the Dissipation of Energy." *Proceedings of the Royal Society of Edinburgh* 8 (1875): 325–34.
———. "On the Age of the Sun's Heat." In *Popular Lectures and Addresses*, vol. 1, 356–75. London: MacMillan and Co., 1891. Originally published in *Macmillan's Magazine* 5 (March 5, 1862): 388–93.
———. "The Sorting Demon of Maxwell." In *Popular Lectures and Addresses*, vol. 1, 144–48. London: MacMillan and Co., 1891.
Thurston, John B. "Devaluing the Human Brain." *Saturday Review of Literature* (April 23, 1949): 24–25.
Tillman, Frank, and Bertrand Russell. "Language, Information, and Entropy." *Logique et analyse* 30 (1965): 126–40.
Trotsky, Leon. *Soviet Economy in Danger*. New York: Pioneer Publishers, 1933.
Tu, Yuhai. "The Nonequilibrium Mechanism for Ultrasensitivity in a Biological Switch: Sensing by Maxwell's Demons." *Proceedings of the National Academy of Sciences* 105, no. 33 (August 19, 2008): 11737–41.
Turing, Alan M. "Computing Machinery and Intelligence." *Mind* 59, no. 236 (October 1950): 433–60.
Tyson, Donald. *The Demonology of King James: Includes the Original Text of Daemonologie and News from Scotland*. Woodbury, MN: Llewellyn, 2011.
Vaihinger, Hans. *Die Philosophie des Als ob: System der theoretischen, praktischen, und religiösen Fiktionen der Menschleit auf Grund eines idealistischen positivismus*. Berlin: Von Reuther & Reichard, 1911.
Voltaire. *Le Siècle de Louis XIV*. Berlin: C.-F. Henning, 1751.
Waddington, Conrad Hal. "The Basic Ideas of Biology." In *Towards a Theoretical Biology*, edited by Conrad Hal Waddington, 1–32. Chicago: Aldine Publishing Co., 1968.
———. "Evolutionary Adaptation." *Perspectives in Biology and Medicine* 2, no. 4 (1959): 379–401.
Waldrop, M. Mitchell. "Soar: A Unified Theory of Cognition?" *Science* 241 (July 15, 1988): 296–98.

———. "Toward a Unified Theory of Cognition." *Science* 241, no. 4861 (July 1, 1988): 27–29.
Wallace, Alfred Russel. *The Action of Natural Selection on Man*. New Haven, CT: C. Chatfield & Co., 1871.
———. *Contributions to the Theory of Natural Selection: A Series of Essays*, 2nd ed. London: Macmillan and Co., 1871.
Watson, David Lindsay. *Scientists Are Human*. London: Watts, 1938.
Watt, Ian. "Order and Disorder in the Arts and Sciences: A Historical Retrospect." In *Disorder and Order: Proceedings of the Stanford International Symposium* (September 14–16, 1981), edited by Paisley Livingston, 34–40. Saratoga, CA: Anma Libri, 1984.
Weinberg, Alvin M. "Are Breeders Still Necessary?" *Proceedings of the American Philosophical Society* 130, no. 3 (September 1986): 343–53.
———. "Global Effects of Man's Production of Energy." *Science* 186, no. 4160 (October 18, 1974): 205.
———. "A Nuclear Power Advocate Reflects on Chernobyl." *Bulletin of the Atomic Scientists* 43, no. 1 (August/September 1986): 57–60.
———. "On the Relation between Information and Energy Systems: A Family of Maxwell's Demons." *Interdisciplinary Science Reviews* 7 (1982): 47–52. Reprinted as "A Family of Maxwell's Demons." In Weinberg, *Nuclear Reactions: Science and Trans-Science*. New York: American Institute of Physics, 1992.
———. "Social Institutions and Nuclear Energy." *Science* 177, no. 4043 (July 7, 1972): 27–34.
———. "Social Institutions and Nuclear Energy: II." *Journal of Nuclear Science and Technology* 32, no. 11 (November 1995): 1071–80.
Weinert, Friedel. *Demons of Science: What They Can and Cannot Tell Us about Our World*. Switzerland: Springer, 2018.
Weizenbaum, Joseph. *Computer Power and Human Reason: From Judgment to Calculation*. San Francisco: W. H. Freeman, 1976.
Wells, H. G. "The Time Machine." *New Review* 12 (January–June 1895): 98–112, 207–21, 329–43, 453–22, 577–88.
Whiting, Harold. "Maxwell's Demons." *Science* 6, no. 130 (July 31, 1885): 83.
Wiener, Norbert. *Collected Works with Commentaries*, edited by P. Masani, vol. 4. Cambridge, MA: MIT Press, 1985.
———. "Cybernetics." *Scientia* 87 (1952): 233–35.
———. *Cybernetics, or Control and Communication in the Animal and the Machine*. New York: John Wiley & Sons, 1948; 2nd ed., Cambridge, MA: MIT Press, 1961.
———. "Entropy and Information." *Proceedings of the Symposia of Applied Mathematics of the American Mathematical Society* 2 (1950). Also published in *Mathematical Review* 11 (1950). Reprinted in Norbert Wiener, *Collected Works with Commentaries*, edited by P. Masani, vol. 4. Cambridge, MA: MIT Press, 1985.
———. *The Human Use of Human Beings: Cybernetics and Society*. London: Free Association, 1989.
———. "Men, Machines, and the World About." In *Medicine and Science*, edited by Iago Galdston. New York: International Universities Press, 1954.
———. "Time, Communication, and the Nervous System." *Annals of the New York Academy of Science* 50 (1948): 197–220.
Wigner, Eugene Paul. "Remarks on the Mind-Body Question." In *The Scientist Speculates* (1961), edited by I. J. Good; reprinted in *Symmetries and Reflections: Scientific Essays of Eugene P. Wigner*, 171–84. Bloomington: Indiana University Press, 1967.

Williams, George C. *Adaptation and Natural Selection: A Critique of Some Current Evolutionary Thought*. Princeton, NJ: Princeton University Press, 1966.
Yourgrau, Palle. *Death and Nonexistence*. Oxford: Oxford University Press, 2019.
Zermelo, Ernst. "Ueber einen Satz der Dynamik und die mechanische Wärmetheorie." *Annalen der Physik* 293, no. 3 (1896): 485–94.
Zöllner, Friedrich. "VI. On Space of Four Dimensions." *Quarterly Journal of Science* 8 (April 1878): 227–37.
———. "Thomson's Daemonen und die Schatten Plato's." In *Wissenschaftliche Abhandlungen*, 710–32. Leipzig: L. Staakmann, 1878.
———. *Transcendental Physics: An Account of the Experimental Investigation from the Scientific Treatises of Johann Carl Friedrich Zöllner*. Boston: Banner of Light Publishing, 1901.
Zurek, Wojciech H. "Algorithmic Randomness, Physical Entropy, Measurements, and the Demon of Choice." In *Feynman and Computation: Exploring the Limits of Computers*, edited by Anthony J. G. Hey, 393–10. Reading, MA: Perseus Books, 1999.
———. "Thermodynamic Cost of Computation, Algorithmic Complexity, and the Information Metric." *Nature* 341 (1989): 119.
Zyga, Lisa. "Maxwell's Demons May Drive Some Biological Systems." *PhysOrg*, September 10, 2008. https://phys.org/news/2008-09-maxwell-demons-biological.html.

Index

Note: Page numbers in italic type indicate illustrations.

Abacus: Magazine for the Computer Professional, 227
Aberdeen Proving Grounds, Maryland, 159
absolute space, 105
acceleration, 109
Adalia (slave ship), 39–40
Adam, 6, 7
Adams, Henry, 75–76, 141, 333n75
Age of Reason, 2, 16. *See also* Enlightenment
Agricultural Adjustment Administration, 142
Aharanov-Bohm effect, 198
AI. *See* artificial intelligence
Aladdin's genie, 116, 178–81, 307
algorithms, 32, 185, 188, 237
altruism, 258, 259, 269
Amazon, 278, 296
American Academy for the Advancement of Science, 141, 293
American Economic Association, 286
American Economic Review (journal), 288
American Journal of Physics, ix, 155
American Physical Society, 178
American Scientist (journal), 165
American Society for Cybernetics, 210
American Telephone & Telegraph (AT&T), 162
Ampère, André-Marie, 342n1
amplification, of small-scale acts, 41–42, 55, 74, 82, 88, 129–31, 146, 148, 151–52, 163, 165, 186, 199, 321
Anales de Química (journal), 245
analytical philosophy, 315–16
Angrist, Stanley W., and Loren G. Hepler, *Order and Chaos,* 167
Annalen der Physik (journal), 80
anthropology, 274–75
Antiaircraft Artillery Board, Camp Davis, North Carolina, 159

Arago, François, 35–36
Argonne National Laboratory, Washington, D.C., 230
Army Research and Development Laboratories, 173
ARPANET, 201
art, as foil to entropy, 251, 261
artificial intelligence (AI): children's comprehension as model for, 202–9; coining of term, 191; dangers of, 237; demons associated with, 7, 185, 188, 190–93, 202–9, 267–71; early developments in, 185–94; fears of, 186; hierarchical organization in, 188, 190, 192–94; learning in, 233, 236–37, 349n162; living systems compared to, 271–72; neural networks and, 190; open-ended architectures for, 237; programming and, 185, 187–88, 190–93, 219–20; Searle on, 219–26; Selfridge and, 188–93; strong AI, 191, 219–22, 225–26; Turing test for, 187, 219. *See also* machine learning
Artificial Intelligence Laboratory, MIT, 202–3
Artzybasheff, Boris, 186
Asimov, Isaac, 260–62; *Life and Energy,* 260; "The Modern Demonology," 260
Assembly of German Naturalists and Physicians, 64
Association for Computing Machinery, 220
astronomy: and Einstein's theory of relativity, 105–7; Laplace's demon and, 30–36, 66, 124; and light's path in the universe, 100; and possible violations of second law of thermodynamics, 74, 85, 87; prediction in, 30–36, 66
atomic bomb, 114, 122–24, 128, 131, 151–56, 164–65, 173, 193, 214, 252, 338n30
atomic energy, 139–41. *See also* nuclear energy

atomistic theories, 80
automata, 7, 21
Ayer, A. J., 12

Babbage, Charles, 29, 38–40, 43–44
Bachelier, Louis, 281
Bar-Hillel, Yehoshua, 342n28
Baudelaire, Charles, 4
BBC (British Broadcasting Corporation), 245
Becquerel, Henri, 113, 116
Beelzebub, 4
Behavioral and Brain Sciences (journal), 220
Bekenstein, Jacob, 214–16; "Black-Hole Thermodynamics," 215
beliefs: computer-held, 191; understanding of reality influenced by, 317–18
Bell, John, 183–84
Bell inequality, 183
Bell Labs, 162, 273–74
Bennett, Charles, 226–31, 234–36
Bentham, Jeremy, 314; "Theory of Fictions," 314
Bergson, Henri, 137, 220, 247–48
Bernstein, Aaron, 95, 97, 101
Bezos, Jeff, 296
Bhushan, Abhay, 201–2
Bigelow, Julian, 159
biology. *See* life/living systems
biology–physics connection, 251–52, 259, 263, 266–67
bits, 163
black holes, 214–18, 322
Bloch, Ernst, 9
block universe, 99–100
Blumenberg, Hans, 322–23
body, Cartesian conception of, 17, 184
Bohm, David, 176, 183–84, 197
Bohm's demon, 176–77
Bohr, Niels, 115, 118, 120, 130, 131–32, 143–44, 252
Bois-Reymond, Emil du, 64–65, 247
Boltzmann, Ludwig, 76–77
Boltzmann's constant, 170
Born, Max, 12, 80, 114, 115, 117, 119–22, 143, 156, 182–83
Born's demon, 200
Boscovich, Roger, 42
bots, 240. *See also* nanobots
Bourdieu, Pierre, 247, 275–76
brain-in-the-vat thought experiment, 16, 267

brain/mind: as computer, 190, 218–19, 267–71; demons in, 218–26; as machine, 190. *See also* thinking
Bretton Woods agreement, 285
Bridgman, Percy, 152–53
Brillouin, Léon, 165–68, 170–71, 177, 182, 194, 215, 227, 253, 262, 279, 284, 338n38, 346n29; *Science and Information Theory*, 177, 194. *See also* Maxwell-Szilard-Brillouin demon
Broad, C. D., 116–17, 337n12
Broglie, Louis de, 120, 177
Brown, Robert, 79, 80
Brownian computers, 232
Brownian motion, 79–92; Campbell-Swinton and, 87; and chance, 91–92; characteristics of, 79, 91, 309; discovery of, 79; economic uses of, 281; Einstein and, 79–81, 88–91; flight patterns compared to, 159; Gouy and, 81–82; harnessing energy of, 79, 82, 90–91, 152; Jevons and, 83–84; Maxwell's demon and, 79, 81, 84–87, 90; Perrin and, 89–91; Random Walk problem identified with, 89; Smoluchowski and, 90–92; substances exhibiting, 83–84
Bush, Vannevar, 122
Butler, Samuel, 352n30
Byron, George Gordon, Lord, 40

calculating machines, 2, 29, 37. *See also* computers
California Institute of Technology (Caltech), 228, 253
Campbell-Swinton, Alan Archibald, 87, 248
Carlyle, Thomas, 41–42
Carnap, Rudolf, 339n73, 342n28
Carnot, Sadi, 61, 66, 82, 86, 87, 113, 115
Carr, Wildon, 87–88
Carroll, Sean, ix–x
Cassirer, Ernst, 65, 150
Catholic Church, 316
causality: critiques of, 41–43, 117–18, 123–24, 135, 138; Einstein's adherence to, 117–18, 120–21, 143–44, 147, 177, 183; Laplace's demon and, 31, 123, 136–39, 150, 163–64, 182–84, 186, 259; laws of nature as basis of, 109–10, 117–18; quantum mechanics and, 117, 120, 135, 144, 200; resistance to abandoning, 117–18, 120–21, 124, 132–33; speed of light and, 97. *See also* determinism

Caves, Carlton, 242–43
cell reproduction, 264
Cervantes, Miguel de, *Don Quixote*, 18–20
chance: Boltzmann's discounting of significance of, 77; Brownian motion and, 91–92; in economic realm, 279–81, 308; Einstein's resistance to, 80, 121, 177; in gambling, 91, 163–64; insignificance of, 65, 68, 140; Laplace's demon and, 163–64; Maxwell's demon and, 260; and origin of life, 263; and perpetual motion machines, 126; quantum mechanics and, 144–46; rational assessments of, 35; reversal of time and, 97; and weather prediction, 68
chaos theory, 263
Charniak, Eugene, 202, 233, 236; *Artificial Intelligence Programming*, 349n162; "Toward a Model of Children's Story Comprehension," 202–3, *204*, 346n59
Charniak's demons, 202
Chemistry World (journal), 245
Chernobyl nuclear power plant, 293
children, as model for AI, 202–9
Chinese room argument, 220–24, 271, 348n118
chromosomes, 246, 250–51
classical observer, 136
Clausius, Rudolf, 69–70, 74
Clifford, William Kingdon, 57
climate change, 289
Club of Rome, 284
CNN (Cable News Network), 245
cogito ergo sum, 16, 26, 187
cognitive psychology, 188, 190, 193, 267–73
Cohen, George, 253
collectivization, 138
communication: information in relation to, 167–68; Maxwell's demon and, 161–63; perfectly efficient, 162; speed of light and, 101–2. *See also* information
communism, 138–39
complementarity, 114
Compton, Arthur, 114, 120, 122–23, 130–31, 143–46, 151, 153–56; *The Freedom of Man*, 131, 143
Compton, Karl, 114, 121–23, 139–43, 153–55
Compton scattering, 122, 130
computers: Babbage's development of, 29, 38–40, 43; beliefs held by, 191; brains as, 190, 218–19, 267–71; economic uses of, 290–91; energy used by, 230–32; Feynman and, 228–29, 231–33, 243; Laplace's demon as precursor of, 29, 38, 185–86; and Maxwell's demon, 198–99, 201, 211, 236; memory of, 229–31, 233–35; miniaturization of, 231; Neumann and, 158; personal, 218, 236, 240; reconceptualization of, 188; software for, 232; speed as key to intelligence of, 231–32; thinking capacity of, 43–44; in World War II, 158. *See also* artificial intelligence (AI); daemons (computer programs); programming
Conant, James, 122
consciousness, 67, 135, 267–68. *See also* intelligence; knowledge
Copenhagen interpretation, 118, 130, 132, 144
Corbató, Fernando, 201
Corbato's daemon, 240
Cornell Aeronautical Laboratory, 189
Crick, Francis, 198, 236, 252–53, 262
criminal responsibility, means of discovering, 39–40, 66, 93, 96, 98–99
Cruft Laboratory, Harvard University, 165
Curie, Marie, 112, 113, 115–16
Curie, Pierre, 113, 116
cybernetic demons: characteristics of, 157, 161; production of, 171; research on, 162; Wiener's use of, 172–73
cybernetics, 158–62, 165, 171, 210–12, 342n1

daemons (computer programs): actions of, 188, 232; and artificial intelligence, 190–93, 202–8, 236–39; defined, 3, 185, 239–40; as UNIX mascot, 241, *241*; uses of, 3, 201, 240–42
Daily Telegraph (newspaper), 245
Daly, Herman, 290
Dante Alighieri, 4, 217
Danto, Arthur, 220, 322
Darling, L., and E. O. Hulburt, "On Maxwell's Demon," 307
Darwin, Charles, 2, 29, 44–48; *On the Origin of Species*, 44, 46–47, 257; *Sketches*, 44–46
Darwin Centennial Celebrations, 254
Darwin's demon: actions of, 250, 254–62, 265–66, 272–73; coining and use of term, 256, 259, 261–62, 265, 268–69; likened to Maxwell's demon, 246, 250, 253–56; as predator, 310, 353n65; purpose of, 257–58, 273

data analysis, 29–30
Dawkins, Richard, 257, 268–69, 354n79; *The Blind Watchmaker*, 272; *The Selfish Gene*, 269
DDT (pesticide), 260
deepfakes, 15
demon of chance, 279–80, 308
demonology, 22–23, 178–82, 260–62, 316, 327n14
demons: AI associated with, 7, 185, 188, 190–93, 202–9, 267–71; atomic bomb associated with, 154–55; belief in, 22; benevolent, 26; Bohm's, 176–77; Born's, 200; in brains, 218–26; characteristics of, 4, 10, 14, 181, 304–10; Charniak's, 202; cognitive psychology and, 267–71; contracts with, 6, 293, 308, 358n14; contributions of, 300–301; devil associated with, 25; in economics, 277–97; Einstein's refusal of, 80, 89, 93–94, 102, 106–10, 174; as fallen angels, 25; Feynman's, 194–96, 232, 243, 245; folk, 26; Gabor's, 168–70, 198, 308, 321, 360n37; gravitational, 94, 106–9; Haugeland's (H-Demon), 223–26, 348n18; heuristic value of, 10, 22, 304–5, 311–12; humanlike characteristics of, 22, 304; in Industrial Revolution, 41; irrationality and superstition associated with, 10–11, 21–22; Landauer's, 230; in literature, 4, 22–23; Loschmidt's, 71–72, 76, 85, 184, 265, 310; Maxwell-Szilard-Brillouin, 262, 265, 266, 309; as means of exploring reality, 15–21; Monod's, 264, 265, 309; old vs. new, 305–10; pagan, 26; Planck's, 284–85; Popper's, 321; prevalence of, in science, ix–x, 9–10, 305, 311, 316–17, 319; in psychology, 212, *213*, 316, 359n15; relationship to one another of, 180–82; in religious contexts, 359n15; role of, in scientific thought, 2–4, 8, 13–14, 107–8, 178–82, 194, 274–75, 298–300, 306, 318–20; rule-following, 265; "savage" vs. scientific, 109–10; scientific opposition to, 140, 177, 200, 235–36, 310–11, 317, 319 (*see also* Einstein's refusal of); Searle's, 218–26, 271, 308, 310; size of, 308–9; in social sciences, 247, 273–76; Szilard's, 125–26, 198, 243; technology associated with, 5–8, 156, 304, 325n6; testing/analysis of, 22–23; in the theater, 20–23; Unruh, 243; usage of term, x; Wheeler's, 214–16, 243; Zermelo's, 85, 184, 284, 310. *See also* daemons (computer programs); Darwin's demon; Descartes's demon; Laplace's demon; Maxwell's demon; quantum demons

Dennett, Daniel, 220, 224–25, 267–68, 270–72; *Darwin's Dangerous Idea*, 271; *Explaining Consciousness*, 267
Descartes, René, 187, 247; and Cervantes's *Don Quixote*, 18; *Discourse on the Method*, 15, 23; *First Meditation*, 16, 23; *Meditations*, 24; *Principles of Philosophy*, 17; religious accusations against, 23–25
Descartes's demon: and AI, 224; heretical features of, 23–25; historical significance of, 2, 7, 15, 267; limitations of, 17, 27; power possessed by, 23–26; reality subverted by, 7, 15–17, 24, 27
determinism: critiques of, 123, 137–38, 155–56, 177, 183; defenses of, 148–49, 163–64, 183–84, 200; Hawking and, 218; Laplacean, 31, 41–42, 123, 138, 149; vitalism vs., 249. *See also* causality; indeterminacy
deus ex machina: explaining relativity by, 108; gravitation as, 106; natural selection as, 254; for perpetual motion machine, 92, 125–26
devil, 6, 22, 25–26, 319
discovery process, 1–2, 8, 11, 299–301
DNA: discovery of, 198, 252; energy expenditure in processes of, 230, 232; information carried by, 246, 253–54; process of, 252–54; Schrödinger and, 250–51. *See also* genes
Dostoevsky, Fyodor, 138
double-slit experiment, 118, 144–46
doubt: Descartes's demon as source of, 15; role of, in science, 15; role of, in thinking, 26–28; sense experience as means of addressing, 20
Doyle, Arthur Conan, 210
Driesch, Hans, 248
Drunkard's Walk, 89, 260, 285, 309
dualism, 17, 184

Eastern mysticism, 184
Eberty, Felix, *The Stars and World History*, 95–96, 111
Eccles, John, 219
ecological economics, 284
economics: chance in, 279–81, 308; computers' impact on, 290–91; demons in, 277–97; ecological, 284; entropy and, 290; environmental, 284; growth in, 277, 281, 284, 287–89, 294–95; information in, 282, 285–86; Laplace's demon in, 278, 279, 283; Maxwell's demon

in, 277–79, 281–82, 285–90, 295–97; neoclassical, 282, 284, 287; optimism vs. pessimism in, 277–78, 286–88, 290–91, 293; physics' connection with, 277–79, 281–90, 293–96; scarcity in, 248, 277, 279, 283, 286, 290, 294–95; shocks in, 283, 285, 293; technological progress and, 287–88
Eddington, Arthur, 93–94, 104–10, 124, 132, 140–41, 284
Eddy, Henry Turner, 73–74, 77
efficient market hypothesis, 280
Ehrenberg, Werner, 197–200
Ehrenberg-Siday-Aharanov-Bohm effect, 198
Ehrenfest, Paul, 120
Eigen, Manfred, 271; *The Laws of the Game*, 264–65; *Steps towards Life*, 266–67
Einstein, Albert, 2; and atomic bomb, 124, 151–52; and Bohm, 176; Bohr's debate with, 120; and Boltzmann, 77; and Brownian motion, 79–81, 88–91; and causality, 117–18, 120–21, 143–44, 147, 177, 183; demons shunned by, 80, 89, 93–94, 102, 106–10, 174; early influences on, 93–95; Eddington and, 104–10; emigration of, to United States, 140; ghosts in work of, 102–5, 119–20; nervous breakdown of, 105; Neumann and, 133; "On the Motion of Small Particles Suspended in Liquids at Rest," 80–81, 88–89; and Pearson's *Grammar of Science*, 93; Poincaré and, 81, 333n4; politics of, 122, 176; popular science published by, 103–5, 110; and probability, 80; and quantum mechanics, 112–13, 115, 117–21, 129–30, 143–44, 146–48, 155, 174–75, 183; and radioactivity, 114; refrigeration patents of, 129; and religion, 94; reputation of, 50, 106, 110, 174; Rothstein and, 173; and speed of light, 102, 104, 105, 110–11, 113, 143; Szilard and, 124–25, 129; theory of relativity, 78, 93, 102–11, 103, 174; and the Vienna Circle, 78; in war years, 103–5
The Einstein Theory of Relativity (documentary), 110
élan vital. See vital force
electrodynamics, 50, 53
Electrolux, 129
electromagnetic waves, 50, 73–74, 95
ELIZA (chatbot), 219, 226
elves, 10
energetics, 77

energy: atomic, 139–41; black holes and, 214–16; and climate change, 289; computer use of, 230–32; conservation of, 13, 67, 116, 154, 180; economic effects of, 287–90; efficient use of, 61, 291–92; expended in forgetting, 210, 234; information exchanged for, 127, 161, 167–69, 229, 231, 234–35, 242–43; nuclear, 293; quantum mechanics and, 113–14, 116; radioactivity and, 113, 115–16, 139–41. See also entropy; perpetual motion machine; thermodynamics
Engels, Friedrich, 41
Enigma machine, 158
Enlightenment, 28, 35–36, 314. See also Age of Reason
entanglement, 114, 143, 183, 244–45
entropy: art as foil to, 251, 261; black holes and, 216; Boltzmann's formula for, 76; economic effects of, 290; as enemy of life, 172; explanation and significance of, 53–54; information in relation to, 127–28, 167–68, 257, 338n38, 359n31; intelligent beings' effect on, 91–92, 125–28, 134–35, 142, 168–70, 244; introduction of concept of, 69; negative, 170–71, 182, 251, 257, 292; noise compared to, 162; quantum mechanics and, 134; scientific enterprise in relation to, 170–71; statistical validity of, 53; violations of, 13, 87, 92, 125–27, 161, 177–78, 210–11, 246, 248–49. See also second law of thermodynamics
environmental economics, 284
epigenetics, 255, 353n66
equilibrium: disturbance of, 88, 141, 266, 292, 295; economic, 279; living beings and, 128, 251, 263; statistical, 145; thermodynamic, 49, 55, 57–59, 63, 66–67, 69, 72, 129, 166, 308, 309; of universe, 97–98
eternal recurrence, 71–73, 85, 140–41, 284
ethics, 302–4. See also morality, genetic perspective on
eugenics, 142–43, 249–50
European Commission, 245
Eve, 6, 7
evolution: gene-level study of, 257–58, 272–73; Maxwell's demon as model for, 250; mechanistic conception of, 29, 46, 247; selection mechanism in, 44–48, 254, 257, 266, 272; Wallace vs. Darwin on, 47–48. See also Darwin's demon; natural selection

evolutionary biology, 254–60
existence, 313–16
experimentation, 11, 19–20, 22–23, 322

fake news, 15
Fama, Eugene, 285–86
Faust, 7, 293, 358n14
Federal Aviation Commission, 142
feedback, 159–61, 255, 266, 352n21
Fermi, Enrico, 150–51
Feyerabend, Paul, 317–20
Feynman, Richard, 2, 50–51, 155, 194–97, 212, 214, 227–33, 243; *Feynman Lectures on Computation*, 228, 229, 231
Feynman's demon, 194–96, 232, 243, 245
file transfer protocol (FTP), 201
first law of thermodynamics, 53, 69, 113, 178, 180–81
Flammarion, Camille, 97–99
flight patterns, 159–61
Foerster, Heinz von, 182, 210–12
Foglio, Phil, design for UNIX t-shirt, *241*
forgetting, energy expended by, 210, 234, 243
four-dimensional spaces, 70, 103, 104, 110
Fourier, Joseph, 37, 56
Frankenstein, 4, 40–41, 154
Franklin, Rosalind, 198
Franz Ferdinand, Archduke of Austria, 103
free will, 29, 36, 58, 70, 107, 130, 131, 146, 155–56, 164, 267
Freud, Sigmund, 316, 358n14
Fyodorov, Nikolai, 138

Gabor, Dennis, 168–70
Gabor's demon, 168–70, 198, 308, 321, 360n37
Galatea, 6
Galton, Francis, 75, 250
gambling, 34, 91, 160–61, 163–64
Gamow, George, 215, 251–52; "Maxwell's Demon," 252
Geiger counters, 147–48, 152
General Motors, 291
genes, 257–60, 269, 272–73. *See also* DNA
genius, 261
Georgescu-Roegen, Nicholas, 277, 281–85, 287–90, 295; *Analytical Economics*, 284; *The Entropy Law and the Economic Process*, 284
Ghirardi, GianCarlo, 176–77

ghouls, 10
giants, 10, 308–9
Gibbs, Josiah Willard, 127; *Elementary Principles of Statistical Mechanics*, 72
Globe Theater, London, 20
gnomes, Hawking's evocation of, 216–17
goblins, 10
God: devil as rival of, 25–26; evolution independent of, 29, 45, 48; Nietzsche on the death of, 72–73; power of, 23, 25–26; quantum mechanics and, 120; secular threats to, 30, 33, 35–36, 48, 50; universe in relation to, 33, 36, 50, 120, 272
Goethe, Johann Wolfgang von, 4, 68; *Faust*, 7, 41, 131–32
Golem, 6, 197
Gospel of John, 20
Gouy, Louis Georges, 81–82, 86
Grandier, Urbain, 317
gravity, 79, 94, 101–2, 106–9
Gray, Asa, 45
Great Depression, 129, 283
growth, economic, 277, 281, 284, 287–89, 294–95

Hahn, Otto, 150
Haizmann, Christoph, 358n14
Haldane, J. B. S., 140
Halley, Edmond, and Halley's Comet, 33–34
Harvard University, 165, 186, 202
Harvey Society, 256
Haugeland, John, 221–23, 267, 348n118
Haugeland's demon (H-Demon), 223–26, 348n18
Hawking, Stephen, 216–18
H-bomb, 214
heat. *See* thermodynamics
heat distribution, mathematical analysis of, 37, 56–57
Heidegger, Martin, 304
Heisenberg, Werner, 114, 120, 131, 132, 148, 151, 155, 177, 182, 259, 319
Hell Machine, 147, 320
Helmholtz, Hermann von, 67–68
Hepler, Loren G., and Stanley W. Angrist, *Order and Chaos*, 167
Hermann, Grete, 114, 148–49, 164, 177, 183
Herschel, John, 106–7
Hertz, Heinrich, 50

hidden variables, in quantum mechanics, 133, 148, 175–77, 183, 199–200, 319
Hilbert, David, 103
Hill, Thomas, 96–97
Hiroshima bombing, 154, 156
history: discipline of, 322–23; knowledge and explanations of, 117; science in relation to, 323
Hitler, Adolf, 139–40, 146
Hofstadter, Douglas, 220, 223–25
Holmes, Samuel Jackson, 249–50
Hopfield, John, 236, 264
Hopkins Marine Station, Stanford University, 272
House Un-American Activities Committee, 176
Hugo, Victor, 9
Hulburt, E. O., and L. Darling, "On Maxwell's Demon," 307
Hume, David, 32–33
humor, 233, 268
Huxley, Aldous, *The Devils of Loudun*, 317–18
Huxley, Thomas, 46–47, 247, 329n56
hypertext transfer protocol (HTTP), 241

IBM. *See* International Business Machines
Icarus, 6
imagination: creatures of, 9; knowledge in relation to, 11–12; role of, in science, 8–13, 298–302, 312, 321–22, 325n7, 326n9
imitation game, 187
immigration, 250
Imperial Chemical Industries, 147
Imperial College London, 168
indeterminacy, 114, 118, 135, 164, 177, 260. *See also* determinism
Industrial Revolution, 41
INFERNOS (Information, Fluctuations, and Energy Control in Small Systems), 245
infinite time, actions in, 84–85, 87, 140–41, 284
information: black holes and, 217–18; communication in relation to, 167–68; in DNA, 246, 253–54; in economics, 282, 285–86; energy required for, 127, 161, 167–69, 229, 231, 234–35, 242–43; entropy in relation to, 127–28, 167–68, 257, 338n38, 359n31; limits on, 133–34; meaning/value of, 168–69, 342n28; quantitative value of, 279; in quantum realm, 135, 244–45. *See also* communication; intelligence; knowledge

Information Systems Branch, Office of Naval Research, University of Illinois at Urbana, 182
Institute for Cognitive Science, University of California–San Diego, 236
Institute of Radiobiology and Biophysics, University of Chicago, 252
intelligence: cybernetics and, 160; demon-based model of, 269–71; entropy countered by, 91–92, 125–28, 134–35, 142, 168–70, 244; as Maxwell's demon, 92, 125–28, 141–42, 351n191; nature of, 218–26; neural networks and, 190; Selfridge's paradigm for, 188–91; Soviet ideal of collective, 138–39; Turing test for determining, 186–87. *See also* artificial intelligence (AI); consciousness; information; knowledge
intelligent design, 256
intentionality, 222, 226
International Business Machines (IBM), 162, 165, 227
International Congress of Physics, 85–86, 120
International Journal of Theoretical Physics, 226
internet, 201, 240, 241
invisible hand, 277
ionization, 121
irrationality, 10, 13, 22, 35, 300, 316
irreversibility. *See* reversible/irreversible laws/effects
isotope separation, 121, 128
Israel, 197

James I, King of England, *Daemonologie*, 21–22
Janet, Paul, 48
Jevons, William Stanley, 83–84
Jinn (genie), 24, 307
Johnstone, James, 248–49
jokes, 233, 268
Jordan, Pascual, 164
Journal de physique, théorique, et appliquée, 81
Journal of Applied Physics, 170

Kant, Immanuel, 28
Katechon, 49
Keller, Evelyn Fox, 253–54
Kendall, Maurice, 279–80
Keynes, John Maynard, 279
kinetic proofreading, 264

knowledge: ambivalence toward, 5–7, 31, 43; difficulties in obtaining, 42–43; energy required for, 167–68; ethics in relation to, 302–4; imagination in relation to, 11–12; limits of, 64–67 (*see also* in quantum realm, incompleteness of); measurement as means of, 133, 134, 136; nature in relation to, 134, 175; in quantum realm, incompleteness of, 113–14, 118–20, 123–24, 133–34, 139; universal, 2, 29–32, 38–40, 46, 51, 65–66, 99, 132, 136–39, 148, 151, 182–84, 186, 218, 259. *See also* consciousness; information; intelligence

Kuhn, Thomas, 317–19

Laing, Richard, 212
Landauer, Rolf, 227–31, 235–36
Landauer's demon, 230
Langevin, Paul, 110
Lanier, Jason, 296
Laplace, Pierre-Simon: accomplishments of, 33–36; failed public service of, 36; on origin of earth, 328n17; *Philosophical Essay on Probabilities*, 30; *Theorie analytique de probabilités*, 38; *Traité de mécanique céleste*, 33, 37
Laplace's demon: biology and, 246, 250–51; Bois-Reymond on, 64–67; and causality/certainty, 2, 29–32, 38–40, 46–47, 51, 65–66, 99, 123, 136–39, 148, 150, 151, 163–64, 182–84, 186, 218, 259; Darwin's work compared to, 46; economic uses of, 278, 279, 283; historical significance of, 29, 31, 36–38, 41–42, 43–44, 51, 64–67, 87–88, 182–83, 320; historiography and, 322–23; junior versions of, 181; limitations of, 113, 119, 123–24, 132, 133, 136–39, 151, 155–56, 284, 339n73; machines' incorporation of, 158; Maxwell's demon compared to, 181, 259; names given to, 31–32; Planck and, 132–33; political uses of, 183; powers of, 2, 29–32, 148, 182, 304, 309, 322; quantum demons as threat to, 112, 146; quantum mechanics and, 164; skepticism about/exorcism of, 41–42, 67, 177, 182, 184, 247–48, 264, 283; and third law of thermodynamics, 178, 180; as time traveler, 99
Lasseter, John, front cover for *Unix System Manager's Manual*, 241
Latour, Bruno, 247, 274–75

Lawrence, Ernest, 150
laws of nature, 2, 10–13, 68, 180
Leibniz, Gottfried, and Leibniz's wheel, 37, 42
Leonard, Andrew, 240, 241
Lerner, A. Y., *Fundamentals of Cybernetics*, 211
Lévi-Strauss, Claude, 301
Lewis, Gilbert Newton, 127–28, 150, 161, 284
Lewontin, Richard, 259–60
Lichtenstein, Léon, 137
life/living systems: AI compared to, 271–72; amplifying effects of, 131; entropy-resisting, 248–49; essential forces of, 220, 247, 251, 256; Laplace's demon and, 246, 250–51; materialistic conception of, 248, 314; Maxwell's demon and, 131, 246–50, 253–56, 260, 262–64; mechanistic conception of, 29, 46, 247, 270, 314; mystery of, 67, 127–28, 131, 246; origins of, 263; quantum mechanics and, 259–60; Schrödinger's speculations on, 246, 250–51; and second law of thermodynamics, 256. *See also* nature
light, 100, 107. *See also* speed of light
Lillie, Ralph, 131, 248, 249
Lincoln, Abraham, 75
Lincoln Laboratory, MIT, 190–92
Lindsay, Peter H., and Donald A. Norman, *Human Information Processing*, 213
literature: demons in, 4, 22–23; exploration of reality in, 21, 22–23; suspicion of, 18, 27
Lloyd, Seth, 244–45
Locke, John, 32
Lockyer, Norman, 247
logic, 123, 137
logical empiricism, 138
logical positivism, 138, 299–300, 302, 316
London, Edinburgh, and Dublin Philosophical Magazine and Journal of Science, 197
Lorentz, Hendrick, 115
Los Alamos National Laboratory, 153–55, 176, 243
Los Angeles Times (newspaper), 233
Louis XIV, King of France, 66
Louis XVIII, King of France, 36
Lovelace, Ada, 29, 40, 43–44, 187
Lucifer, 4, 309
Lucretius, 80
Lyell, Charles, 46

Mach, Ernst, 77–78, 105
machine learning, 159–60, 185, 188–91, 202, 233, 271, 349n162. *See also* artificial intelligence (AI)
machines. *See* mechanics/machines
MacKay, Donald, 191
Magazine of Fantasy and Science Fiction, 260
Malkiel, Burton, *A Random Walk down Wall Street*, 285
management practices, 273–74, 296–97
Manchester Mark I, 186
Manhattan Project, 114, 124, 131, 153–55, 214
mapmaking, 107–8
Margenau, Henry, 136–37
Mark III computer, 186
Marlowe, Christopher, *Doctor Faustus*, 7
Martínez-Alier, Juan, 294–95
Marx, Karl, 41, 43
Marxism, 183
Massachusetts Institute of Technology (MIT), 114, 122–23, 142, 157, 159, 162, 168, 185, 190–92, 201–2, 219, 244
materialism, 16, 36, 183, 184, 248, 270, 314. *See also* secularism
Mathematical Gazette (journal), 127
mathematics: and celestial mechanics, 33; intellectual power of, 32, 38; quantum indeterminism and, 123, 137; and statistics/probability, 30, 35, 38
Maxwell, James Clerk: "Concerning Demons," 52–55; contributions of, 50–51; *Theory of Heat*, 55–57, 71, 75
Maxwell's demon: Aladdin's genie in relation to, 181; Asimov on, 260; biology and, 131, 246–50, 253–56, 260, 262–64; and black holes, 214–15; Brownian motion and, 79, 81, 84–87, 90; computers and, 198–99, 201, 211, 236; construction of, 121; cybernetics and, 160–62, 165, 210–12; economic uses of, 277–79, 281–82, 285–90, 295–97; Ehrenberg on, 197–200; Feynman and, 195–96, 227–32; historical significance of, 49–53, 68–71, 74–87, 125, 320–21; at human scale, 59, 124, 134, 169, 171; intelligent beings as, 92, 125–28, 141–42; Laplace's demon compared to, 181, 259; Lewis and, 127–28; librarians as, 150; limitations of, 90, 113, 114, 133–36, 139, 166, 169, 196, 212, 234, 236, 239, 242–44, 289–92; in literature, 252; machines' incorporation of, 158; management practices likened to, 273–74, 296–97; manuscript on, 52, 53–55; naming of, 53, 57, 59; and nature, 265; operations and powers of, 2, 49–50, 55, 59–64, 74–75, 86, 88, 90, 252, 292; philosophy and, 275; Poincaré and, 76, 81, 84–87; in politics, 75–76, 123, 141–43; quantum mechanics and, 115, 136, 141, 244–45; reversal of entropy by, 125–27, 177–78, 210–11; scientist as, 274–75; and second law of thermodynamics, 51, 53–56, 70, 177–78, 180, 199, 228, 242–43, 248, 289, 294, 296; skepticism about, 69–70, 226–27; social manifestations of, 273–76, 291–94; sorting as chief activity of, 49, 59, 74–75, 141, 150, 163, 231, 249–50, 274–76, 282, 308; Szilard and, 125; time reversed by, 57–58; torchlight of, 166, 228; as a valve, 49, 54–55, 92, 135, 168, 194, 242
Maxwell-Szilard-Brillouin demon, 262, 265, 266, 309
McCarthy, John, 191, 219, 267
McCarthyism, 176
McKinsley, William, 76
measurement, in quantum domain, 133, 134, 136, 178
mechanics/machines: brains as, 190; corporeal, in Cartesian worldview, 17; cybernetics and, 157–59; evolution based on, 29, 44–48; knowledge derived by means of, 31; Laplace's demon and, 158; living systems compared to, 128–29; Maxwell's demon and, 158; reversibility of laws of, 63–64; statistical, 77; threats represented by, 41, 50. *See also* perpetual motion machine
mechanistic conception of life, 29, 46, 247, 270, 314
mechanistic conception of universe, 46, 50, 84, 247, 337n12
media, psychological effects of, 27, 305
Meinong, Alexius, 315
Meitner, Lisa, 150
memory, 210, 229–31, 233–35, 243
Menabrea, Luigi Federico, 43–44
Mephistopheles, 4
metaphysics, 78, 137, 321
metastability, 161–62, 171
Methodos (journal), 180

Metz, André, 137
Meyer, Garry S., "Infants in Children's Stories," 206–9
microbots, 244
microchips, 198–99, 228, 231, 238
microparticles, 79–81
microprocessors, 291
Microsoft, 278, 296
Millikan, Robert, 137
Milton, John, 4
mind. *See* brain/mind
Mind (journal), 186
mind-body relationship, 17, 184
Minsky, Marvin, 185, 202, 203, 220, 233; "Steps toward Artificial Intelligence," 192
miracles, 22, 33–34, 46, 57, 166
Mises, Richard von, 123–24
MIT. *See* Massachusetts Institute of Technology
molecular biology, 252
molecules: as basic components of nature, 46–47, 89, 174; Brownian motion and, 79–92; Laplacean knowledge of, 32; limitations of thermodynamic theory based on, 73; Loschmidt's models of, 71; Maxwell's demon's action upon, 55–58, 60; velocity of, 60
Monod, Jacques, 252–53, 256, 274; *Chance and Necessity*, 262, 263, 265
Monod's demon, 264, 265, 309
monsters, 9–10
morality, genetic perspective on, 257–59, 269. *See also* ethics
Morgan Bank, 286
Morowitz, Harold J., *Entropy for Biologists*, 263
Morse code, 190
Morton, Jack, 273–74
Moscow Trials, 138
MSNBC (cable news outlet), 245
Musk, Elon, 7, 325n6
Mussolini, Benito, 151
mythology, and scientific discovery, 299–302

Nagasaki bombing, 154, 156
nanobots, 244
Napoleon, 33, 36
National Defense Research Council, 153, 159
National Physical Laboratory, Britain, 187, 280
National Radium Conference, 124
National Science Foundation, 177
natural resources, 278, 279, 284, 287–90
natural selection, 46–48, 74, 190–91, 254–55, 257, 261, 266, 271. *See also* Darwin's demon; evolution
nature: classical theories of, 117; Darwin's evolutionary theory of, 46–48; indeterminacy in, 164, 177; knowledge in relation to, 134, 175; Laplacean analysis of, 29, 31; laws of, 2, 10–13, 68, 180; Maxwell's demon and, 265; molecular conception of, 46–47, 89, 174; regularity of, disproved by Maxwell's research, 49–51, 54; uncertainty in, 112, 119, 123, 130–31, 143–44, 151, 164. *See also* life/living systems; reality; universe
Nature (journal), ix, 57, 70, 89, 140, 149, 182, 242–43, 247, 293–94
Nazis, 123, 129, 132, 149, 151, 165
necromancers, 23
negentropy/negative entropy, 170–71, 182, 251, 257, 292
Nemeth, Evi, 238–39
neoclassical economics, 282, 284, 287
Nernst, Walther, 65
Neumann, John von: and atomic bomb, 122, 153–54, 338n30; and biology, 252; and computers, 158; emigration of, to United States, 140; and information, 343n28; *Mathematical Foundations of Quantum Mechanics*, 133; and quantum mechanics, 114, 133–36, 148, 175–76, 183; student of, 151
neural networks, 190, 268, 270
Neurath, Otto, 137–38, 339n73
neuroscience, 267–73
neutrons, 150–51
Newell, Allen, 193, 220, 233, 237, 267; "The Chess Machine," 193
New Hacker's Dictionary, 239
New Scientist (journal), 242, 243, 268, 293
Newton, Isaac, 35–36, 50, 106–7, 123
New York Review of Books (magazine), 225
New York Times (newspaper), ix, 116, 139–41, 173, 178, 238
Nietzsche, Friedrich, 72–73
Nobel Prize, 80, 89, 114, 116, 119, 120, 122, 130, 137, 168, 184, 219, 253, 256, 263, 264, 275, 279, 287
Noether, Emmy, 148
noise, 162, 309

nonexistents, 313–16
nonlinear dynamics, 159, 281
nonlocality, 114, 118, 135, 143, 147–48
Norman, Donald A., and Peter H. Lindsay, *Human Information Processing*, 213
nuclear chain reactions, 124, 150–51
nuclear energy, 293. *See also* atomic energy
nuclear magnetic resonance demons (NMR demons), 245

Odum, Howard, 287, 291–92
open-ended programming, 185–86, 237
open-source code, 203
operationalism, 152
Oppenheimer, Robert, 122, 132, 153, 176, 338n30
order: art as exemplar of, 251, 261; as essential feature of life, 249, 250–51, 264–65; social, 275–76
osmosis, 74–75
Ostwald, Wilhelm, 77

Pais, Abraham, 175
Pandemonium computer-programming model, 188–93
Pandora (goddess), 6
Papert, Seymour, 202, 346n59
parallel distributed procession (PDP), 232–33
parallel processing, 188, 190, 192, 232, 268
Pascal, Blaise, and Pascaline, 37, 42
Pasteur, Louis, 326n15
Pauli, Wolfgang, 131–32, 256, 343n28
PDP Research Group, 236
Pearson, Karl, 89, 94, 250, 282, 283; *The Grammar of Science*, 75, 93, 282
Penrose, Oliver, *Foundations of Statistical Mechanics*, 209–10
perpetual motion machine: Brownian motion and, 82–83, 87, 90; computers and, 229; Feyerabend and, 359n31; Gabor and, 168–69; impossibility of, 53; intervention of intelligent beings as means to, 91–92, 125–26; Maxwell's demon and, 49, 51, 55; perfect communication compared to, 162; radioactivity and, 116, 122, 154; Szilard's quest for, 124–26; Thomson brothers' quest for, 61–62
Perrin, Jean, 89–91
personal computers, 218, 236, 240
personhood, 17. *See also* subjectivity

philosophy: Descartes and, 7, 15–17, 23–28; epistemological concerns of, 32–33, 42–44; and existence, 313–16; and Laplace's demon, 136–38; Maxwell's demon and, 275; and mystery of life, 67, 127–28, 131; and the nature of intelligence, 218–26; quantum mechanics and, 116–17, 135, 136–37, 150, 173–75; and reality/truth, 21, 23–24, 32–33; Socrates and, 4; Vienna Circle and, 78, 137–38
Philosophy of Science (journal), 179–80
photoelectric cells, 130, 169–70
photons, 127, 130
Physical Review (journal), 121, 140
Physical Review Letters (journal), 243
physics–biology connection, 251–52, 259, 263, 266–67
Physics Bulletin (journal), 216
physics–economics connection, 277–79, 281–90, 293–96
Pierce, John R., 245; *Symbols, Signals, and Noise*, 194, *194*, 320
pilot waves, 119–20, 176
Pinker, Steven, 268–71
Pittendrigh, Colin, 255–59, 272–73; "Reflections of a Darwinian Clock-Watcher," 272
Planck, Max, 2, 284; contributions of, 71, 77, 112; and determinism, 183; and Laplace's demon, 132–33; and Maxwell's *Theory of Heat*, 71; and quantum mechanics, 71, 114, 115, 132
Planck scale, 131
Planck's constant, 149
Planck's demon, 284–85
Plato, 309; *The Republic*, 70, 331n56
plutonium, 113
Podolsky, Boris, 143
Pohle, Joseph, 101
Poincaré, Henri: and Brownian motion, 81; Einstein and, 81, 333n4; and eternal recurrence, 284; and Maxwell's demon, 76, 81, 84–87; *Science and Hypothesis*, 76, 333n4; and speed of light, 97–98, 101–2; and thermodynamics, 333n14
Polidori, John, *The Vampyre*, 40
politics: Laplace's demon in, 183; Maxwell's demon in, 75–76, 123, 141–43. *See also* social sciences, demons in

pollution, 60–61, 235, 242, 248, 279, 286, 288
polypeptide fibers, 262
Popper, Karl, 319–22, 352n38, 359n31; *The Logic of Scientific Discovery*, 319; "Metaphysical Epilogue," 321; *Postscript to the Logic of Scientific Discovery*, 320
Popper's demon, 321
prediction. *See* statistics/probability
Prigogine, Ilya, 263–64; *La Nouvelle alliance* (with Isabelle Stengers), 263–64
Princeton University, 121, 140, 176, 214
probability. *See* statistics/probability
Proceedings (journal of Royal Society of Edinburgh), 57
Proceedings of the Institute of Radio Engineers (journal), 162, 192
Proctor, Richard, *Other Worlds than Ours*, 100–101
programming: and AI, 185, 187–88, 190–93, 219–20; context as goal of, 202–3, 233; of demons, 185, 188, 193, 203–8, 236–38, 240; open-ended programming, 185–86, 237. *See also* daemons (computer programs)
Project MAC, 201
Prometheus, 6
proteins, 262
psychology: demons in the field of, 212, 213, 316, 359n15; and reality, 316
Pygmalion, 6

quantitative sociology, 30
quantum biology, 164
quantum demons: actions of, 13, 112–14, 123, 130–31, 146, 149–50, 151, 173; Curie and, 115; disguised actions of, 114, 145–46, 156; double-slit experiment and, 118, 144–46; Maxwell's demon compared to, 115; predictability of, 112; Rothstein and, 181
quantum Maxwell's demons (QMD), 244–45
quantum mechanics: actions of, 156; and causality, 117, 120, 135, 144, 200; and chance, 144–46; connections in nature according to, 113; Einstein and, 112–13, 115, 117–21, 129–30, 143–44, 146–48, 155, 174–75, 183; epistemological consequences of, 113–14, 118–20, 123–24; hidden variables in, 133, 148, 175–77, 183, 199–200, 319; historical development of, 114–24; information and, 135, 244–45; interpretive disagreements over, 114–15; Laplace's demon and, 164; and living systems, 259–60; logic and mathematics affected by, 123, 137; mass–energy relationship in, 114; Maxwell's demon and, 115, 136, 141, 244–45; measurement and, 133, 134, 136; Neumann and, 114, 133–36, 148, 175–76; philosophical implications of, 116–17, 135, 136–37, 150, 173–75; Planck and, 71, 114, 115, 132; Schrödinger and, 146–48; statistical nature of, 119–21, 123, 183–84, 200; universe conceived according to, 116, 119, 136, 145–46, 150; waves and particles in, 118–20
Quarterly Journal of Science, 70

radioactivity, 112–16, 124, 139–41, 147–48, 150–56
radium, 113, 114, 116
RAND Corporation, 193
random motion, 159–60
randomness, in evolution, 254
Random Walk, 89, 285, 309. *See also* Drunkard's Walk
Raphael (artist), 251
Reagan, Ronald, 238
reality: Cervantes's *Don Quixote* and, 18–20; demons as means of exploring, 15–21; Descartes's demon and illusions of, 7, 15–17, 24, 27; Einstein's conception of, 102, 115, 119; influence of beliefs on, 317–18; literary explorations of, 21, 22–23; observer's effect on, 134, 175; the unreal in relation to, 313–16. *See also* nature
refrigeration, 129
Reichenbach, Hans, 163–64, 299–300, 302
relativity: demons' role in, 13; Einstein and, 78, 93, 102–11, 103, 174; Einstein's conception of, 175; illustrations and examples of, 110; limitations of, 114; Poincaré and, 81; space affected by, 102–5; special, 102; time affected by, 102–5
relativity demons, 181
La Repubblica (newspaper), 245
reversible/irreversible laws/effects: living beings, 265; Maxwell's demon and, 84; mechanics, 63–64, 76, 86, 127, 229; quantum mechanics, 176; thermodynamics, 134; universe, 71–72, 196, 265. *See also* time: reversal of

Revius, Jacobus, 23, 24
Revue de métaphysique et de morale (journal), 84
Röntgen Society, 87
Roosevelt, Franklin D., 122–23, 124, 141, 151–53
Roosevelt, Theodore, 76
Rorty, Richard, 220
Rosen, Donn Eric, 265–66, 353n66
Rosen, Nathan, 143
Rosenblatt, Frank, 189
Rothstein, Jerome, 173–75, 178–82, 184, 343n43; *Communication, Organization, and Science*, 178
Russell, Bertrand, 315, 343n28; *Principia Mathematica* (with Alfred North Whitehead), 193
Rutherford, Ernest, 116–17

Sagan, Carl, 10, 11
Salk Institute, 274
Samuel, Arthur, 271
Samuelson, Paul, 277, 279–82, 284–86, 292–96; "Microscopic Time Asymmetry of Maxwell's Demons," 292; "Scientific Correspondence: The Law Beats Maxwell's Demon," 294
Satan, 4
Sathanus, 4
scarcity, 248, 277, 279, 283, 286, 290, 294–95
Schank, Roger, 207, 219–20, 223
Schmeck, Harold M., Jr., "A Scientist Gives Demons Their Due," 178–79, *179*
Scholem, Gershom, 197
Schrödinger, Erwin, 120, 146–48, 246, 250–51, 253, 259; *What Is Life?* 250
Schrödinger's cat, 147–48
science: as belief system, 317; demons' prevalence in, ix–x, 9–10, 305, 311, 316–17, 319; demons' role in, 2–4, 8, 13–14, 107–8, 178–82, 194, 274–75, 298–300, 306, 318–20; Descartes's contribution to birth of, 16; discovery process in, 1–2, 8, 11, 299–301; entropy resulting from, 170–71; and ethics, 302–4; Feyerabend and Kuhn on, 317–20; harmful applications/outcomes of, 5–6, 41, 156, 199, 303–4; history in relation to, 323; imagination's role in, 8–13, 298–302, 312, 321–22, 325n7, 326n9; nature of, 78, 109, 170; Popper on, 319–22; progress of, 12–13; witchcraft compared to, 317–18. *See also* technologies
Science (journal), 74, 127, 130, 141, 173, 236, 249, 292

Scientia (journal), 103
Scientific Advisory Committee, U.S. Department of War, 153–54
Scientific American (journal), ix, 192, 197, 234, 242, 293–94
Searle, John, 219–26; "Mind, Brains, and Programs," 219
Searle's demon, 218–26, 271, 308, 310
second law of thermodynamics: black holes and, 216; Brownian motion and, 82, 86, 91; desire to circumvent, 72; explanation and significance of, 53–54, 74; intelligent beings' effect on, 91–92, 125–26, 128–29, 134–35, 142, 168–70, 244, 351n191; living systems and, 256; Maxwell's demon and, 51, 53–56, 70, 177–78, 180, 199, 228, 242–43, 248, 289, 294, 296; political application of, 333n75; radioactivity and, 152; statistical validity of, 53, 89; violations of, 51, 55–56, 74, 77–78, 82, 86, 125, 152, 165–66, 169, 171, 248, 248–49. *See also* entropy
secularism: cognitive psychology and, 270; evolution and, 29; God's role threatened by, 29, 30, 33, 35–36; mechanistic theories as feature of, 29, 30, 33. *See also* materialism
selfish genes, 257, 259, 269, 354n79
Selfridge, Oliver, 185, 188–93, 202; "Pandemonium: A Paradigm for Learning," 188–90, *189*; "Pattern Recognition by Machine," 192
semiconductors, 228
senses, reliability of, 21, 24, 27, 78, 326n9
Serres, Michel, 247, 275
Shakespeare, William, 4, 20–21; *Hamlet*, 20, 261, 326n9; *Macbeth*, 21; *The Tempest*, 21
Shannon, Claude, 162–63, 168, 343n28
Shelley, Mary, *Frankenstein*, 4, 7, 40–41
SHRDLU (computer program), 219, 224
Siday, Reymond, 197
Simon, Herbert, 220
skepticism: role of, in science, 15; superstition questioned by, 28
Slade, Henry, 70
small-scale acts. *See* amplification, of small-scale acts
smart weapons, 159–60
Smith, Adam, 277
Smoluchowski, Marion von, 90–92, 105, 125, 168, 351n191, 359n31

Smyth, Henry De Wolf, 121–22, 154
Smyth, Samuel Phillips Newman, 250, 351n10
social physics, 30
social sciences, demons in, 247, 273–76. *See also* politics
Socrates, 4
Soddy, Frederick, 116
software, 232, 269–70. *See also* daemons (computer programs); programming
Solid State Electronics Branch, Electron Division, Signal Corps Engineering Laboratories, 173
Solow, Robert, 286–87
Solvay Congress, 115, 120
Somerville, Mary, 40
Sommerfeld, Arnold, 105
sorting: as chief activity of Maxwell's demon, 49, 59, 74–75, 141, 150, 163, 231, 249–50, 274–76, 282, 308; as human activity, 306–8
Soviet Union, 138–39
Soyka, Ed, front cover for *Abacus*, 227
space: absolute, 102–3, 105; curvature of, 107–9; Einstein's theory of relativity and, 102–5; empty, 102
special relativity, 102
spectral-line shift, 105
speed of light: communication possibilities based on, 101–2; fascination with, 73; as limit, 102, 104, 105, 110–11, 113, 143, 283; quantum physics and, 113; and time reversal, 93–94, 97, 101; time travel involving, 93–101, 103–4. *See also* light
Spengler, Oswald, 117
spiritualism, 48, 70–71
Stalin, Joseph, 138
Stanford University, 207, 219, 272
statistical mechanics, 77
statistics/probability: as aid in discerning truth, 30, 32–33, 35; applications of, 30, 34; Einstein and, 80; in infinite amount of time, 84–85, 140–41, 284; Laplace and, 30, 32, 34–35, 37; quantum mechanics and, 119–21, 123, 183–84, 200; and thermodynamics, 113; thermodynamics and, 53, 59, 76–78, 91; uncertainty linked to, 130–31
Stengers, Isabelle, with Ilya Prigogine, *La Nouvelle alliance*, 263–64

Stewart, Balfour, *The Unseen Universe* (with Peter Guthrie Tait), 57
Stimson, Henry, 154
stock market, 279–82, 285–86, 293–95
strong AI, 191, 219–22, 225–26
subatomic particles, 112, 114–20, 123, 144
subjectivity, 26. *See also* personhood
Summers, Lawrence, 293
sun, energy of, 62–63
superstition: demons associated with, 10; rational examination of, 22, 28, 33, 212, 305, 310–11
Systematic Zoology (journal), 266
Szilard, Leó, 284; and biology, 252, 253; and effect of observer on reality, 343n30; Einstein and, 129, 151–52; emigrations of, 140, 151; Gabor and, 168–69; and information, 279; and Maxwell's demon, 125, 134, 161, 338n38; Neumann and, 133, 134; and nuclear energy/atomic bomb, 124, 151–55; and perpetual motion machine, 124–28; at Salk Institute, 274. *See also* Maxwell-Szilard-Brillouin demon
Szilard's demon/engine, 125–26, 198, 243
Szilard's exorcism, 125, 134–35, 161, 176

Tait, Peter Guthrie, 52–54, 63–64, 69–70; *The Unseen Universe* (with Balfour Stewart), 57
Talmud, 308
Talos, 6
technologies: demons associated with, 5–8, 156, 304, 325n6; diversity of, 4–5; economics and, 287–88; and ethics, 303–4; harmful applications/outcomes of, 5–6, 156, 199, 303–4, 325n6; progress of, 12–13. *See also* science
Technology Review (journal), 140
telegraph, 97, 100
television, 87
Tennessee Valley Authority, 142
theater, 20–23
Theophilus of Adana, 6
thermodynamics: black holes and, 214–16; Clausius's contribution to, 69; demons and, 178; economics in relation to, 282; Maxwell's contribution to, 51; Maxwell's demon and, 13, 51, 53–54, 69–71, 74, 76; Planck's contribution to, 71; Poincaré and, 333n14; politics and,

333n75; radiant vs. kinetic theories of, 73–74; statistical, 76–78, 91, 113; violations of, 88–89, 92, 134, 174, 189. *See also* first law of thermodynamics; second law of thermodynamics; third law of thermodynamics
thinking: computers' capacity for, 43–44; doubt's role in, 26–28. *See also* artificial intelligence (AI); brain/mind
third law of thermodynamics, 178, 180
Thomas (apostle), 20
Thomson, James, 61–62
Thomson, William (later Lord Kelvin), 53, 55, 57–63, 70, 71, 74, 86, 331n56
thought experiments, 8–9, 319–20
time: absence of, in block universe, 99–100; Einstein's theory of relativity and, 102–5; recurrence of, 71–73; reversal of, 57–58, 93–94, 97, 101. *See also* infinite time
Time (magazine), 186
time travel, 93–101, 103–4
tipping points. *See* amplification, of small-scale acts
Torrey Pines Mesa laboratory, 274
transistors, 274
Trans World Airlines, 291
trial-and-error reasoning, 11
Triglandius, Jacques, 23, 24
Trotsky, Leon, 138–39
T. Rowe Price, 286
Truman, Harry S., 123, 154
truth: Descartes's search for a method of discerning, 15–16, 26–27; ethics in relation to, 302–4; probability and, 32–33; science and, 302–3
Tukey, John, 163
Turing, Alan, 158, 185; "Computing Machinery and Intelligence," 186–87
Turing Award, 220
Turing test, 187, 219

uncertainty: atomic bomb and, 155; in genetics, 260; and limits of knowledge, 120, 139; mathematics and logic in relation to, 123; measurement linked to, 133, 134; in nature, 112, 119, 123, 143–44, 151, 164; as quantum principle, 114, 118, 120–21, 135, 182; Schrödinger and, 148; significance of, 156; statistical, 130–31

unconscious thought, 191
unicorns, 10
Universal Exposition (St. Louis, 1904), 86, 101
universe: block model of, 99–100; Cartesian conception of, 17, 27; consciousness's effect on, 135; consequential effects of small actions in, 74, 87; death of, 50, 62–63, 66, 72; Einstein's conception of, 91, 102–10, 120; God in relation to, 33, 36, 50, 120, 272; Laplacean analysis of, 2, 29, 31, 33, 46–47, 51, 88, 138, 184; materialistic conception of, 184; Maxwell's demon and, 84–87; mechanistic conception of, 46, 50, 84, 247, 337n12; mysteries of, 166–67, 214, 236; quantum mechanics and, 116, 119, 136, 145–46, 150; time travel stories and, 95–100. *See also* nature
University of California, Berkeley, 126, 219
University of Chicago, 122, 252
UNIX, 238–41
Unix System Administration Handbook, 238
Unix System Manager's Manual, 241
UNIX t-shirt design, 241
Unruh, William, 243
Unruh demon, 243
uranium, 113, 128, 148, 151–52, 165
US Atomic Energy Commission, 183, 230
US Department of Defense, 202

Vaihinger, Hans, 315–16
vampires, 10
Vienna Circle, 78, 137–38
Vietnam War, 259
virtual reality, 7, 17, 24
vital force/vitalism, 220, 247, 249, 251, 256, 309
Voltaire, 42, 328n41

Waddington, Conrad Hal, 254–55, 266
Wallace, Alfred Russel, 47–48
Walras, Léon, 283
Washington Conference on Theoretical Physics, 251–52
Washington Post (newspaper), 245
Watson, James, 198, 252–53
Wauchope, R., 39
wave-particle duality, 118–19
weather prediction, 68
Weinberg, Alvin, 288–89, 291, 293
Weinert, Friedel, 325n3

Weizenbaum, Joseph, 219, 226; *Computer Power and Human Reason*, 226
Wells, H. G., *The Time Machine*, 99–100
Western Electric, 162
Weyer, Johann, *De Praestigiis Daemonum et Incantationibus ac Venificiis* (On the tricks of demons and on spells and poisons), 316, 358n14
Weyl, Herman, 235–36
Wheeler, John Archibald, 214, 243
Wheeler's demon, 214–16, 243
Whitehead, Alfred North, and Bertrand Russell, *Principia Mathematica*, 193
Whiting, Harold, 74
Wiener, Norbert, 137, 157–63, 168, 171–74, 281, 355n1; *Cybernetics*, 157, 162, 194; *The Human Use of Human Beings*, 171
Wigner, Eugene, 151–52, 184, 288
Wigner's friend, 345n87
Wilkins, Maurice, 198
will. *See* free will

Williams, George, 259, 272, 354n79; *Adaptation and Natural Selection*, 269
Winograd, Terry, 202, 219, 346n59
witchcraft, 317–18
witches, 23
World War I, 103, 128
World War II, 151–55, 158–59
World Wide Web, 241

X-ray tube, 198

Yale University, 219
Yourgrau, Palle, 314

Zeitschrift für Physik (journal), 119, 125
Zermelo, Ernst, 85
Zermelo's demon, 85, 184, 284, 310
Zeus, 6
Zöllner, Johann Carl Friedrich, 70–71, 331n56
Zurek, Wojciech, 243–44, 351n191